Springer Series in Advanced Manufacturing

Series Editor

Duc Truong Pham, University of Birmingham, Birmingham, UK

The **Springer Series in Advanced Manufacturing** includes advanced textbooks, research monographs, edited works and conference proceedings covering all major subjects in the field of advanced manufacturing.
The following is a non-exclusive list of subjects relevant to the series:

1. Manufacturing processes and operations (material processing; assembly; test and inspection; packaging and shipping).
2. Manufacturing product and process design (product design; product data management; product development; manufacturing system planning).
3. Enterprise management (product life cycle management; production planning and control; quality management).

Emphasis will be placed on novel material of topical interest (for example, books on smart manufacturing and sustainable manufacturing) as well as new treatments of more traditional areas.

As advanced manufacturing usually involves extensive use of information and communication technology (ICT), books dealing with advanced ICT tools for advanced manufacturing are also of interest to the Series.

Springer and Professor Pham welcome book ideas from authors. Potential authors who wish to submit a book proposal should contact Anthony Doyle, Executive Editor, Springer, e-mail: anthony.doyle@springer.com.

INDEXED IN SCOPUS

D. T. Pham · Natalia Hartono
Editors

Intelligent Engineering Optimisation with the Bees Algorithm

Springer

Editors
D. T. Pham
Department of Mechanical Engineering
University of Birmingham
Birmingham, UK

Natalia Hartono
Department of Engineering
University of Exeter
Exeter, UK

ISSN 1860-5168 ISSN 2196-1735 (electronic)
Springer Series in Advanced Manufacturing
ISBN 978-3-031-64935-6 ISBN 978-3-031-64936-3 (eBook)
https://doi.org/10.1007/978-3-031-64936-3

This Springer imprint is published by the registered company Springer Nature Switzerland AG
The registered company address is: Gewerbestrasse 11, 6330 Cham, Switzerland

Preface

The Bees Algorithm (BA) is a nature-inspired optimisation method that mimics the foraging behaviour of honey bees. It was first proposed by Pham, Ghanbarzadeh, Koç, Otri, Rahim and Zaidi in 2005 and has since been applied to a wide range of engineering problems, such as machine scheduling, computer vision, mechanical design, robot path planning, process control and many more. The BA has also been modified by various researchers to enhance its performance and adaptability.

This book provides an overview of some of the latest developments and applications of the BA in engineering optimisation. It consists of two parts: the First Part covers improvements to the BA, while the Second Part showcases applications of the BA in various engineering domains.

The First Part comprises four chapters. In the chapter "Enhanced Bees Algorithm Implementing Early Neighbourhood Search with Efficiency-Based Recruitment", Packianather, Imanguliyev and Pham introduce an enhanced version of the BA that implements strategies to improve the exploration and exploitation abilities of the algorithm. The chapter "Improving the Bees Algorithm Using Gradual Search Space Reduction" by Albakr and Pham proposes a technique that uses regional elimination to avoid premature convergence and stagnation. In the chapter "Local Optimal Issue in Bees Algorithm: Markov Chain Analysis and Integration with Dynamic Particle Swarm Optimisation Algorithm", Liu, Ang, Ng and Chaw study the convergence properties of the BA and its integration with a dynamic particle swarm optimisation algorithm. This is followed by the chapter "Development of the Bees Algorithm Toolkit for Optimisation in LabVIEW" by Sahin and Pham which describes a suite of BA optimisation tools embedded in LabVIEW, a popular system for data acquisition and control applications.

The Second Part consists of 16 chapters discussing various engineering applications of the BA. The chapter "Geometrical Optimisation of Smart Sandwich Plates Using the Bees Algorithm" by Duc and Quan presents the optimisation of the geometry of smart sandwich plates on Pasternak-type elastic foundations. The authors used the BA to maximise the natural frequency and the critical buckling load of the sandwich plates.

In the chapter "Integrating the Bees Algorithm with WSAR for Search Direction Determination and Application to Constrained Design Optimisation Problems", Baykasoglu and Senol propose integrating the BA with a weighted superposition attraction–repulsion approach for search direction determination. The authors describe the application of the resulting method to four constrained design optimisation problems.

How welding is performed can affect the final geometry of the welded component. The chapter "Bees Algorithm-Based Optimisation of Welding Sequence to Minimise Distortion of Thin-Walled Square Al–Mg-Si Alloy Tubes" by Wu and Wang discusses an integrated optimisation approach combining the BA with an artificial neural network model to find welding sequences that minimise distortion of thin-walled alloy tubes.

In the chapter "Hybrid Genetic Bees Algorithm (GBA) for Continuous and Combinatorial Optimisation Problems", Tran, Vu, Truong and Nguyen propose a hybrid Genetic BA for training a multi-layer perceptron and solving a two-stage flow shop scheduling problem and a just-in-time job shop scheduling problem. The authors also compare the results obtained with those of the Genetic Algorithm and Particle Swarm Optimisation.

The surface finish of 3D-printed parts depends on the parameters of the 3D printing process. The chapter "Optimisation of Surface Roughness in 3D Printing Using the Bees Algorithm" by Kamaruddin, Ridzuan and Sukindar describes how the BA was employed to find the parameter combination to maximise the surface quality of parts printed by fused deposition modelling.

The chapter "The Bees Algorithm for Robotics-Enabled Collaborative Manufacturing" by Xu, Yang, Ji, Liu and Liu examines the optimisation of collaborative manufacturing (and remanufacturing). The authors used the improved multiobjective discrete BA and the improved discrete BA-Pareto algorithm to balance a robotised disassembly line, select the optimum collaborative manufacturing service aggregation for a robotised assembly system and plan the task sequence for a human–robot collaborative disassembly operation to minimise human fatigue.

The chapter "Bees Algorithm for Hyperparameter Search with Deep Learning to Estimate the Remaining Useful Life of Ball Bearings" by Kumar, Kumar, Gupta, Theinnoi and Pham shows how a two-parameter version of the BA was used to find the optimal set of hyperparameters for a Convolutional Neural Network combined with a Long Short-Term Memory. The resulting deep-learning system was able accurately to predict the remaining useful life of ball bearings in rotating machinery.

In the chapter "Bees Local Phase Quantisation Feature Selection for RGB-D Facial Expression Recognition", Mousavi and Ilanloo discuss the automatic recognition of facial expressions (anger, joy, surprise, disgust and fear) from colour and depth images. After extracting local phase quantisation features from the images, they employed the BA to select the appropriate number of features to maximise the final recognition accuracy.

The chapter "Optimisation of Convolutional Neural Network Parameters Using the Bees Algorithm" by Packianather and Alamri describes the combining of the BA and Bayesian Optimisation to improve the classification accuracy of a convolutional neural network. The engineering application considered by the authors is the detection of cracks in images of concrete.

The chapter "Ergonomic Risk Assessment Combining the Bees Algorithm and Simulation Tools" by Singh, Caterino, Rinaldi, Fera, Macchiaroli and Pham proposes a methodology for integrating BA optimisation and simulation to reduce the workload for workers on production and assembly lines. Simulation is first employed to obtain ergonomics data for different workers. Then, the BA is used as a tool for optimising the job rotations of the workers among the workstations.

Song, Guo, Li, Wu, Yang and Chen examine the problem of building a team of experts for a project in their chapter "A Knowledge Transfer-Based Bees Algorithm for Expert Team Formation Problem in Internet Company." The authors propose a BA that performs 'knowledge transfer' by searching multiple subpopulations and moving individuals from dominant subpopulations to weaker subpopulations.

The chapter "Green Vehicle Routing Optimisation Using the Bees Algorithm" by Satpathy, Das, Hartono and Pham describes the application of the BA to find the optimal set of vehicle tours to service a subset of customers while incorporating stops at fuel stations to refuel. This is the first study to use the BA to solve this so-called Green Vehicle Routing problem.

In the chapter "Utilising the Bees Algorithm for UAV Path Planning—A Simultaneous Collision Avoidance and Shortest Path Approach", Dasgupta, Kumar, Vaibhav and Ismail present the BA as a tool for UAV path planning. The aim is to determine the safest trajectory for an unmanned aerial vehicle (UAV), ensuring it reaches its destination from the starting point via the shortest path while avoiding obstacles along the way.

The chapter "A Tabu-Based Bees Algorithm for Unmanned Aerial Vehicles in Maritime Search and Rescue Path Planning" by Guo, Song, Li, Ou, Xing and Zhang also deals with path planning for UAVs. The context is maritime search and rescue. The authors adopt a reverse 'tabu' strategy to reduce the possibility of invalid search and speed up convergence.

The chapter "Pedestrian-Aware Cyber-Physical Optimisation of Hybrid Propulsion Systems Using a Fuzzy Adaptive Cost Map and Bees Algorithm" by Li, Liu, Wang, Gu, Zhou, Wen, Pham and Xu proposes a system to reduce hybrid vehicle emissions on pedestrians. The system uses a combination of the BA and a fuzzy adaptive cost map to optimise the rule-based power-split parameters in a Plug-in Hybrid Propulsion system.

Finally, the chapter "Surrogate Model-Assisted Bees Algorithm for Global Optimisation of Microwave Filters" by Lan, Qian, Castellani, Wang, Pham and Wang deals with a black-box optimisation problem where the objective function is unknown. The authors first constructed an approximate model of the unknown objective function. They then used the BA to generate candidate solutions. Only those solutions deemed promising by the surrogate model were subsequently investigated further using more accurate but computationally expensive means.

Our aim in compiling this book is for it to serve as a source of inspiration for readers who wish to explore the potential of the BA as an engineering optimisation tool. In addition to the small sample featured in the book, there are many more potential engineering applications of the BA. We warmly encourage researchers and practitioners to try the BA when they next require a simple and effective means to solve complex optimisation problems.

Birmingham, UK D. T. Pham
Exeter, UK Natalia Hartono

Acknowledgements

Our warm thanks go to the authors for their contributions to the book and, more importantly, to the development, application and dissemination of the Bees Algorithm. The field of nature-inspired optimisation is richer through their work.

Numerous people have added value to the book, in particular, the committee members of the International Workshop on the Bees Algorithm and its Applications, by critically reviewing and helping to improve submissions. We acknowledge the input of Prof. Doriana M. D'Addona, Dr. Mei Choo Ang, Dr. Luca Baronti, Dr. Marco Castellani, Dr. Mario Caterino, Prof. Ahmed Haj Darwish, Dr. Hector de la Torre Gutierrez, Dr. Fabio Fruggiero, Dr. Asrul Harun Ismail, Dr. Kaiwen Jiang, Dr. Shafie Kamaruddin, Dr. Ebubekir Koç, Dr. Feiying Lan, Dr. Ji Li, Dr. Joey Lim, Dr. Jiayi Liu, Dr. Ernesto Mastrocinque, Dr. Michael Packianather, Prof. F. Javier Ramirez, Dr. Murat Sahin, Dr. Shahnorbanun Sahran, Mr. Hamid Furkan Suluova, Prof. Wenjun Xu, Dr. Baris Yuce, Dr. Sultan Zeybek and Mr. Feifan Zhao. Dr. Rory Adams and Dr. Michael Packianather, the keynote speakers at the workshop, set the scene for the book and deserve special credit.

The book was produced with the patient and expert support of Springer's Executive Editor Mr. Anthony Doyle and his colleagues Mr. Deivasigamani Arumugam and Mr. Vishnu Muthuswamy, to whom we express our sincere appreciation.

Contents

Bees Algorithm Development

Enhanced Bees Algorithm Implementing Early Neighbourhood Search with Efficiency-Based Recruitment

Michael S. Packianather, Azar Imanguliyev, and D. T. Pham

Abstract This chapter discusses modifications to the basic Bees Algorithm to enhance its performance. The proposed changes involve adding an early neighbourhood scanning step to determine promising areas from which to start the local search operation and making the recruitment process dynamic by varying the number of bees recruited to exploit a patch according to how efficiently it has been exploited. The enhanced version of the Bees Algorithm was applied to well-known complex continuous benchmark functions with high dimensions to find their optima. The results obtained from this study show that the new Bees Algorithm performed better than the basic version. The results were also compared with those of other swarm-based optimisation methods.

Keywords Basic Bees algorithm (Basic BA) · Enhanced BA with early neighbourhood scanning and efficiency-based recruitment (BA_{SE}) · Optimisation

1 Introduction

Population growth and resource depletion create tough competition in many areas of life. To address these issues, industries try to maximise their production, and optimisation techniques have become an important tool for efficient operation.

M. S. Packianather (✉)
School of Engineering, Cardiff University, Queen's Buildings,The Parade, Cardiff CF24 3AA, UK
e-mail: PackianatherMS@cardiff.ac.uk

A. Imanguliyev
Baku Engineering University, Khirdalan City, Hasan Aliyev Street, 120 AZ0101, Absheron, Baku, Azerbaijan
e-mail: azimanquliyev@beu.edu.az

D. T. Pham
Department of Mechanical Engineering, School of Engineering, College of Engineering and Physical Sciences, The University of Birmingham, Birmingham B15 2TT, UK
e-mail: d.t.pham@bham.ac.uk

D. T. Pham and N. Hartono (eds.), *Intelligent Engineering Optimisation with the Bees Algorithm*, Springer Series in Advanced Manufacturing,
https://doi.org/10.1007/978-3-031-64936-3_1

Optimisation is a process of seeking the values of variables to find an optimal solution for the optimisation problem that needs to be maximised or minimised. There are various types of optimisation techniques available in the literature. These techniques can be classified in many different ways. One such method is to classify based on their variables which could be continuous or discrete. Another classification method is according to whether an optimisation technique is deterministic and stochastic. To solve problems in polynomial time, deterministic optimisation techniques are used. On the other hand, there are optimisation problems that cannot be solved in polynomial time. Stochastic optimisation techniques are utilised to solve these types of problems. Many stochastic optimisation techniques, such as the Genetic Algorithm [8], Evolutionary Programming [2], Particle Swarm Optimisation [7], The Ant Colony Optimisation [6], and the Bees Algorithm [10, 12, 13, 16–18, 21, 22], were inspired by nature.

2 Optimisation Techniques

The mathematical technique concerned with finding the "best" solution for a problem, where the "best" refers to the fittest solution in the solution space, is called optimisation. In many fields, such as physics, chemistry, medicine, manufacturing or economic analysis, various optimisation techniques have been used. However, there is no optimisation technique that is suitable for every problem [20]. A block diagram of the optimisation process is given in Fig. 1 [5].

Moving from the real-world problem to the algorithm, the model or solution technique is called analysis. Here, the main task is eliminating non-crucial details and focusing on important elements. Moving from the algorithm, model, and solution technique to the computer implementation is called numerical methods, and from computer implementation back to the algorithm, model, and solution technique is called verification.

Finally, moving from the algorithm, model, and solution technique to real-world problems is called validation and sensitivity analysis. In this step, the obtained results are compared with the real world, and in case of failure, the process goes to the next cycle.

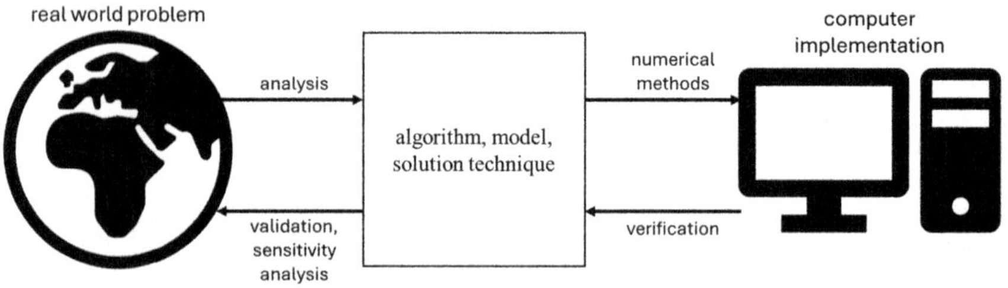

Fig. 1 Block diagram of the optimisation process [adapted from Chinneck [5]]

The goal of optimisation is to maximise or minimise the objective function concerning constraints and search space. An example is given below:

Given:

Function:

f (X) defined as f: A $\rightarrow \Re^n$

X $\in$ A and A is subset of n dimensional Euclidian space $\Re^n$

Constraints:

Inequality constraints: $a_i(X) \leq 0 \quad i = 1, 2 \ldots m$.

Equality constraints: $b_i(X) = 0, \quad i = 1, 2 \ldots p$.

Sought:

Maximisation: max $\in$ A such that f (max) $\geq$ f (X) for all X $\in$ A.

Minimisation: min $\in$ A such that f (min) $\leq$ f (X) for all X $\in$ A.

where f(X) is called the "objective function", f is called the "search space" or "parameter space", and each element of A is called the "candidate solution" [4]. Candidate solutions are tested in the objective function to find an "optimal solution". An optimal solution is the maximised or minimised solution of an objective function.

3 The Basic Bees Algorithm

The Bees Algorithm is a stochastic optimisation technique inspired by the foraging behaviour of honey bees. The Bees Algorithm has both global exploration and local exploitation strategies that increase the success rate of the algorithm in finding the global optimum. To demonstrate its performance, the Bees Algorithm was implemented on several single and multiobjective functions. The basic Bees Algorithm has undergone many improvements since it was introduced in 2005 by Pham and colleagues. Most of the improvements were focused on the neighbourhood search site, such as an abandonment strategy, population and neighbourhood size change strategies. The other improvements were focused on parameter tuning and hybridisation of the basic Bees Algorithm with other well-known optimisation techniques, such as Ant Colony and Particle Swarm Optimisation techniques.

Although several modifications were introduced to the Bees Algorithm, there is still opportunity for further improvements. For example, the Bees Algorithm has certain weaknesses that were not studied properly, such as a poor initialisation stage, the absence of the memory and the number of parameters. Moreover, new neighbourhood search strategies can also be developed to make the Bees Algorithm more.

competitive. In this study, an improved neighbourhood search with recruitment of bees is proposed [11]. The working mechanisms of the basic Bees Algorithm will be described below.

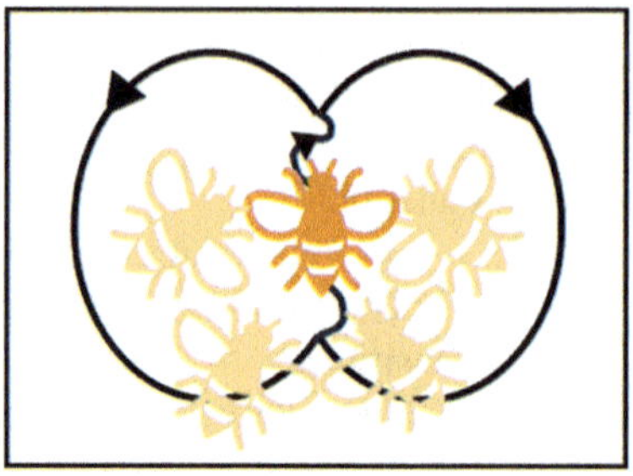

Fig. 2 Waggle dance of the scout bee [adapted from Bandgar et al. [3]]

3.1 Foraging Behaviour of Honey Bees

The Bees Algorithm is also one of the swarm intelligence-inspired algorithms that was developed by researchers at the Manufacturing Engineering Centre (MEC) in Cardiff University under the supervision of Pham [13] after observing bees foraging for nectar. A colony of honey bees explores a wide area around their hive to find a food source (nectar). Bees assigned for initial exploration are called scout bees. Scout bees can fly up to 11 km to find better flower patches [9, 14]. When a scout bee finds a food source, its job is to share information regarding the discovered patch with bees waiting in the hive. After delivering nectar to the hive, scouts go to a special area (dance floor) in front of the hive and perform eight shape movements, also known as the 'waggle dance' [14]. The waggle dance contains information about the direction, distance and quality of the flower patch found by the bee [19]. A waggle dance is shown in Fig. 2. The relation between the duration of the dance and distance from the hive is a positive correlation, indicating that as the distance increases, the waggle run duration also increases [15]. Information related to the direction of the source is given in Fig. 3.

After the performance of the "dancers", the colony decides the number of bees that need to be assigned for the food source. More bees go to more promising patches for harvesting. Recruited bees monitor food levels on every patch and share this information with the colony when they return to the hive. Therefore, bees concentrate on better patches all the time, which makes the food gathering process much faster and more efficient. This behaviour of honey bees was computationally modelled as a search algorithm.

3.2 The Algorithm

The parameters and the pseudo-code of the algorithm are given below.

Number of scout bees (n),

Number of sites selected out of n visited sites (m),

Number of best sites out of m selected sites (e),

Number of bees recruited for best e sites (nep),

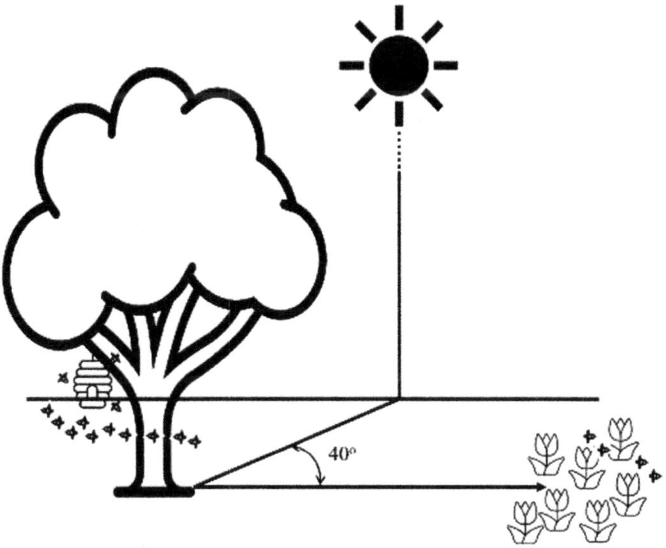

Fig. 3 Relation between dance and the direction of the food source [adapted from Seeley [14]]

Number of bees recruited for the other (m-e) selected sites (nsp),
Patch size around a selected best location (ngh).
The steps of the basic Bees Algorithm are given in Fig. 4.

Start
  Initialise population with random solutions.
  Evaluate fitness value of the population.

  While (stopping criterion not met)

    For each best patch
      Select the best m patches for neighbourhood search.
      Recruit bees for selected patches (more bees for best patches).
      Evaluate their fitness.
      Select the fittest bee value from each patch.
    End

  Assign remaining bees to search randomly and evaluate their fitness.

  End

Fig. 4 Steps of the basic Bees Algorithm

According to Fig. 4, the basic Bees Algorithm has six main steps. The first step is placing the "n" scout bees in a search space. In the following steps, the fitness values of the visited patches are evaluated. The patches with the highest fitness values are chosen as "selected sites" for neighbourhood search as step 3. In step 4, the algorithm performs a neighbourhood search on selected areas by assigning more scout bees to the elite sites 'e' and fewer scout bees to the nonelite best sites 'm-e'. In step 5, the scout bees around the best sites with the highest fitness values are selected as representative bees to form a new population. The remainder of the bees are assigned for random search to find potential solutions in step 6. This process continues until one of the stopping criteria has been met.

The proposed improvements to the basic Bees Algorithm will be presented in the next section.

4 Proposed Enhancements to the Bees Algorithm with Early Neighbourhood Scanning and Efficiency-Based Recruitment Strategies

In the literature, there are several optimisation algorithms with different search abilities and each of them has its own strengths and weaknesses. Considering the Bees Algorithm, which is the focus of this study, it has a random initialisation stage. Such initialisation has both advantages and disadvantages. The results produced by the algorithm are subject to this random initialisation process. This can be overcome by starting the search from a more promising location.

This study presents two modifications to the basic Bees Algorithm. These are early neighbourhood scanning and efficiency-based recruitment. The aim of these changes is to enhance the performance of the initialisation stage and make the neighbourhood search more effective, which will empower the overall performance of the algorithm on high-dimensional problems. The steps and flow chart for the proposed version of the enhanced Bees Algorithm called BA_{SE} are given in Figs. 5 and 6.

4.1 The Early Neighbourhood Scanning Strategy

The Early Neighbourhood Scanning Strategy-based Bees Algorithm starts with a random initialisation of n scout bees in search space, the same as in the basic Bees Algorithm. The initialisation stage in the basic BA considers a list of all random points visited by each scout bee. This may not contain enough information about the surrounding space. To obtain a better view of the vicinity of the point and to start a neighbourhood search from more promising patches, an early neighbourhood scanning stage is introduced during the first scouting process to preselect some of the

Start
Initialise population with random solutions.
Do (early neighbourhood search for each random solution)
Evaluate fitness values of each neighbourhood.
End

While (stopping criterion not met)
  For each best patch
    Select sites for neighbourhood search.
    Recruit bees for selected sites (using normal strategy + efficiency based enhancement) and evaluate fitnesses.
    Select the fittest bee from each patch.
  End
  Assign remaining bees to search randomly and evaluate their fitnesses.
End

Fig. 5 Steps of BA_{SE}

promising patches. During the first scouting process, the bees will go to random positions in the search space, and their positions will be evaluated and ranked according to their fitness values ready for neighbourhood search. Then, the neighbourhood search is carried out with a minimum number of scout bees to avoid affecting the computational time of the algorithm too much by increasing the number of iterations. The proposed solution, which starts the search from more promising patches, will inevitably lead to the discovery of better sites from where the local search will be carried out because if the algorithm starts its search from an advantageous position, it will have a better opportunity to converge to the global optimum.

In addition, another improvement is proposed below, which is efficiency-based recruitment for each best patch to increase the performance of the algorithm.

4.2 *Efficiency-Based Recruitment Strategy*

Efficiency characterises how well the time, cost and effort used for the implementation of a task or job compare to that achieved by alternative methods. This term has widely varying meanings and applications in different disciplines. In particular, for engineering, it can be generalised such that efficiency is a capability of producing a specific outcome effectively with a minimum amount or quantity of waste, cost, or unnecessary work.

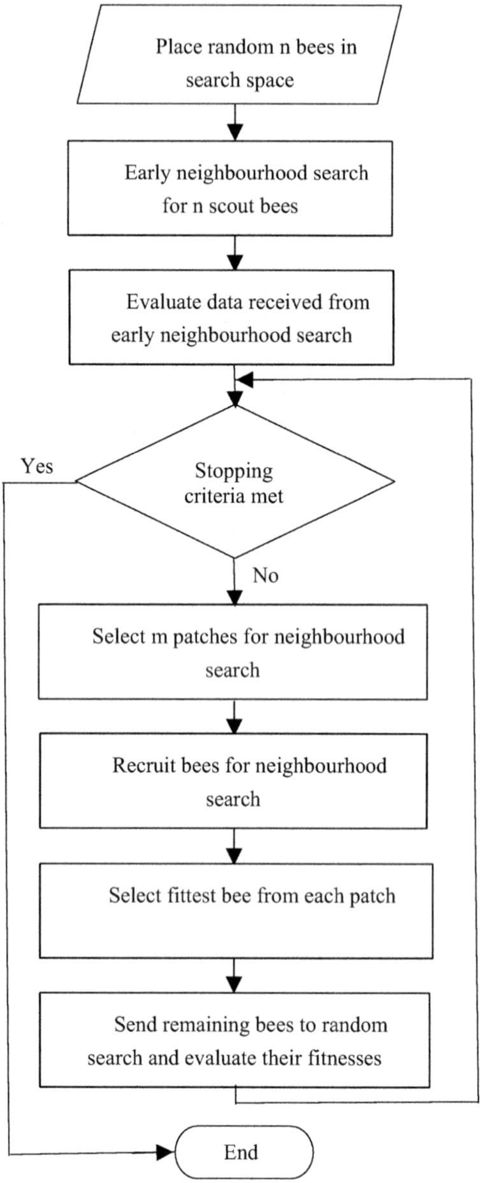

Fig. 6 Flow chart of BA_{SE}

Efficiency can be shown as a percentage of what ideally could be achieved. The efficiency of any work in its simplest form can be calculated with the formula below:

$$Efficiency = \frac{Output}{Input} \times 100 \tag{1}$$

Efficiency-based recruitment for neighbourhood search strategy is the second step of the proposed Bees Algorithm. In this stage, the fitness values of each patch are evaluated to choose the "m" best patches to start the neighbourhood search. The neighbourhood search process is performed as in the basic Bees Algorithm with the addition of efficiency-based recruitment.

The number of recruited bees for the neighbourhood search changes dynamically according to the efficiency of the related sites where the number of bees around elite (nep) and nonelite best sites (nsp) are computed based on Eqs. 2 and 3, respectively.

$$nep_i = nep_{i-1} + nab \tag{2}$$

$$nsp_i = nsp_{i-1} + nab \tag{3}$$

where "*nab*" is the number of added bees according to the efficiency calculation and "*i*" is the number of iterations.

The number of added bees, *nab*, is computed based on the efficiency rate (ER) of the best sites after a predefined number of iterations β. The 'ER' for each selected patch is calculated based on the fitness value F and difference in fitness values ΔF, as given in Eq. 4.

$$ER^j = \frac{\Delta F_i^j - \Delta F_{min}}{\Delta F_{max} - \Delta F_{min}} \tag{4}$$

where "*i*" is the iteration number, "*j*" is the site number, $\Delta F_i^j = F_i^j - F_{i-\beta}^j$, and $\Delta F_{min} = min\{F_i - F_{i-\beta}\}$, $\Delta F_{max} = max\{F_i - F_{i-\beta}\}$, ER_j is the efficiency rate of patch "*j*". ΔF_{min} is the difference of less productive fitness values in the β iteration, and ΔF_{max} is the difference of most productive fitness values in the β iteration. Each patch is ranked according to their efficiency rate. Therefore, the number of recruited bees around each site changes according to the ranks (grouped A to E), as given in Table 1.

Finally, the remaining scout bees are assigned randomly to carry out a global search. The process will run until the stopping criteria are met. The stopping criteria for the proposed version of the algorithm are as follows:

Table 1 The patch range and needed numbers of bees

Range of *ER*	Group Type	Required Bees
$0.0 <= ER <= 0.2$	E	0 Bees
$0.2 <= ER <= 0.4$	D	+ 1 Bees
$0.4 <= ER <= 0.6$	C	+ 2 Bees
$0.6 <= ER <= 0.8$	B	+ 3 Bees
$0.8 <= ER <= 1.0$	A	+ 4 Bees

- The global optimum was found with an acceptable error rate (ErrR) (in this study, the error rate was chosen as ErrR < 0.0001).
- Maximum number of Evaluations (in this study, this value is chosen as 5,000,000).
- Number of repetitions of the global optimum (in this study, this value is chosen as 100).

5 Experiments, Results and Discussion

To measure the performance of the enhanced Bees Algorithm, some well-known continuous type benchmark problems were selected. Each of these functions has different characteristics, so the obtained results illustrate the strengths and weaknesses of the algorithm in different situations. The algorithm was run a hundred times for each function. The results were compared with the basic Bees Algorithm (BA) and other well-known optimisation techniques, such as Particle Swarm Optimisation (PSO), Evolutionary Algorithm (EA) and Artificial Bee Colony (ABC).

The Bees algorithm requires a number of parameters to be set manually for each benchmark function. Furthermore, the number of recruited bees for early neighbourhood scanning and 'β' for efficiency-based recruitment must be predefined in the proposed version of the Bees Algorithm. In this study, the number of recruited bees for early neighbourhood scanning and 'β' were defined as 2 and 10, respectively. The other parameters to run the proposed algorithm to solve different benchmark problems are given in Table 2 [1].

The performance of the proposed algorithm was assessed according to the accuracy, and the average evaluation numbers and results were compared to well-known optimisation techniques. These are given in Tables 3 and 4. Experimental results for PSO, EA and ABC were extracted from Ahmad [1].

The accuracy of the algorithms was computed based on the average absolute differences of the best results of a hundred runs. According to this approach, the more accurate results are closer to zero.

Goldstein-Price 2D: The expected optimum result for the function is 3. The average result obtained from the basic BA for a hundred runs was found to be 3.0005. The result received from BA_{SE} on the same problem was found to be 3.0007. The basic BA used an average of 504 evaluation numbers to find that result, whereas the average of the new algorithm evaluation numbers was 21,496. Both algorithms produced similar average global optima. In this case, the performance of the basic BA was not improved on the given problem by applying the proposed enhancements, and consequently, the average number of evaluations used by the BA_{SE} was significantly greater than the number of evaluations used by the basic BA.

Schwefel 2D: The expected optimum for the function is -837.97. The average global optimum obtained from one hundred runs of the basic BA and BA_{SE} were -838.1192 and -837.9704, respectively. The basic BA used an average of 250,049 evaluation numbers to find that optimum, whereas the average for the new algorithm's evaluation was 338,600. Please note that this test function has complex topography,

Table 2 The parameters to run the BA_{SE} on different benchmark functions

Bees Algorithm parameters	Function									
	Goldstein & Price (2D)	Schwefel (2D)	Schaffer (2D)	Rosenbrock (10D)	Sphere (10D)	Ackley (10D)	Rastrigin (10D)	Martin & Gaddy (2D)	Easom (2D)	Griewank (10D)
n	10	10	100	15	10	100	100	10	100	100
m	3	2	4	8	7	8	3	5	4	40
nsp	2	5	10	10	20	10	20	10	10	10
e	1	1	2	5	1	1	1	1	2	20
nep	13	6	30	30	30	20	40	30	30	30
ngh	0.005	0.5	3	0.0015	0.05	0.7	0.01	0.1	0.5	1.5

Adapted from Ahmad [1] with permission

Table 3 Accuracy of the proposed algorithm compared with other well-known optimisation techniques

No	Functions	PSO		EA		ABC		BA		BA_{SE}	
		Average Absolute Difference	Standard Deviation	Average Absolute Difference	Standard Deviation	Average Absolute Difference	Standard Deviation	Average Absolute Difference	Standard Deviation	Average Absolute Difference	Standard Deviation
1	Goldstein & Price (2D)	0.0000	0.0000	0.0000	0.0000	0.0000	0.0001	0.0005	0.0006	0.0007	0.0008
2	Schwefel (2D)	4.7376	23.4448	4.7379	23.4448	0.0000	0.0000	0.1492	0.7679	0.0004	0.0057
3	Schaffer (2D)	0.0000	0.0000	0.0009	0.0025	0.0000	0.0000	0.0096	0.0018	0.0009	0.0029
4	Rosenbrock (10D)	0.5998	1.0436	61.5213	132.6307	0.0965	0.0880	0.0003	0.0003	0.0002	0.0003
5	Sphere (10D)	0.0000	0.0000	0.0000	0.0000	0.0000	0.0000	0.0003	0.0003	0.0001	0.0001
6	Ackley (10D)	0.0000	0.0000	0.0000	0.0000	0.0000	0.0000	0.0294	0.0477	0.0001	0.0028
7	Rastrigin (10D)	0.1990	0.4924	2.9616	1.4881	0.0000	0.0000	0.005	0.02	0.0002	0.0003
8	Martin & Gaddy (2D)	0.0000	0.0000	0.0000	0.0000	0.0000	0.0000	0.0000	0.0003	0.0000	0.0003
9	Easom (2D)	0.0000	0.0000	0.0000	0.0000	0.0000	2.0096	0.3	0.23	0.0000	0.0003
10	Griewank (10D)	0.0008	0.0026	0.0210	0.0130	0.0052	0.0078	0.3158	0.1786	0.0049	0.0019

Table 4 Average evaluation of the proposed algorithm compared with other well-known optimisation techniques

No	Functions	PSO		EA		ABC		BA		BA_{SE}	
		Avg. evaluations	Std Dev	Avg. evaluations	Std Dev	Avg. evaluations	Std Dev	Avg. evaluation	Std Dev	Avg. evaluations	Std Dev
1	Goldstein & Price (2D)	3262	822	2002	390	2082	435	504	211	21,496	36,855
2	Schwefel (2D)	84,572	90,373	298,058	149,638	4750	1197	250,049	680	338,600	0
3	Schaffer (2D)	28,072	21,717	219,376	183,373	21,156	13,714	121,088	174,779	112,430	66,120
4	Rosenbrock (10D)	492,912	29,381	500,000	0	497,728	16,065	935,000	0	148,193	116,904
5	Sphere (10D)	171,754	7732	36,376	2736	13,114	480	285,039	277,778	95,644	89,997
6	Ackley (10D)	236,562	9119	50,344	3949	18,664	627	910,000	0	236,299	123,325
7	Rastrigin (10D)	412,440	67,814	500,000	0	207,486	57,568	885,000	0	53,935	44,779
8	Martin & Gaddy (2D)	1778	612	1512	385	1498	329	600	259	15,888	16,554
9	Easom (2D)	16,124	15,942	36,440	28,121	1542	201	5280	6303	1120	1345
10	Griewank (10D)	290,466	74,501	490,792	65,110	357,438	149,129	4,300,000	0	316,443	97,830

so it is difficult to converge to the global optimum, but the Bees Algorithm with both global and local search found the optimum with no problem. However, BA_{SE} with early neighbourhood scanning and enhanced local search was more accurate on the given task.

Schaffer 2D: The expected optimum for the function is 0. The average of a hundred global optimums of the basic BA was 0.0096. The corresponding result obtained from the proposed version of the Bees Algorithm (a hundred runs) was 0.0009. The basic BA used an average of 121,088 evaluation numbers to find that result, whereas the average for the BA_{SE} was 112,430. Hence, BA_{SE} performed more accurately and faster than the basic BA on this optimisation problem.

Rosenbrock 10D: The expected answer is 0. The average global optimum obtained from the basic BA (a hundred runs) was 0.0003. The result received from BA_{SE} was 0.0002. The basic BA used an average of 935,000 evaluations to find that result, whereas the average for the new algorithm was 148,193. The basic BA found a less accurate global optimum, whereas the proposed version of the BA produced a more accurate result. The performance of the proposed algorithm for the given problem was increased. This is related to the extra initialisation during the early phases of the proposed algorithm.

Sphere 10D: The expected optimum result for this function is 0. The average result obtained from a hundred runs of the basic BA on the given function was 0.0003. The corresponding result obtained by the proposed version of the Bees Algorithm for 100 runs was 0.0001. The basic BA used an average of 285,039 evaluations to find that result, whereas the average of the proposed algorithm's evaluation numbers was 95,644. In this case, the performance of the basic BA was increased significantly by BA_{SE} for the given function.

Ackley 10D: The expected global optimum for the function is 0. The average optimum result obtained from 100 runs of the basic BA on the given function was 0.0294. The corresponding result received from the proposed version of the Bees Algorithm was 0.0001. The basic BA used an average of 910,000 evaluations to find that result, whereas the average number of evaluations needed by the new algorithm was 236,299. According to the experimental results, the performance of BA_{SE} on the Ackley function was significantly better than that of the basic BA. Again, it is because of the complex search space of the function that makes it difficult for algorithms to converge to the global optimum with the standard approach. Thus, the introduced enhancements empowered the Bees Algorithm to find a more accurate solution.

Rastrigin 10D: The global optimum of the function is 0. The average optimum obtained from the basic BA for a hundred runs was 0.005, and the basic BA used an average of 885,000 evaluations to find that value. The corresponding result received from the proposed version of the Bees Algorithm was more accurate (0.0002), and the average of the new algorithm's evaluations was only 53,935. The Rastrigin function is highly multimodal, which makes it very difficult for global optimisation algorithms to find an optimum. Even the basic BA, with local and global search strategies, was not very accurate on the Rastrigin function. However, the BA_{SE} with improved local (efficiency-based recruitment) and global search (early neighbourhood scanning) strategies was successful on this problem.

Martin & Gaddy 2D: The expected optimum for this function is 0. The experimental result obtained from the basic BA (a hundred runs) was 0.000 with 600 evaluations. BA_{SE} found the same result; however, the number of evaluations was too high (15,888). This is related to the structure of the new algorithm because it is not necessary to perform extra calculations (efficiency rate, early neighbourhood scanning) for such "easy" functions where it will only increase the number of evaluations.

Easom 2D: The expected optimum result for this function is -1. The basic BA performance on the given function was not satisfactory. An average of one hundred results obtained from the basic BA was 0.7, while the basic BA used an average of 5280 evaluations. On the other hand, the results obtained from BA_{SE} were better. The respective results received from BA_{SE} were 1 and 1120 (evaluations). Easom is another hard optimisation problem. BA_{SE} performed better than the basic BA on this function.

Inverted Griewank 10D: The expected global optimum for this function is 10. The average result obtained from the basic BA for a hundred runs was 9.6842. The corresponding result received from the proposed version of the Bees Algorithm was 9.9951. The basic BA used an average of 4,300,000 evaluations to find that result, whereas the average of the new algorithm's evaluation numbers was 316,443. The experimental results obtained from the basic BA on the given function were not satisfactory. Although the average global optimum was close to the expected optimum, the number of evaluations to obtain that result was extremely high. However, the BA_{SE} performed significantly well on the Griewank function. Both the average optimum and number of evaluations for the proposed Bees Algorithm were better than the basic BA's corresponding results. The proposed version of the algorithm's number of evaluations is still too high.

A further t test was utilised to perform statistical analysis of the algorithm where the confidence level was selected to be 95% ($\alpha < 0.05$). Based on the observed results (Table 5), the proposed algorithm is statistically significantly better than the basic BA on all benchmark functions.

Overall, the results illustrate that the BA_{SE}'s performance on complex high-dimensional functions is better than that on lower-dimensional ones. Although the proposed algorithm finds an accurate global optimum, the number of evaluations needed to obtain that result is relatively high. This is due to extra computation performed in the proposed version of the algorithm.

According to the 'no free lunch' theorem, if an algorithm performs well on a certain class of problems, then it necessarily pays for that with degraded performance on the set of all remaining problems [20].

Table 5 The statistical analysis between the proposed BA_{SE} and the basic BA

No	Function	Significance between the basic BA and the improved Bees Algorithm (BA_{SE})	
		Significant ($\alpha < 0.05$)	α
1	Goldstein & Price (2D)	Yes	0.0004
2	Schwefel (2D)	Yes	3.698 E-18
3	Schaffer (2D)	Yes	6.472 E-52
4	Rosenbrock (10D)	Yes	0.0045
5	Sphere (10D)	Yes	1.9650 E-06
6	Ackley (10D)	Yes	7.150 E-08
7	Rastrigin (10D)	Yes	0.0085
8	Martin & Gaddy (2D)	Yes	0.0010
9	Easom (2D)	Yes	0.0024
10	Griewank (10D)	Yes	0.019

6 Conclusion

In this study, two novel enhancements have been presented for the Bees Algorithm. The basic Bees Algorithm was improved both with early neighbourhood scanning in the initialisation stage and efficiency-based recruitment in the neighbourhood search stage. The proposed algorithm has been successfully applied to continuous type benchmark functions and compared with other well-known optimisation techniques.

To test the performance of the proposed algorithm, the following approaches were utilised: accuracy analysis, average evaluation and t test.

According to the accuracy analysis and the average evaluation, the proposed algorithm performed better on higher-dimensional rather than lower-dimensional functions. Finally, the statistical significance of the proposed algorithm was computed with a t test, and the results were compared with the basic Bees Algorithm. Based on t test results, it can be claimed that the proposed algorithm is statistically significantly better in performance than the basic Bees Algorithm.

Acknowledgements The authors would like to thank the Azerbaijan Republic, ASTUTE 2020 project funded by the ERDF and MANUELA project funded by the EU H2020 for their support.

References

1. Ahmad S (2012) A study of search neighbourhood in the Bees algorithm. PhD Thesis, Institute of Mechanical and Manufacturing Engineering, Cardiff University
2. Back T, Fogel DB, Michalewicz Z (1997) Handbook of evolutionary computation. Computational Intelligence Library, Taylor and Francis, New York

3. Bandgar PS, Pondkule KA, Khyade VB (2018) Fascinating communication in honey bees. Int J Curr Microbiol App Sci 7:3704–3718
4. Blondin J (2009) Particle swarm optimisation: a tutorial. Availaible from: http://cs. armstrong. edu/saad/csci8100/pso tutorial. pdf, 34
5. Chinneck JW (2000) Practical optimisation: a gentle introduction. Carleton University
6. Dorigo M, Maniezzo V, Colorni A (1996) Ant system: optimisation by a colony of cooperating agents. IEEE Trans Syst, Man Cybern, Part B (Cybern) 26(1):29–41
7. Eberhart RC, Kennedy J (1995) A new optimiser using particle swarm theory. In: Proceedings of 6th international symposium on micromachine and human science, pp 39–43
8. Goldberg DE (1989) Genetic algorithms in search, optimisation, and machine learning. Addison-Wesley Publishing Company, Inc., Massachusetts, USA
9. Gould JL, Gould CG (1988) The Honey Bee. Scientific American Library, 144 New York
10. Hartono N, Ramírez FJ, Pham DT (2022) Optimisation of robotic disassembly plans using the Bees algorithm. Robot Comput-Integr Manuf 78:102411
11. Imanguliyev A (2013) Enhancements for the Bees algorithm. PhD Thesis, Cardiff University
12. Liu J, Zhou Z, Pham DT, Xu W, Yan J, Liu A, Ji C, Liu Q (2018) An improved multiobjective discrete bees algorithm for robotic disassembly line balancing problem in remanufacturing. Int J Adv Manuf Technol 97(9):3937–3962
13. Pham DT, Ghanbarzadeh A, Koç E, Otri S, Rahim S, Zaidi M (2005) The Bees algorithm—technical report. Manufacturing Engineering Centre, Cardiff University, Cardiff
14. Seeley TD (1995) The social physiology of honey Bee colonies: the wisdom of the hive. Harvard University Press, London
15. Seeley TD, Visscher PK, Passino KM (2006) Group decision making in honey bee swarms. Am Sci 94(3):220–229
16. Shouran M, Anayi F, Packianather M, Habil M (2021) Load frequency control based on the bees algorithm for the Great Britain power system. Designs 5(3):50
17. Shouran M, Anayi F, Packianather M (2022) Design of sliding mode control optimised by the Bees algorithm for LFC in the Great Britain power system. Mater Today: Proc 52:937–943
18. Shouran M, Anayi F, Packianather M, Habil M (2022) Different fuzzy control configurations tuned by the Bees Algorithm for LFC of two-area power system. Energies 15(2):657
19. Talbi E (2009) Metaheuristics: from design to implementation. Wiley
20. Wolpert DH, Macready WG (1997) No free lunch theorems for optimisation. IEEE Trans Evol Comput 1(1):67–82
21. Xu W, Tang Q, Liu J, Liu Z, Zhou Z, Pham DT (2020) Disassembly sequence planning using discrete Bees algorithm for human-robot collaboration in remanufacturing. Robot Comput-Integr Manuf 62:101860
22. Zeybek S, Pham DT, Koç E, Seçer A (2021) An improved bees algorithm for training deep recurrent networks for sentiment classification. Symmetry 13(8):1347

Improving the Bees Algorithm Using Gradual Search Space Reduction

Turki Albakr and D. T. Pham

Abstract The Bees Algorithm is a well-known metaheuristic optimisation method that has been applied in many disciplines with noticeable success. For example, it has been applied to machine scheduling, training artificial neural networks (ANNs) for pattern recognition, and the design of mechanical structures. There have been many attempts to improve the Bees Algorithm's performance to tackle some of its weaknesses with more focus on the local search stage. This research attempts to improve the Bees Algorithm with more attention directed to stages other than the local search. The suggested method employs an adapted notion of the regional elimination method to achieve the abandonment and reduction of the search space within the Bees Algorithm. To assess the performance, the proposed method was tested on 24 benchmark functions, and it was applied to two engineering problems. The results obtained indicate a statistically significant improvement.

Keywords Engineering design · Optimisation · Bees algorithm · Bracketing · Metaheuristic algorithm · Region elimination · Swarm intelligence · ABC · PSO

1 Motivation

The Bees Algorithm belongs to a relatively new category of algorithms called swarm intelligence (SI) algorithms that mimic the social behaviour of natural creations living together in swarms. SI algorithms are nontraditional algorithms that were proposed to solve types of problems that are infeasible to solve using deterministic methods due to their level of complexity and massive search space. The SI algorithms involve

T. Albakr (✉)
Computing and Informatics College, Saudi Electronic University, Riyadh, Saudi Arabia
e-mail: t.albaker@seu.edu.sa

D. T. Pham
Department of Mechanical Engineering, College of Engineering and Physical Sciences, The University of Birmingham, Birmingham B15 2 TT, UK
e-mail: d.t.pham@bham.ac.uk

D. T. Pham and N. Hartono (eds.), *Intelligent Engineering Optimisation with the Bees Algorithm*, Springer Series in Advanced Manufacturing,
https://doi.org/10.1007/978-3-031-64936-3_2

a great deal of stochastic features. These algorithms experience slow convergence rates, becoming trapped in local optima and long computational times, particularly for hard problems [1, 2]). This is due to the stochastic nature of these algorithms and the high number of iterations that must be performed. The bee algorithm is not an exception [3–6]. However, similar to many other population-based SI algorithms, BA experiences some problems due to its stochastic nature. For example, being trapped in local optima and the slow convergence to the optimum due to lack of search guidance. Meanwhile, while BA performs reliably on noisy problems [7], its performance will degrade down on smooth unimodal functions [7, 8]. Moreover, BA contains a great number of control parameters that need to be tuned, which implies that the process of parameter tuning by itself is an optimisation problem. Hence, this research aims to enhance the Bees Algorithm. Since the proposal of BA in 2005, it has been subject to many improvement attempts. One of the attempts proposed by the original developer of the BA, Pham et al. (2008c), incorporates neighbourhood shrinking and abandonment procedures into the basic BA (BBA). Hussein et al. (2014) also attempted to improve the initialisation phase of the BA by incorporating Levy flight in the neighbourhood search. Yuce et al. (2013) proposed the use of an adaptive neighbourhood search that shrinks and enlarges according to fitness values for both the shrinking and abandonment procedures. Another attempt, by Shatnawi et al. (2013), introduced the BA with two types of memory-based lists (local and global); here, the researchers claim a minimum 59% improvement over the BA. The attempt, proposed by Ghanbarzadeh (2007), aims to increase the exploitability of the neighbourhood search area with a large patch size parameter and decrease accordingly. Neighbourhood search area around the selected patch if no improvement is achieved. We can conclude from the above survey of BA enhancement attempts that the majority of these efforts focused more on the local search space. This indicates a gap in improvements applied to global search. Additionally, the careful inspection of the BA indicates that the local search in the BA works as a refinement medium for the global search.

2 Bees Algorithm

The Bees Algorithm (BA) was originally developed by Pham and his team in 2005 [9]. BA is one of the important nature-inspired algorithms motivated by the foraging behaviour of bees in nature. BA begins with the initialisation stage where scout bees are randomly distributed to search for the best sites. Those visited sites are ranked according to their fitness. The best sites are selected to search in their vicinity according to a certain neighbourhood distance. Next, more intensified neighbourhood searches are taking place, and more bees are recruited for neighbourhood searches around the top of the fittest sites or the elite sites; however, fewer bees are recruited for sites with lower fitness. The global search will start when the remaining scout bees are assigned to search randomly again in the search area. The new population consists of fittest bees from the local search and the global search. The search process proceeds until the stopping criteria are satisfied. The stopping criteria are usually reached when either the exact optimum or near the optimum solution is found or when a certain number of iterations are reached.

The steps for the Bees Algorithm in its basic form:

1. Initialise the scout population with random solutions.
2. Evaluate the fitness of the population.
3. While (stopping criterion not met)/Forming the new population.
4. Select sites for the neighbourhood search.
5. Recruit bees for selected sites (more bees for fittest sites) and evaluate fitness.
6. Select the fittest bee from each patch.
7. Assign remaining bees to search randomly and evaluate their fitness.
8. End While.

3 Gradual Search Space Reduction Method

The proposed method in this research is inspired by region elimination methods [10, 11]. Originally, the region elimination method was used to solve unimodal problems where there is only one global optimum and no local optima. In this research, the original region elimination method was adapted within BA to work for problems with multiple local and global optimums. Rather than routinely extracting the sample from the exact search space. The consecutive elimination and reduction of the search space helps to achieve smarter and focused sample extraction. However, as the global optimum location varies with different optimisation problems, the elimination and reduction of the search space must vary. Hence, different scenarios of search space elimination and reduction are introduced. To explain how the proposed method works, first, the original region elimination method will be presented.

3.1 Regional Elimination

The core concept of the region elimination method is to consecutively eliminate parts of the search space until the exact minimum is found. The following are the steps of this method:

1. A search space region is specified, for example (a, b).
2. Two points are selected, x 1 and x 2, where x 1 < x 2.
3. f(x1) is evaluated, and
   - If f(x1) < f(x2) (Fig. 1c, the minimum cannot exist beyond x2 in the period (x2, b), then segment (x2, b) is abandoned from the search space.
   - If f(x1) > f(x2) (Fig. 1b, the minimum cannot exist before x1 in period (a, x1). Therefore, segment (a, x 1) is abandoned from the search space [10]. See Fig. 1 below.

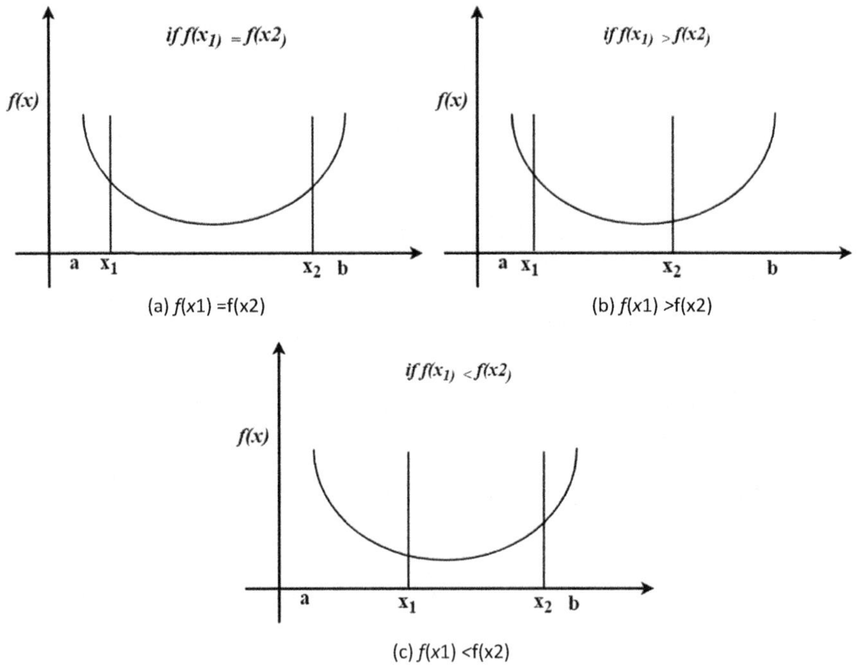

Fig. 1 Region elimination method

3.2 *The Bees Algorithm with Search Space Reduction (BASR)*

To adapt the region elimination to this proposed method, some essential factors are introduced. The region elimination method was originally designed for optimising unimodal functions, whereas the proposed method considers multimodal problems where more than one local or global optimum might exist. Moreover, the region elimination method is a deterministic numerical optimisation method used to find the exact minimum. However, given that the bee algorithm is a metaheuristic approximation method, the minimum or near minimum within the stipulated error margin will be sufficient. Hence, the following modifications are applied:

- The initial search space region (a, b) is specified equal to the whole search space specified for the domain of the problem being tested, where b > a.
- The number of points selected should be equal to the number of the initial sample specified by parameter n of the bee algorithm (x1, x 2..., xn)
- The elimination interval

$$L = \Delta x = (b - a)/n \tag{1}$$

where n is the number of initial samples. Then,

$$F(x) = X = (x1, x2 \ldots xn) \tag{2}$$

is evaluated, and if

$$F(x) - F_{optm} > 0.001 \tag{3}$$

where F(x) is the calculated optimum, F_{optm} is the standard optimum for the problem being tested, and 0.001 is the stipulated error margin. The segment L is eliminated from the search space from both ends. The new search space is calculated as follows:

$$(b - L, a + L)$$

The process will continue until the search yield values close to the optimum within the stipulated error margin where $F(x) - F_{optm} \leq 0.001$; then, the search process is terminated. If too many unsuccessful searches are attempted, the width of the search area will almost equal zero, and the process will then be restarted from the original search space (a, b). If the maximum number of function evaluations is reached (500,000) without finding a solution within the stipulated error margin, the search is stopped.

Furthermore, to better use the global search, bees are assigned to search in different areas of the search space. There are five different search scenarios for the search space:

- searching the whole search space, which decreases gradually from both ends of the search space by the elimination factor L Eq. (2)
- searching after the first quarter (1/4) of the search space quarter from the two different ends of the search space, which gradually decreases from both ends by the elimination factor L.
- searching the area between the center and the right (b) end of the search space, then it is gradually decreased from both sides by the elimination factor L.
- searching only the area between the center and the left end (a) of the search space and is gradually decreased from both sides by the elimination factor L.
- searching the whole search space, which gradually decreases from the left end (a) only by elimination factor L.

The Steps for BASR algorithm:

1. Initialise the population with random solutions using the search space reduction technique
2. Evaluating the fitness of the population
3. While (stopping criterion not met) forming a new bee population
4. Select best and elite sites for neighbourhood search
5. Recruit bees around best and elite sites and evaluate fitness
6. Select the fittest bee from each site

7. Assign remaining bees to search randomly using the search space reduction technique with five different scenarios and evaluate their fitness
8. If (stopping criterion not met) reduce the search space by L
9. End while

3.3 Experiment Setup

To evaluate the performance of the proposed algorithm, twenty-four continuous benchmark functions are selected. The criteria for selecting benchmark functions for testing are designed to account for the varying topography of the search space, complexity, and modality. This is essential to ensure testing objectivity and reliability [12]. Complexity includes separability and dimensionality, where separability refers to the interdependency between function parameters. Nonseparable functions are mostly harder to solve than separable functions [12]. The dimensionality of a function demonstrates its number of decision variables or parameters. As the number of dimensions in a function increases, it becomes harder to optimise. Modality is a property related to the number of peaks in the search space. A function is unimodal if it has only one optimum; otherwise, with many global and/or local optima, it is regarded as multimodal [12]. The test functions used in this research are listed and classified according to their properties in Table 1. Although more than half of them are two-dimensional, they are predominantly multimodal and nonseparable, which adds to the difficulty in solving them.

The three key performance metrics used in this investigation are the accuracy of the fitness solution, the success rate, and the number of functions evaluated (NFE). The accuracy of the fitness solution refers to how close the optimum is to the standard optimum. It can also be described as the quality of the solution. The success rate refers to the number of times the algorithm was able to converge to the optimum within the maximum number of function evaluations permitted in all the runs. The number of function evaluations (NFE) denotes the number of times the benchmark function.

being tested were executed in every individual run. It can also be described as the speed with which the algorithm converges to the optimum. This is obvious, as the algorithm will need less time to converge if the function under testing is executed fewer times. The methodology of this research primarily considers the quality of the solution or accuracy and then looks at the success rate and the number of function evaluations (NFE). Hence, the main objective of his research should mainly contribute to the quality of the solution. The results of the test are compared to those of the basic bee algorithm (BBA) and two other well-known algorithms, PSO and ABC. for PSO, the relatively new version SPSO2011 will be used, and for ABC, the quick ABC variant qABC will be used [3, 13]. The algorithm should continue to run until the stopping criteria defined below are met:

Table 1 List of test functions and their properties

Functions	Differentiability	Separability	Scalability	Modality
Sphere (10D)	Yes	No	Yes	Unimodal
Rosenbrock (10D)	Yes	No	Yes	Unimodal
Quartic (30D)	Yes	Yes	Yes	Unimodal
Ackley (10D)	Yes	No	Yes	Multimodal
Schaffer (2D)	Yes	No	Yes	Multimodal
Easom (2D)	Yes	Yes	No	Multimodal
Rastrigin (10D)	Yes	Yes	Yes	Multimodal
Shekel (4D)	Yes	Yes	Yes	Multimodal
Langerman (10D)	Yes	No	Yes	Multimodal
Griewank (10D)	Yes	No	Yes	Multimodal
Branin (2D)	Yes	No	No	Multimodal
Sumpow (10D)	Yes	Yes	Yes	Unimodal
Bukin6 (2D)	No	No	No	Multimodal
Crossit (2D)	No	No	No	Multimodal
Drop (2D)	No	Yes	No	Multimodal
Shubert (2D)	Yes	Yes	No	Multimodal
Beale (2D)	Yes	No	No	Multimodal
McCorm (2D)	Yes	No	No	Multimodal
Camel6 (2D)	Yes	No	No	Multimodal
Bohal (2D)	Yes	No	No	Multimodal
Colville (2D)	Yes	No	No	Multimodal
Powersum (2D)	Yes			Unimodal
Salomon (2D)	Yes	No	Yes	Multimodal
Alpine (2D)	Yes	Yes	No	Multimodal

- either the global optimum found with acceptable error rate (ER) (chosen here as $ER <= 0.001$) or
- maximum number of function evaluations reached (stipulated in this research as 5,000,000).

3.4 Results and Discussion

The assessment of the performance of the proposed BASR algorithm involves a comparison with three other algorithms, namely, BBA, qABC and SPSO2011. The parameter settings for these algorithms are shown above in Tables 2, 3, 4 and 5. The analysis of performance according to the figures in Table 6 shows that the proposed BASR algorithm performed better than all other algorithms in at least 20 functions. The mean and standard deviation figures in Table 7 confirm the conclusion about the performance data, with the proposed algorithm achieving a better mean in 19 functions and a better standard deviation in 17 functions. However, in two functions, BASR reached the maximum NFE without finding the optimum within the stipulated error margin. Additionally, in the Shekel function, BASR achieved a better mean value even though qABC performed better and achieved a more accurate optimum than BASR. Despite the above consideration, BASR performance remains remarkable, especially if we consider the result in functions with complex topography and multipocketed landscapes, such as Rosenbrock, Ackley, Shaffer, Rastrigin, Griewank, and Drop, where BASR achieves a better optimum in all 50 runs. To assess the statistical significance of the performance of BASR, we need to look at the p value calculated using the Mann–Whitney significance test. Looking at Table 8, we can obviously observe that the figures confirm the good performance of BASR with a better p value in at least 18 functions, which comprise three-quarters of the functions. These findings support the conclusion that the gradual search space reduction technique used in the proposed BASR algorithm helped improve the performance of the Bees algorithm.

3.5 Success Rate and Number of Function Evaluation

Looking at Table 9, we can observe that the proposed BASR algorithm achieved the highest SR average among all the algorithms. It was able to achieve 100% SR in all functions except in three functions. Turning to Table 10, we notice another aspect of improved performance by BASR against qABC and BBA. However, the first glance at the performance of BASR against SPSO2011 indicates less improved performance. Although the figures show lower NFE figures by SPSO2011, which suggests a better performance, it cannot be taken as an indication of satisfactory performance since SPSO2011 ended the search earlier, while its obtained optimum does not fall within the stipulated error margin of 0.001. In fact, for the functions (Easom, Shekel, Langerman, Crossit, Drop, Shubert, and McCorm), the optimum was located far from the error margin. This phenomenon is referred to as premature convergence. This is also reflected in the mean and standard deviation figures in Table 11. Hence, if SPSO2011 figures were excluded, we would be able to notice the improved performance of BASR. Nevertheless, the above findings have yet to be confirmed by the statistical significance test using the Mann–Whitney test.

The p value statistical figures in Table 12 for BASR against BBA and qABC confirm that BASR improved significantly in terms of NFE. BASR was better than BBA in 19 test functions and better than qABC in 17 functions. This means that BASR performed faster than these two algorithms. However, with regard to SPSO2011, the p value figures were lower, suggesting less significant performance. BASR is better than SPSO2011 in only 14 functions. Although these p value figures still suggest better performance by BASR, this cannot be taken as an indication of satisfactory

Table 2 List of parameter values for testing BASR

No.	Functions	n	m	nsp	e	nep	ngh	Stlim
1	Sphere (10D)	11	4	10	2	30	0.003	10
2	Rosenbrock (10D)	11	4	10	2	30	0.003	10
3	Quartic(30D)	13	4	10	2	30	0.005	10
4	Ackley (10D)	11	4	10	2	30	0.00001	10
5	Schaffer (2D)	11	4	10	2	30	0.01	10
6	Easom (2D)	100	4	10	2	30	0.0009	10
7	Rastrigin (10D)	11	4	10	2	30	0.003	10
8	Shekel (4D)	1000	4	10	2	30	0.001	10
9	Langerman (10D)	1000	4	10	2	30	0.09	10
10	Griewank (10D)	17	10	10	2	30	0.001	10
11	Branin (2D)	15	10	10	2	30	0.01	10
12	Sumpow (10D)	20	4	10	2	30	0.0001	10
13	Bukin6 (2D)	10	4	10	2	30	0.00001	10
14	Crossit (2D)	15	4	10	2	30	0.05	10
15	Drop (2D)	11	4	10	2	30	0.01	10
16	Shubert (2D)	1000	4	10	2	30	0.0005	10
17	Beale (2D)	45	4	10	2	30	0.045	10
18	McCorm (2D)	10	4	10	2	30	0.05	10
19	Camel6 (2D)	100	4	10	2	30	0.005	10
20	Bohal (2D)	11	4	10	2	30	0.05	10
21	Colville (2D)	10	4	10	2	30	0.01	10
22	Powersum (2D)	10	4	10	2	30	0.01	10
23	Solomon (2D)	10	4	10	2	30	0.001	10
24	Alpine (2D)	13	4	10	2	30	0.001	10

performance by SPSO2011, as discussed earlier. It is just an aspect of the premature convergence phenomenon where algorithms abruptly stop processing without finding any satisfactory solution.

3.6 Testing with Engineering Problems

To further evaluate the performance of the proposed BASR algorithm, two engineering problems were selected to apply the proposed algorithm. These engineering problems are gear train problems, nonconstrained problems and tension/compression springs, constrained problems.

Table 3 List of parameter values for testing BBA

No.	Functions	n	m	nsp	e	nep	ngh	Stlim
1	Sphere (10D)	11	4	10	2	30	0.003	10
2	Rosenbrock (10D)	10	4	10	2	30	003	10
3	Quartic (30D)	13	4	10	2	30	005	10
4	Ackley (10D)	10	4	10	2	30	0.01	10
5	Schaffer (2D)	10	4	10	2	30	0.01	10
6	Easom (2D)	10	4	10	2	30	0.01	10
7	Rastrigin (10D)	10	4	10	2	30	0.03	10
8	Shekel (4D)	10	4	10	2	30	0.01	10
9	Langerman (10D)	10	4	10	2	30	0.09	10
10	Griewank (10D)	10	10	10	2	30	0.1	10
11	Branin (2D)	10	10	10	2	30	0.001	10
12	Sumpow (10D)	10	4	10	2	30	0.1	10
13	Bukin6 (2D)	10	4	10	2	30	0.0001	10
14	Crossit (2D)	10	4	10	2	30	0.1	10
15	Drop (2D)	10	4	10	2	30	0.01	10
16	Shubert (2D)	10	4	10	2	30	0.1	10
17	Beale (2D)	10	4	10	2	30	0.05	10
18	McCorm (2D)	10	4	10	2	30	0.5	10
19	Camel6 (2D)	10	4	10	2	30	0.05	10
20	Bohal (2D)	10	4	10	2	30	0.05	10
21	Colville (2D)	10	4	10	2	30	0.0005	10
22	Powersum (2D)	10	4	10	2	30	0.1	10
23	Solomon (2D)	10	4	10	2	30	0.01	10
24	Alpine (2D)	13	4	10	2	30	0.001	10

Table 4 qABC parameter settings

Parameter	Value
Population size	10
cycles (max number)	100
Employed bees ne	5
Onlooker bees ne	4
Random scouts	1
Stagnation limit for site (abandonment stlim)	200
ra	1

Table 5 SPSO2011 parameter settings

Parameter	Value
Population size	10
PSO cycles (max number) T	500,000
Connectivity	Default (3)
Maximum velocity	Default 1.1
C1	Default 1.1
C2	200
Wmax	Default 0.7
Wmin	Default 0.7

3.6.1 Gear Train Problem

The gear train design problem is a real-world problem that aims to minimise the gea r ratio to be particularly close to 1/6.931 [14]. The design variables here represent the number of teeth for every gear, restricted to values between 12 and 60 (Fig. 2).

To reduce the error value, the gear ratio should be as close as possible to 1/6.931 [15]. To calculate the error value:

$$Error\ value(\%) = \frac{Gear\ Ratio - \left(-\frac{1}{6.931}\right)}{\frac{1}{6.931}} \times 100\% \tag{4}$$

$$Gear\ Ratio = \frac{x_3 x_2}{x_1 x_4} \tag{5}$$

the objective function formulated as:

$$\min_x f(x) = \frac{1}{6.931} - \frac{x_3 x_2}{x_1 x_4} \tag{6}$$

The result of the experiment was taken out of 20 runs, and the algorithm was forced to conduct a minimum of 100 iterations to acquire a good result. The stopping criteria used are the same criteria used to evaluate the algorithm with benchmark functions, or 500,000 NFE. The selection of parameter values for the proposed algorithms was partially decided from the literature review of other algorithms based on BA and partially through trial and error. The results obtained by the test were compared with some of those found in the literature review for ABC [16], CS [17] Gandomi et al. 2013) and PSO-GA [18] (Table 13).

According to Table 14, the gear ratio solution achieved by PSO-GA, BASR, ABC, and CS obtains a gear train value of 0.14428 and an error value of 0.001%. This was the best result found when reviewing the literature on solving the gear train problem. The fact that the proposed algorithm achieved a level of performance within the best achieved thus far is an indication of a satisfactory performance by the proposed algorithm BASR. However, with regard to the speed of convergence, ABC was the

Table 6 Performance Figures for BASR, BBA, qABC and SPSO2011

No.	Functions	Result of 50 Runs		Result of 50 Runs		Result of 50 Runs	
		BASR	BBA	BASR	SPSO2011	BASR	qABC
1	Sphere (10D)	50	0	50	0	50	0
2	Rosenbrock (10D)	50	0	41	9	50	0
3	Quartic (30D)	50	0	49	1	50	0
4	Ackley (10D)	50	0	50	0	50	0
5	Schaffer (2D)	50	0	50	0	50	0
6	Easom (2D)	32	18	50	0	19	31
7	Rastrigin (10D)	50	0	50	0	50	0
8	Shekel (4D)	34	16	50	0	20	30
9	Langerman (10D)	27	23	50	0	36	14
10	Griewank (10D)	50	0	50	0	50	0
11	Branin (2D)	39	11	0	50	19	31
12	Sumpow (10D)	45	5	38	12	37	13
13	Bukin6 (2D)	38	12	48	2	42	8
14	Crossit (2D)	28	22	50	0	31	19
15	Drop (2D)	50	0	50	0	50	0
16	Shubert (2D)	44	6	50	0	14	36
17	Beale (2D)	33	17	28	22	34	16
18	McCorm (2D)	29	21	50	0	28	22
19	Camel6 (2D)	38	12	50	0	28	22
20	Bohal (2D)	50	0	50	0	50	0
21	Colville (2D)	38	12	25	25	47	3
22	Powersum (2D)	28	22	39	11	48	2
23	Solomon (2D)	50	0	50	0	50	0
24	Alpine (2D)	50	0	50	0	50	0
Total:		BASR=24 BBA=0		BASR=22 None = 1 SPSO2011 = 1		BASR=20 qABC=4	

fastest, as it needed only 60 NFEs to achieve its best result, which is the lowest among all, while CS and BASR needed 5000 and 9725 NFEs, respectively.

3.6.2 Tension/Compression Spring

The aim of the tension/compression spring problem is to reduce the spring steel wire volume, which means decreasing the weight of the spring [19]. More details of the tension/compression spring are included in Appendix B (Fig. 3).

The objective function of the tension/compression problem is:

Table 7 Mean and standard deviation for BASR, BBA, qABC and SPSO2011

No.	Functions	BBA		qABC		SPSO2011		BASR	
		Mean	Std.Dev.	Mean	Std.Dev.	Mean	Std.Dev.	Mean	Std.Dev.
1	Sphere (10D)	8.19E-04	1.27E-04	6.24E-04	2.49E-04	7.98E-04	1.44E-04	4.1E-32	1.31E-32
2	Rosenbrock (10D)	1.8E+00	9.23E-01	4.09E-01	8.85E-01	8.88E+00	4.36E+01	9.55E-04	3.72E-05
3	Quartic (30D)	3.7E+00	7.34E-01	5.23E-01	2.88E-01	1.18E-01	6.7E-02	4.18E-03	2.14E-02
4	Ackley (10D)	1.38E+01	1.25E+00	3.11E-03	3.05E-03	3.7E-01	6.35E-01	1.42E-14	2.25E-15
5	Schaffer (2D)	2.57E-04	2.61E-04	9.65E-04	8.04E-04	2.69E-04	2.6E-04	0.0E+00	0.0E+00
6	Easom (2D)	4.56E-04	2.64E-04	2.73E-04	3.28E-04	1.0E+00	2.74E-06	3.41E-04	2.29E-04
7	Rastrigin (10D)	1.33E+01	3.84E+00	6.22E-04	2.78E-04	1.93E+01	1.01E+01	0.0E+00	0.0E+00
8	Shekel (4D)	6.67E-04	1.96E-04	2.34E-02	1.61E-01	9.32E+00	3.44E-01	5.01E-04	2.26E-04
9	Langerman (10D)	2.74E-01	1.79E-01	3.5E-01	1.63E-01	7.06E-01	9.64E-05	3.21E-01	1.5E-01
10	Griewank (10D)	5.68E-01	7.78E-02	4.65E-02	2.19E-02	1.38E-01	8.28E-02	0.0E+00	0.0E+00
11	Branin (2D)	6.43E-04	1.96E-04	2.63E-04	2.76E-04	3.58E-07	0.0E+00	3.85E-04	2.95E-04
12	Sumpow (10D)	6.86E-04	2.44E-04	3.82E-04	1.12E-04	5.48E-04	2.73E-04	2.17E-04	3.05E-04
13	Bukin6 (2D)	1.59E-02	4.93E-03	2.81E-02	1.61E-02	7.35E-02	3.97E-02	1.02E-02	4.07E-03
14	Crossit (2D)	4.23E-04	2.71E-04	5.65E-04	3.01E-04	4.43E-02	4.49E-02	3.77E-04	2.57E-04
15	Drop (2D)	4.06E-04	2.48E-04	1.12E-02	2.17E-02	2.25E-01	1.14E-01	0.0E+00	0.0E+00
16	Shubert (2D)	5.24E-03	4.89E-03	2.69E-04	3.13E-04	7.88E+01	4.48E+01	4.93E-04	2.6E-04
17	Beale (2D)	4.82E-04	2.74E-04	1.98E-03	3.49E-03	1.58E-02	1.07E-01	4.21E-04	2.72E-04
18	McCorm (2D)	5.26E-04	2.92E-04	5.06E-04	2.83E-04	1.58E-01	1.32E-01	4.68E-04	2.6E-04
19	Camel6 (2D)	4.87E-04	2.86E-04	4.01E-04	3.26E-04	4.44E-01	2.85E-01	2.63E-04	2.94E-04
20	Bohal (2D)	4.68E-04	2.92E-04	3.2E-04	2.78E-04	4.12E-04	2.65E-04	0.0E+00	0.0E+00
21	Colville (2D)	9.28E-04	7.65E-05	3.43E-02	3.1E-02	8.08E-04	1.76E-04	8.45E-04	1.45E-04
22	Powersum (4D)	6.84E-04	2.26E-04	7.13E-03	8.58E-03	8.31E-04	2.21E-04	6.42E-04	2.1E-04
23	Solomon (2D)	5.49E-04	2.57E-04	4.8E-02	4.88E-02	2.64E-03	1.39E-02	1.06E-15	3.07E-16
24	Alpine (2D)	6.25E-04	2.41E-04	5.58E-04	2.84E-04	6.69E-04	2.35E-04	1.07E-16	3.93E-17

$$\min_x f(x) = ((x_3 + 2)x_2x_1^2)$$

Subject to the following constraints:

$$g_1(x) = 1 - \frac{(x_2^3 x_3)}{71785x_1^4} \le 0$$

$$g_2(x) = \frac{(4x_2^2 - x_1x_2)}{12566x_2x_1^3 - x_1^4} + \frac{1}{12566x_1^2} - 1 < 0_g$$

$$g_3(x) = 1 - \frac{(140.45x_1)}{x_2^2 x_3} \le 0$$

$$g_4(x) = \frac{(x_2 + x_1)}{1.5} - 1 \le 0$$

Engineering design problems are complex and nonlinear with many different variables and constraints. These variables and constraints must be satisfied to achieve

Table 8 P values using the Mann–Whitney test (a = 0.05) for accuracy acquired by BASR over ABC, BBA, and PSO

No.	Functions	BASR-qABC		BASR-SPSO2011		BASR-BBA	
		p value	Significant	p value	Significant	p value	Significant
1	Sphere (10D)	7.07E-18	Yes	7.07E-18	Yes	7.07E-18	Yes
2	Rosenbrock (10D)	7.07E-18	Yes	7.75E-09	Yes	7.07E-18	Yes
3	Quartic (30D)	1.08E-17	Yes	6.34E-17	Yes	7.07E-18	Yes
4	Ackley (10D)	1.86E-18	Yes	1.82E-18	Yes	1.86E-18	Yes
5	Schaffer (2D)	3.31E-20	Yes	3.31E-20	Yes	3.31E-20	Yes
6	Easom (2D)	7.40E-03	qABC	9.12E-20	Yes	2.40E-02	Yes
7	Rastrigin (10D)	3.31E-20	Yes	3.29E-20	Yes	3.31E-20	Yes
8	Shekel (4D)	1.43E-01	No	7.07E-18	Yes	2.92E-04	Yes
9	Langerman (10D)	4.22E-04	Yes	7.07E-18	Yes	6.97E-01	No
10	Griewank (10D)	3.31E-20	Yes	3.31E-20	Yes	3.31E-20	Yes
11	Branin (2D)	1.43E-02	qABC	3.31E-20	PSO	8.87E-06	Yes
12	Sumpow (10D)	3.95E-05	Yes	1.58E-07	Yes	4.71E-10	Yes
13	Bukin6 (2D)	5.89E-08	Yes	5.85E-15	Yes	1.84E-07	Yes
14	Crossit (2D)	2.13E-03	Yes	7.07E-18	Yes	4.18E-01	No
15	Drop (2D)	3.31E-20	Yes	3.31E-20	Yes	3.31E-20	Yes
16	Shubert (2D)	2.99E-05	qABC	7.07E-18	Yes	8.11E-13	Yes
17	Beale (2D)	1.25E-04	Yes	6.03E-02	No	2.43E-01	No
18	McCorm (2D)	5.74E-01	No	7.07E-18	Yes	3.26E-01	No
19	Camel6 (2D)	6.72E-02	No	7.07E-18	Yes	9.14E-05	Yes
20	Bohal (2D)	3.31E-20	Yes	3.31E-20	Yes	3.31E-20	Yes
21	Colville (2D)	1.35E-13	Yes	4.18E-01	No	3.90E-04	Yes
22	Powersum (2D)	2.03E-14	Yes	5.90E-07	Yes	2.63E-01	No
23	Solomon (4D)	6.06E-18	Yes	7.07E-18	Yes	7.07E-18	Yes
24	Alpine (2D)	7.07E-18	Yes	7.07E-18	Yes	7.07E-18	Yes
Total Performance		BASR:18 qABC:3 None:3		BASR:21 SPSO2011:1 None:2		BASR:19 BBA:0 None: 5	

an optimised solution [17]. The variables were forced to remain within the domain of the objective function. With regard to the constraints, at the end of every iteration, only the feasible solutions that satisfy all the constraints will be considered. Twenty runs were performed, and every run was forced to achieve at least 100 iterations. To evaluate the performance of the three proposed algorithms, their performances were compared with a collection of algorithms with remarkable performance as indicated by the literature. They are the following: evolutionary strategies [20], PSO [21], ABC [16], society and civilisation algorithm (SCA) [22] and Unified Particle Swarm Optimisation (UPSO) [14]. Table 15 shows the parameters used to apply the proposed algorithms on the tension/compression spring.

When studying the figures for Table 16, it indicates that BASR achieved the best result among all the other algorithms; however, in terms of mean and standard

Table 9 Success rate figures for the BASR, BBA, qABC and SPSO2011 algorithms

No.	Functions	qABC Success Rate	SPS2011O Success Rate	BBA Success Rate	BASR Success Rate
1	Sphere (10D)	100%	100%	100%	100%
2	Rosenbrock (10D)	0%	74%	0%	100%
3	Quartic	38%	0%	0%	96%
4	Ackley (10D)	24%	74%	0%	100%
5	Schaffer (2D)	66%	100%	100%	100%
6	Easom (2D)	100%	100%	100%	100%
7	Rastrigin (10D)	98%	0%	0%	100%
8	Shekel (4D)	98%	100%	100%	100%
9	Langerman (10D)	4%	100%	30%	18%
10	Griewank (10D)	0%	0%	0%	100%
11	Branin (2D)	100%	0%	100%	100%
12	Sumpow (10D)	100%	100%	100%	100%
13	Bukin6 (2D)	6%	4%	0%	0%
14	Crossit (2D)	100%	100%	100%	100%
15	Drop (2D)	74%	100%	100%	100%
16	Shubert (2D)	100%	100%	14%	100%
17	Beale (2D)	76%	98%	100%	100%
18	McCorm (2D)	100%	100%	100%	100%
19	Camel6 (2D)	100%	100%	100%	100%
20	Bohal (2D)	100%	100%	100%	100%
21	Colville (2D)	12%	100%	100%	100%
22	Powersum (2D)	18%	100%	100%	100%
23	Solomon (4D)	48%	96%	100%	100%
24	Alpine	100%	100%	100%	100%
SR Average		65%	77%	69%	92%

deviation, it was among the lowest, suggesting that its performance was not stable and consistent throughout all the runs. Nevertheless, it came second in terms of speed, where it needed 17,500 NFE to converge for its best result. Although this is not an optimal performance, it still indicates that the method used has the potential to achieve better performance (Table 17).

4 Conclusion and Future Work

In this research, a new method, BASR, is presented to improve the performance of the Bees algorithm. It used an adaptation of region elimination methods. It was evaluated on 24 benchmark functions and two engineering problems. Its performance on benchmarks was remarkably accurate and fast with an acceptable level of consistency. However, the suggested method performed well on engineering problems, although it was not the best among all methods with a low level of consistency. Nevertheless,

Table 10 NFE figures for BASR, BBA, qABC and SPSO2011 algorithms

No.	Functions	Result of 50 Runs		Result of 50 Runs		Result of 50 Runs	
		BASR	qABC	BASR	SPSO2011	BASR	BBA
1	Sphere (10 D)	50	0	50	0	50	0
2	Rosenbrock (10D)	50	0	44	6	50	0
3	Quartic (30D)	40	10	48	2	50	0
4	Ackley (10D)	50	0	49	0	50	0
5	Schaffer (2D)	50	0	50	0	50	0
6	Easom (2D)	19	31	0	50	44	6
7	Rastrigin (10D)	50	0	50	0	50	0
8	Shekel (4D)	7	43	0	50	25	25
9	Langerman (10D)	9	41	0	50	8	42
10	Griewank (10D)	50	0	50	0	50	0
11	Branin (2D)	30	20	50	0	49	1
12	Sumpow (10D)	49	1	50	0	50	0
13	Bukin6 (2D)	26	24	0	50	3	46
14	Crossit (2D)	39	11	2	48	37	4
15	Drop (2D)	50	0	0	50	50	0
16	Shubert (2D)	8	42	0	50	50	0
17	Beale (2D)	47	3	44	6	12	33
18	McCorm (2D)	46	4	0	50	35	12
19	Camel6 (2D)	17	33	0	50	23	19
20	Bohal (2D)	50	0	50	0	50	0
21	Colville (2D)	46	4	6	44	50	0
22	Powersum (2D)	50	0	38	12	45	5
23	Solomon (4D)	50	0	50	0	50	0
24	Alpine	50	0	50	0	50	0
Total:		BASR:19		BASR:14		BASR:20	
		qABC:5		SPSO2011:10		BBA:3 None:1	

concept search space manipulation has the potential to provide satisfactory performance in the optimisation of problems with a continuous search space due to its ability to achieve a focused search.

Table 11 Mean and Std.Dev. figures for BASR, BBA, qABC and SPSO2011

No.	Functions	BBA		qABC		SPSO2011		BASR	
		Mean	Std.Dev.	Mean	Std.Dev.	Mean	Std.Dev.	Mean	Std
1	Sphere (10D)	9180.34	4550.16	33446.14	9643.9	5884	437.89	102	0
2	Rosenbrock (10D)	500042.2	223747	500048.9	25.9	209694	190711.6	55705.44	2036.26
3	Quartic (30)	500068	225933.7	377869.9	180710.9	500000	0	160903.4	151059.8
4	Ackley (10D)	500039	235690.6	468309.6	70444.77	139692	213573	102	0
5	Schaffer (2D)	5066.22	2820.48	233117.2	235650	2122	544.53	102	0
6	Easom (2D)	4060.24	1890.02	2650.48	4249.69	100	0	1972.66	814.79
7	Rastrigin (10D)	500068.9	235704.7	180555.7	92942.62	500000	0	102	0
8	Shekel (4D)	62742.04	33644.05	43292.62	96619.86	100	0	51442.88	11641.98
9	Langerman (10D)	385919.2	237557	491350.5	60942.64	100	0	431486.8	150993.1
11	Griewank (10D)	500045.5	235691.8	500049.6	27.24	500000	0	110	0
10	Branin (2D)	4115.46	2048.44	1052.04	885.35	500000	0	857.66	388.14
12	Sumpow (10D)	951.2	438.41	150	0	1502	327.1	75.68	47.38
13	Bukin6 (2D)	500012.5	235700.7	490635.5	57616.16	490634	65562	500031.9	14
14	Crossit (2D)	400.84	207.09	274	115.86	100	0	164.24	69.92
15	Drop (2D)	27392.88	21770.94	236593.4	196388.3	100	0	102	0
16	Shubert (2D)	460845.8	223514.6	5662.18	9603.01	100	0	10468.8	3658.4
17	Beale (2D)	654.84	468.63	180170.2	213103.4	11656	69765.16	954.96	349.25
18	McCorm (2D)	1319.68	690.69	13572.06	17897.62	100	0	598.82	332.08
19	Camel6 (2D)	1025.76	540.44	790	369.86	120	44.72	874.4	300.75
20	Bohal (2D)	93133.46	69487.08	1872.26	1145.41	4992	742.12	102	0
21	Colville (2D)	90739.72	43316.04	453006.8	141528.2	14160	4506	31410.62	17993.65
22	Powersum (2D)	74731.52	50147.5	463048.3	94270.83	23094	23524.03	11361.32	8803.96
23	Solomon (2D)	38934.28	29375.34	273926.7	238028.5	26946	96573.06	102	0
24	Alpine (2D)	3546.6	1893.11	2344	8793.92	2912	1601.83	102	0

Table 12 P values using the Mann–Whitney test (a = 0.05) for NFE acquired by BASR over ABC, BBA, and PSO

No.	Functions	BASR -qABC		BASR -SPSO2011		BASR -BBA	
		p value	Significant	p value	Significant	p value	Significant
1	Sphere (10D)	3.31E-20	Yes	1.72E-20	Yes	3.28E-20	Yes
2	Rosenbrock (10D)	6.84E-18	Yes	3.05E-11	Yes	6.69E-18	Yes
3	Quartic (30 D)	4.58E-08	Yes	1.73E-17	Yes	6.21E-18	Yes
4	Ackley (10D)	3.30E-20	Yes	1.59E-20	Yes	3.11E-20	Yes
5	Schaffer (2D)	3.30E-20	Yes	1.72E-20	Yes	3.30E-20	Yes
6	Easom (2D)	1.21E-02	qABC	2.50E-20	PSO	2.11E-13	Yes
7	Rastrigin (10D)	3.31E-20	Yes	1.59E-23	Yes	3.16E-20	Yes
8	Shekel (4D)	2.10E-10	qABC	2.55E-20	PSO	0.297714	No
9	Langerman (10D)	2.65E-08	qABC	2.46E-20	PSO	1.46E-09	BBA
11	Griewank (10D)	3.30E-20	Yes	1.59E-23	Yes	3.04E-20	Yes
10	Branin (2D)	9.96E-02	No	2.44E-20	Yes	3.41E-16	Yes
12	Sumpow (10D)	5.98E-21	Yes	1.24E-18	Yes	1.69E-18	Yes
13	Bukin6 (2D)	2.17E-01	No	3.63E-20	PSO	3.88E-13	BBA
14	Crossit (2D)	1.42E-07	Yes	8.39E-18	PSO	5.54E-09	Yes
15	Drop (2D)	3.31E-20	Yes	1.59E-23	PSO	3.31E-20	Yes
16	Shubert (2D)	7.67E-10	qABC	2.38E-20	PSO	6.34E-18	Yes
17	Beale (2D)	1.36E-14	Yes	3.55E-12	Yes	3.23E-05	BBA
18	McCorm (2D)	1.14E-12	Yes	2.32E-20	PSO	4.92E-08	Yes
19	Camel6 (2D)	2.02E-02	qABC	2.86E-19	PSO	0.1711787	No
20	Bohal (2D)	3.19E-20	Yes	1.74E-20	Yes	3.31E-20	Yes
21	Colville (2D)	7.63E-14	Yes	2.96E-12	PSO	3.77E-15	Yes
22	Powersum (2D)	7.06E-18	Yes	1.07E-03	Yes	8.59E-12	Yes
23	Solomon (2D)	3.30E-20	Yes	1.78E-20	Yes	3.31E-20	Yes
24	Alpine (2D)	3.11E-20	Yes	1.76E-20	Yes	3.28E-20	Yes
Total		BASR = 17 qABC = 5		BASR = 14 SPSO2011 = 10 None = 0		BASR = 19 BBA = 2 None = 3	

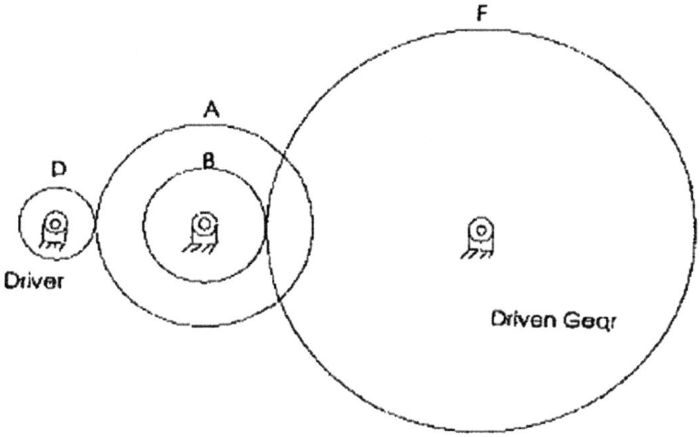

Fig. 2 Gear train design scheme

Table 13 BASR Parameter Setting for Gear Train Problem

Parameter	Parameter value
Number of scout bees	20
Number of selected bees, m	4
Number of elite bees, e	2
Number of forager bees for elite bee, nep	30
Number of foragers for selected bee, nsp	10
Neighbourhood size, ngh	0.1

Table 14 Performance figures for the gear train problem

Item	ABC	PSO-GA	CS	BASR
Mean	3.641339 E-10	1.285602 E-10	1.9841E-9	1.28E-10
Std. Dev	5.525811E-10	8.7787 E-10	3.5546 E-9	2.819994E-10
Best Solution	2.700857 E-12	2.70085 E-12	2.7009 E-12	2.700857E-12
X1	49	43	43	49
X2	16	16	16	19
X3	19	19	19	16
X4	43	49	49	43
Gear Ratio	0.144281	0.14428096	0.144281	0.144281
Error %	0.001	0.001	0.001	0.001

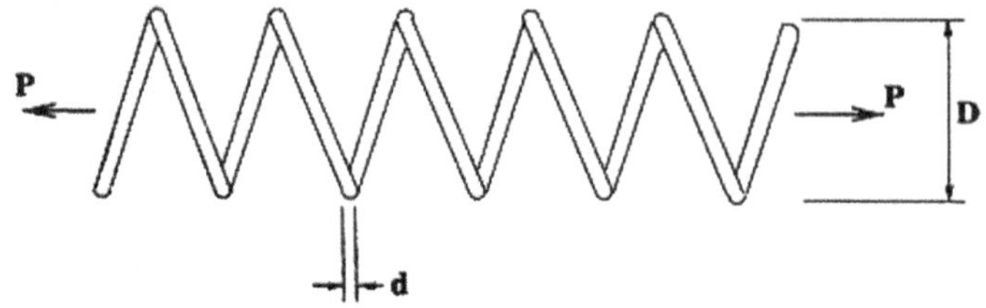

Fig. 3 Tension/compression spring

Table 15 BASR parameter values for the tension/compression spring

Parameter	Parameter value
Number of scout bees	100
Number of selected bees, m	4
Number of elite bees, e	2
Number of forager bees for elite bee, nep	30
Number of foragers for selected bee, nsp	10
Neighbourhood size, ngh	0.0001

Table 16 Comparison of the figures of the tension/compression spring problems with different algorithms involved in the testing

Tension/ compression spring	ABC	SCA	$\mu + \lambda - ES$	UPSOm	PSO	BASR
Best	0.012665	0.012669	0.012689	0.01312	0.012665	0.008838
Mean	0.012709	0.012923	0.013165	0.0229478	0.012702	0.035154
Std. Dev	0.0128	0.00059	0.00039	0.0072	4.1E-05	0.012017
Evaluations	30,000	25,167	30,000	1,00,000	15,000	17,889

Table 17 List of variable and constraint values achieved by BASR

Variables and constraints values	BASR
X1	0.0685
X2	0.5667
X3	12.204
g1	– 0.4
g2	– 0.341
g3	– 1.456
g4	– 0.576

References

1. Liang JJ, Qin AK, Suganthan PN, Baskar S (2006) Comprehensive learning particle swarm optimizer for global optimization of multimodal functions. IEEE Trans Evol Comput 10(3):281–295
2. Beheshti Z, Shamsuddin SMH (2013) A review of population-based meta-heuristic algorithms. Int J Adv Soft Comput Appl 5(1):1–35
3. Kamsani SH (2016) Improvements on the Bees algorithm for continuous optimisation problems (Doctoral dissertation). PhD thesis. University of Birmingham
4. Pham DT, Darwish HA (2010) Using the Bees algorithm with Kalman filtering to train an artificial neural network for pattern classification. Proc Inst Mech Engineers J Syst Control Eng, pp 885–892
5. Alfi A, Khosravi A (2012) Constrained nonlinear optimal control via a hybrid BA-SD. Int J Eng-Trans C: Asp 25(3):197–204
6. Yuce B, Pham DT, Packianather MS, Mastrocinque E (2015) An enhancement to the Bees algorithm with slope angle computation and Hill Climbing algorithm and its applications on scheduling and continuous type optimisation problem. Prod Manuf Res 3(1):3–19
7. Pham DT, Castellani M, Fahmy A (2008) Learning the inverse kinematics of a robot manipulator using the Bees algorithm. In: 2008 6th IEEE international conference on industrial informatics. IEEE, pp 493–498
8. Pham DT, Pham QT, Ghanbarzadeh A, Castellani M (2008) Dynamic optimisation of chemical engineering processes using the bees algorithm. IFAC Proc 4(2):6100–6105
9. Pham DT, Ghanbarzadeh A, Koc E, Otri S, Rahim S, Zaidi M (2005) The Bees algorithm technical note. Tech. rep. Manufacturing Engineering Centre, Cardiff University, UK
10. Bhattacharjya RK (2009) Region elimination method. In: CE 602: Optimization Method [Lecture notes]. Guwahati: Indian Institute of Technology. http://www.iitg.ac.in/rkbc/CE602/ce602.htm
11. Deb K (2012) Optimization for engineering design: algorithms and examples. New Delhi, India: PHI Learning Pvt. Ltd
12. Jamil M, Yang X (2013) A literature survey of benchmark functions for global optimisation problems. J Numer Optim 2:150–194
13. Karaboga D, Gorkemli B (2012) A quick artificial bee colony-qABC-algorithm for optimization problems. In: 2012 International symposium on innovations in intelligent systems and applications. Trabzon, Turkey, pp 1–5
14. Parsopoulos KE, Vrahatis MN (2005) Unified particle swarm optimization for solving constrained engineering optimization problems. In: 1st International conference on natural computation, ICNC 2005, Changsha, China. Springer, Berlin, Germany, pp 582–591
15. Kannan BK, Kramer SN (1994) Augmented Lagrange multiplier based method for mixed integer discrete continuous optimization and its applications to mechanical design. J Mech Des, Trans ASME 116(2):405–411
16. Akay B, Karaboga D (2012) Artificial bee colony algorithm for large-scale problems and engineering design optimization. J Intell Manuf 23(4):1001–1014
17. Gandomi AH, Yang XS, Alavi AH (2013) Cuckoo search algorithm: a metaheuristic approach to solve structural optimization problems. Eng Comput 29(1):17–35
18. Garg H (2016) A hybrid PSO-GA algorithm for constrained optimization problems. Appl Math Comput 274:292–305
19. Guo CX, Hu JS, Ye B, Cao YJ (2004) Swarm intelligence for mixed-variable design optimization. J Zhejiang University SCIENCE A 5(7):851–860
20. Mezura-Montes E, Coello CA (2005) Useful infeasible solutions in engineering optimization with evolutionary algorithms. In: MICAI 2005: Advances in Artificial Intelligence. Springer, Berlin, Germany, pp 652–662
21. He S, Prempain E, Wu QH (2004) An improved particle swarm optimizer for mechanical design optimization problems. Eng Optim 36(5):585–605

22. Ray T, Liew KM (2003) Society and civilization: an optimization algorithm based on the simulation of social behavior. IEEE Trans Evol Comput 7(4):386–396

Local Optimal Issue in Bees Algorithm: Markov Chain Analysis and Integration with Dynamic Particle Swarm Optimisation Algorithm

JianBang Liu, Mei Choo Ang, Kok Weng Ng, and Jun Kit Chaw

Abstract Inspired by the foraging behaviour of bees in nature, the bees algorithm (BA) is a biomimetic optimisation method designed according to how bees work together in choosing food sources and collecting high-quality honey. However, there are only a few mathematical studies on the convergence and optimisation problem of BA in the literature, and most of the BA implementations are dependent on trial-and-error. Thus, this study attempts to use Markov chain theory to enhance the understanding and analysis of BA convergence from the perspectives of the neighbourhood contraction strategy and site abandonment strategy. To improve the optimisation performance of the BA, this study established a combination model of the BA and a dynamic particle swarm optimisation algorithm on the local search process of the BA and the effect of the task allocation of scout bees for target location. Finally, the simulation experiment was conducted in MATLAB using the combination model. The simulation result shows that the probability of the combination model falling into the local optimum is much lower than that of the conventional BA based on the good convergence performance for the optimisation problem. It can enrich the theoretical base of the BA as a resource for convergence performance and further optimisation analysis.

Keywords Bees algorithm · Convergence performance · Markov chain theory · Dynamic particle swarm optimisation algorithm · Optimisation problem

J. Liu · M. C. Ang (✉) · J. K. Chaw
Institute of Visual Informatics, Universiti Kebangsaan Malaysia, 43600 UKM Bangi, Selangor, Malaysia
e-mail: amc@ukm.edu.my

K. W. Ng
Department of Mechanical, Materials and Manufacturing Engineering, Faculty of Science and Engineering, University of Nottingham Malaysia, Jalan Broga, 43500 Semenyih, Selangor, Malaysia

J. Liu
College of Mathematics and Computer, Xinyu University, 338004 Xinyu, China

D. T. Pham and N. Hartono (eds.), *Intelligent Engineering Optimisation with the Bees Algorithm*, Springer Series in Advanced Manufacturing,
https://doi.org/10.1007/978-3-031-64936-3_3

1 Introduction

The behaviour of bee colonies in nature is very enlightening to human beings. A bee colony contains thousands of individual bees, but such a large colony can flexibly adapt to various living environments in a cluster intelligent way and reasonably allocate and divide tasks according to dynamic living conditions. Many researchers have established models for the intelligent behaviour of bee colonies and tried to use the intelligent behaviour of bee colonies to solve the actual optimisation problem. Seely first proposed the self-organisation model of bee colonies [1]. Passino, Seeley and Visscher [2] model and simulate the nesting and location behaviour of bee colonies and emphasise that the swarm intelligent decision-making ability of bee colonies is certain, which has potential inspiration for people to solve practical optimisation problems. The bees system (BS) is the first algorithm inspired by bee colony behaviour [3], which is mixed with a genetic algorithm to improve the optimisation performance of the algorithm. The artificial bee colony algorithm (ABC) proposed by Karaboga [4] is one of the algorithms inspired by bee behaviour. BA is a swarm intelligence optimisation algorithm proposed by Pham, Ghanbarzadeh, Koc, Otri, Rahim and Zaidi [5] to simulate the foraging behaviour of bees.

Since the BA was proposed, many experts in various fields have also adopted BA to solve relevant optimisation problems. Tolabi, Ali, Ayob and Rizwan [6] combined fuzzy control and BA to optimise the distributed generation problem. Akpinar and Baykasoğlu [7] proposed multiple swarm BA to optimise the balance problem of hybrid model assembly lines. The results of simulation experiments showed that the optimisation effect was better than that of other methods. Considering the advantages such as simplicity and the high efficiency of BA, Sarailoo, Rahmani and Rezaie [8] applied a swarm intelligence method in nonlinear predictive control for an optimisation problem of a water tank system control. Zarei, Atabati and Kor [9] introduced a novel BA that used random adaptive neighbourhood search to improve the original BA by studying the impact of the reduction of recruited bees on the performance of the algorithm and designing the relocation of bees accordingly. After conducting experiments on 23 standard test functions, it was shown that the method significantly improved the convergence performance of the original BA, and the results on 12 standard test functions were better than the conventional BA. Deghbouch and Debbat [10] considered that the local search process of BA lacked efficient exploitation due to the random assignment of search agents throughout neighbourhoods, which lowered algorithm accuracy and resulted in delayed convergence. Based on this, the researchers mixed the grasshopper optimisation algorithm (GOA) and the BA. By utilising the strength of the GOA to enhance the exploitation phase of the BA, this shortcoming was alleviated.

From the perspective of the optimisation problem, numerous optimisation algorithms and techniques have been developed to address a variety of optimisation issues. Rudolph [11] analysed the convergence performance of the canonical genetic algorithm (CGA) for optimisation problems and demonstrated, by means of homogeneous finite Markov chain theory, that the algorithm would never converge to global optimisation regardless of the parameters, process design, and function. Sun, Yang, Gao, Zhang, Li and Li [12] presented the ABC and Markov random field (MRF) strategies in terms of determining an ABC–MRF-cluster model to address the oversimplification of computing models and underutilisation of spatial information in hyperspectral remote sensing image processing. Sagayam and Hemanth [13] developed a hybrid 1-D hidden Markov model (1-D HMM) with ABC optimisation to improve performance metrics for the Cambridge hand gesture dataset, including recognition rate and mistake rate. Kıran and Gündüz [14] introduced a hybridisation of particle swarm optimisation (PSO) and ABC techniques based on the recombination process as the global best and neighbour food source for onlooker bees, which strengthened the global and local search capabilities to address the continuous optimisation problem. Zhang, Wang and Ji [15] offered that the global best solutions derived by the PSO and ABC were utilised for recombination, and the resulting solution was distributed to the PSO and ABC populations as the global best and neighbour food source for onlooker bees, respectively. However, there is currently a lack of studies on BA convergence performance and optimisation analysis based on Markov chain theory and dynamic particle swarm optimisation algorithms, respectively.

The solution search speed of BA is very fast, which has obvious advantages over many existing methods in route optimisation problems. However, the limitation easily falls into a local optimisation problem affecting the optimisation effect of the BA. Therefore, this study analysed the convergence of BA based on Markov chain theory. The iterative sequence of the scout bee individual and evolutionary generation in the solution space has Markov characteristics, and the process of searching for the optimal solution by the BA is analysed in the form of Markov state transition. The evolutionary generations of scout bees are grouped to simplify the one-step transition probability matrix and obtain the Markov static distribution, which demonstrates the convergence properties of the algorithm. This chapter analysed the influence of the neighbourhood contraction strategy and site abandonment strategy on the convergence performance of the BA. Finally, considering the drawback of BA, the combination model of BA and the dynamic particle swarm optimisation algorithm was established to avoid improving the ability of the algorithm to escape from the local optimal solution while destroying the convergence performance of the algorithm based on the simulation experiment. The experimental results show that the optimisation effect of the combination model is much better than that of the conventional BA.

2 Method

2.1 Introduction to BA

The BA is an optimisation algorithm inspired by the foraging behaviour of bees in nature. In the algorithm, the location of the food source represents a set of solutions to the optimisation problem. A small search area is formed by the location of the food source and its neighbourhood, which is called a flower cluster. In the process of searching for food, the bees first send scout bees to search. The scout bees constantly fly from one food source to another to search for all possible food sources. When the income of the food source (such as the content of sugar) exceeds a certain amount, the scout bees return to the hive, remove the honey, and dance a magical "swing dance" on the "dance board". This kind of "swing dance" is a necessary tool for bees to exchange information. It conveys to the whole bee colony the location of the found food source, the distance to the hive, and the degree of return. After dancing, the scout bees lead the following bees waiting outside the hive to fly back to the corresponding food source to collect honey; food sources with high income will recruit more following bees to collect honey. In the process of collecting honey, the bees calculate the income degree of the food source, return to the hive and dance the "swing dance" to convey the information of the income degree of the current food source to the bees. If the income of the food sources is still high, more following bees will be recruited to collect honey. Otherwise, the current food source will be abandoned [16].

Step 1. Initialise ns to detect the location of bees and conduct a random search;

Step 2. The ns value and the fitness value of scout bees were calculated and arranged in descending order;

Step 3. Among the fitness values in descending order, the best bee was selected;

Step 4. Best bees return to the hive to recruit bees for local search;

Step 5. The fitness values of scout bees were calculated and arranged in descending order;

Step 6. Among the fitness values in descending order, some bees were selected as elite bees;

Step 7. Elite bees return to the hive to recruit bees for local search. At the same time, some best bees act as scout bees and continue to search randomly in the search space;

Step 8. Return to step 1, iterate repeatedly until the convergence result is reached, and end the task allocation process.

Figure 1 shows the basic flow of BA; Table 1 summarises the BA configuration parameters.

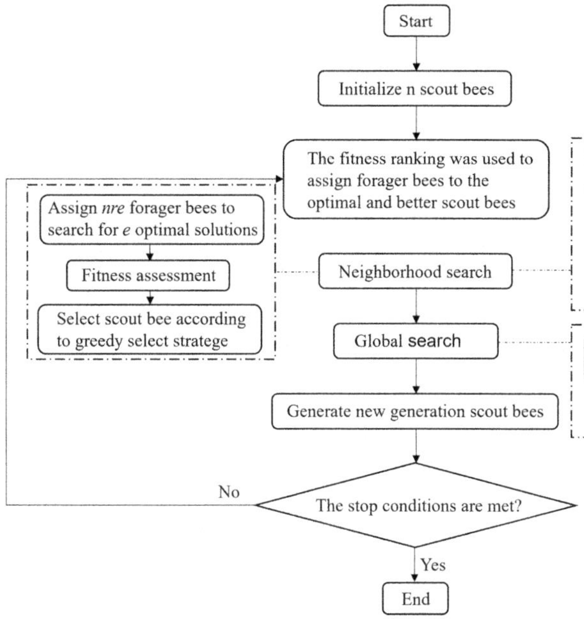

Fig. 1 Flowchart of BA

Table 1 Terminology table of parameter descriptions

Parameter name	Description
n	Number of scout bees
m	Number of better scout bees
e	Number of optimal scout bees
nre	Number of forager bees following the optimal scout bees
nrb	Number of forager bees following the better scout bees
ngh	Initial size of neighbourhood
nsc	Neighbourhood shrinkage coefficient
stlim	Limit of stagnation cycles for site abandonment
criteria	Algorithm stopping criteria

2.2 Application and Limitations of BA

Since the BA was proposed, many previous studies on improving the BA have been carried out to improve the optimisation performance of the algorithm. To create a computational architecture for producing branded design concepts that can satisfy a given functional need, ABC established a shape grammar and integrated it with BA [17]. A computational method for producing branded design concepts that can satisfy functional criteria is developed when a form language is paired with an optimisation method such as an evolutionary algorithm [18]. In addition, the neighbourhood contraction strategy [19] and site abandonment strategy [20, 21] have been used in many improved versions of the BA. The neighbourhood contraction strategy requires the BA to set the initial size *ngh* of the neighbourhood slice to a relatively large value. If the solution can be continuously improved in the neighbourhood search process, the neighbourhood size will remain unchanged. If the improvement cannot be obtained, the neighbourhood size will be adjusted according to formula (1). According to the neighbourhood contraction strategy, the neighbourhood slice in the BA has a large coverage, and the algorithm has a strong exploration ability in the solution space to find the possible area of the optimal solution. With the operation of the algorithm, the neighbourhood is gradually reduced, and the algorithm will focus on a more detailed search for the best solution to improve the search accuracy.

$$ngh_{n+1} = 0.8 * ngh_n \tag{1}$$

The site abandonment strategy is mainly used to address the objective function with multiple locally optimal solutions. Once the scout bee of the original BA falls into the local optimal solution, it will be difficult to leave the local optimal solution and reach the global optimal solution. The site abandonment strategy simulates the process in which the food source mined in nature is discarded by the bee colony. To do so, the parameter *stlim* (limit of stagnation cycles for site abandonment) needs to be introduced. If the search in a neighbourhood slice fails to improve the solution in successive *stlim* iterations, the neighbourhood search is considered to have reached a local optimal position and that any neighbourhood search in the future will not be able to find a solution with higher fitness. At this time, the solution sampled by the scout bees is regarded as the local optimal solution, then the search activity in the neighbourhood slice is stopped, and the scout bee is forced to leave the current position and arrive at any position in the solution space at random. Together with global search, this strategy maintains the diversity of detection bees and enhances the ability of the algorithm to escape from the local optimal solution.

In addition, other improvement strategies also include the use of a fuzzy selection strategy to reduce the configuration parameters of BA [22]; for example, BA, which uses pheromones to transmit information and lead foraging bees to improve the search speed of the algorithm [23]; BA, which uses mutation, creep, crossover, interpolation and extrapolation search operators and a new bee internal competition strategy to improve the optimisation performance of the algorithm [24]; and neighbourhood search of the Gaussian distribution, which is used to enhance the stability of the algorithm and reduce the configuration parameters [25]. BA has been successfully applied in engineering and computer science. Table 2 lists some related applications of BA.

Table 2 Applications of BA

Optimisation field	Continuous optimisation	Combinatorial optimisation
Application	Machine design [26]	PCB device assembly [19, 27]
	Control system coordination [28]	Task scheduling [29]
	Pattern classifier training [30]	Task allocation of swarm robots [31]
	Chemical treatment problem [32]	Timetable problem [33]
	Clustering problem [34]	Manufacturing cell forming [35]
	Environmental/economic energy scheduling problem [36]	
	Gene regulatory network [37]	
	Fuel efficiency of semi rail air-cushion vehicle [38]	
	Robot PID or manipulation control [20]	
	Quality attribute optimisation of manufacturing network [39]	

3 Convergence Performance of BA

The problem that optimisation algorithms usually need to solve is whether the algorithm can finally converge, whether it can converge to the real global optimal solution, or how much probability to find the global optimal solution. BA is a probabilistic heuristic algorithm that combines neighbourhood development search and global exploration search and is characterised by multipoint parallel search. There is information exchange between all kinds of scout bees and forager bees in each evolutionary iteration, so the whole bee colony search path of BA is more complex.

An ideal optimisation algorithm should converge to the global optimal solution with probability 1. For most problems, how to determine the convergence is also related to the search accuracy. For example, the search accuracy is needed to be ε ($\varepsilon > 0$); then, when Eq. 2 is satisfied, the algorithm is considered to have converged to the global optimal solution.

$$|f(x) - f(\xi)| \leq \varepsilon \tag{2}$$

$$fitness(x) = f(x) \tag{3}$$

where ξ is the actual global optimal solution of the objective function.

3.1 Markov Chain

Markov chains are stochastic processes with Markov properties in probability theory and mathematical statistics that exist in discrete index sets and state spaces [40]. Let $\{x(t), t \geq 0\}$ be a series of random variables with values on the finite state space $I = \{i\}$, which is called the state space. If for any $t \geq 1$, $i_k \in I$ $(k \leq t + 1)$. There is an equation as follows:

$$p\{x(t+1) = i_{t+1}|x(t) = i_t, \ldots, x(0) = i_0\} = p\{x(t+1) = i_{t+1}|x(t) = i_t\} \quad (4)$$

where means the probability of $x(t + 1)$ is only related to $x(t)$ but independent of $x(0), x(1), \ldots, x(t-1)$, so $\{x(t), t \geq 0\}$ is called a Markov chain, which greatly simplifies the calculation of conditional probability.

The one-step conditional transition probability of a Markov chain is $p_{i,j}(t) = p\{x(t+1) = i|x(t) = j\}$ where $i, j \in I$. When the transition probability $p_{i,j}(t)$ of a Markov chain $\{x(t), t \geq 0\}$ is independent of t for any $i, j \in I$, the Markov chain is said to be homogeneous, and $p_{i,j}(t)$ is recorded as $p_{i,j}$. Then, $p_i = p\{x(0) = i\}(i \in I)$ is called the initial probability, and $\{p_i, i\}$ is called the initial distribution of the Markov chain. The initial probability represents the probability that the system is in state i when $t = 0$.

3.2 BA and Markov Chain Formalism

According to the optimisation accuracy ε, the objective function solution space Ω can be divided into finite subspaces. The accuracy difference between subspaces is an integer multiple ε. The number of subspaces was set as Nx, and the higher the optimisation accuracy (ε is closer to 0), the greater the value of Nx. The solution space Ω can be expressed as $\{x_1, x_2, \ldots, x_{Nx}\}$.

Considering that the scout bees are distributed in the solution space, the state of a scout bee in the $x(t > 0)$ evolutionary generation is recorded as $x(t) = x_i$ in the BA. The state space is $X = \{x_i|x_i \in \Omega\}$. The scout bee is in one of the Nx possible states in the solution space of the objective function.

From the perspective of the state transition of scout bees, for any two states $x_{i_t}, x_{i_{t+1}} \in X$, the probability of the scout bee transferring from state x_{i_t} to $x_{i_{t+1}}$ after an evolutionary generation from the t-th iteration is recorded as $p\big(x(t+1) = x_{i_{t+1}}|x(t) = x_{i_t}\big)$. The state transition probability of this step is calculated according to the role of the scout bee in the BA, as shown in the following formula:

$$p\left(x(t+1) = x_{i_{t+1}} | x(t) = x_{i_t}\right) = \begin{cases} p_e\left(x(t+1) = x_{i_{t+1}} | x(t) = x_{i_t}\right) \text{ Optimal scout bees} \\ p_b\left(x(t+1) = x_{i_{t+1}} | x(t) = x_{i_t}\right) \text{ Better scout bees} \\ p_r\left(x(t+1) = x_{i_{t+1}} | x(t) = x_{i_t}\right) \text{ Random scout bees} \end{cases} \quad (5)$$

The optimal and better scout bees participate in the neighbourhood search process. The one-step transfer probability of the optimal and better scout bees can be calculated by the following formulas:

$$x^f = x^s rngh \quad (6)$$

$$p_e\left(x(t+1) = x_{i_{t+1}} | x(t) = x_{i_t}\right) = \frac{nre}{\|ngh\|} q_1\left(x_{i_{t+1}}, x_{i_t}\right) \quad (7)$$

$$p_b\left(x(t+1) = x_{i_{t+1}} | x(t) = x_{i_t}\right) = \frac{nrb}{\|ngh\|} q_1\left(x_{i_{t+1}}, x_{i_t}\right) \quad (8)$$

where x^f represents the forager bee led by the scout bee x^s; r is a random number subject to a uniform distribution in the range of [0,1]; ngh returns the number of subspaces in the neighbourhood search range specified by ngh in the search space; *and* $q_1\left(x_{i_{t+1}}, x_{i_t}\right)$ is the decision variable, which is determined by the following equation:

$$q_1\left(x_{i_{t+1}}, x_{i_t}\right) = \begin{cases} 1, \ x_{i_{t+1}} \in \left\| x_{i_t} - ngh, x_{i_t} + ngh \right\| \ \& \ fittness\left(x_{i_{t+1}}\right) \geq fittness\left(x_{i_t}\right) + \varepsilon \\ 0, \ other \end{cases} \quad (9)$$

where $\left\| x_{i_t} - ngh, x_{i_t} + ngh \right\|$ is the size of the hypercube space composed of domain slices centered on x_{i_t}. This equation means that if the state $x_{i_{t+1}}$ is within the coverage of the x_{i_t} domain and its fitness is better than that of x_{i_t}, the optimal or better scout bees will transfer from x_{i_t} to $x_{i_{t+1}}$ with a positive probability; otherwise, the transfer probability is 0. The calculation of the one-step transition probability of a random scout bee is shown in the following formula:

$$p_r\left(x(t+1) = x_{i_{t+1}} | x(t) = x_{i_t}\right) = \frac{1}{Nx} q_2\left(x_{i_{t+1}}, x_{i_t}\right) \quad (10)$$

where this equation indicates that the random scout bees search in the whole solution space, in which the decision variable $q_2\left(x_{i_{t+1}}, x_{i_t}\right)$ is determined by the following formula:

$$q_2\left(x_{i_{t+1}}, x_{i_t}\right) = \begin{cases} 1, \ fittness\left(x_{i_{t+1}}\right) fittness\left(x_{i_t}\right) + \varepsilon \\ 0, \ other \end{cases} \quad (11)$$

where this equation means that the random scout bees will transfer from x_{i_t} to $x_{i_{t+1}}$ with a positive probability if the fitness of state $x_{i_{t+1}}$ is better than that of x_{i_t} for random scout bees.

Theorem 1 *The state sequence {x(t), t ≥ 0} of a scout bee in the evolution process constitutes a finite homogeneous Markov chain in BA. The proving was as follows:*

(1) The state space of scout bees $X = \{x_i | x_i \in \Omega\}$ is finite, and the number of contained states is *Nx*. It satisfies finiteness;
(2) $\forall x_i$ X and the positive integer $i_0, i_1, \ldots, i_{t+1} \le Nx$;

$$p\big(x(t+1) = x_{i_{t+1}} | x(t) = x_{i_t}, x(t-1) = x_{i_{t-1}}, \ldots, x(0) = x_{i_0}\big)$$
$$= p\big(x(t+1) = x_{i_{t+1}} | x(t) = x_{i_t}\big) \tag{12}$$

where it shows that there is no aftereffect in the state transfer of scout bees.
(3) $\forall x_i$ X, $1 \le r$ and the positive integer $i_t, i_{t+1} \ldots, x_{t+r} \le Nx$;

$$p\big(x(t+r) = x_{i_{t+r}} | x(t) = x_{i_t}\big) = \sum\nolimits_{x_{i_{t+1}} \in X} \cdots$$
$$\sum\nolimits_{x_{i_{t+r-1}} \in X} p(x(t+r) = x_{i_{t+r}} | x(t+r-1) = x_{i_{t+r-1}}) \cdots$$
$$p\big(x(t+1) = x_{i_{t+1}} | x(t) = x_{i_t}\big)a = p^{(r)}\big(x_{i_{t+r}} | x_{i_t}\big) \tag{13}$$

where the state transition probability of the *r*-step of scout bees is only related to step *r* but not to the initial evolutionary generation *t*, which satisfies homogeneity.

From the above equations, it is known that the state sequence $\{x(t), t \ge 0\}$ of the scout bees constitutes a finite homogeneous Markov chain, so the state transition probability of the scout bees $p\big(x(t+1) = x_{i_{t+1}} | x(t) = x_{i_t}\big)$ can be simplified as $p\big(x_{i_{t+r}} | x_{i_t}\big)$.

From the perspective of the evolutionary generation state, a scout bee evolutionary generation contains n scout bees in the BA, so the n scout bee states form an evolutionary generation state, expressed as $u = (x^1, x^2, \ldots, x^n)$, while $x^i \in X\,(i = 1, 2, \ldots, n)$ represents the *i*-th state of scout bees. The set of possible states of all evolutionary states that constitutes the evolutionary state space is expressed as $U = \{u | u = \big(x^1, x^2, \ldots, x^n\big), x^k \in X)$.

From the perspective of the state transition of evolutionary generation, for any two states $u_{i_t}, u_{i_{t+1}} \in U$, the probability of the evolutionary generation of the scout bees from the state u_{i_t} to $u_{i_{t+1}}$ after the *t*-th iteration is recorded as $p\big(u(u+1) = u_{i_{t+1}} | u(t) = u_{i_t}\big)$. This step state transition probability is calculated according to the one-step state transition probability of the scout bees, and the formula is as follows:

$$p\big(u(t+1) = u_{i_{t+1}} | u(t) = u_{i_t}\big) = \prod\nolimits_{i=1}^{n} p\big(x^i(t+1) = x_{i_{t+1}} | x^i(t) = x_{i_t}\big) \tag{14}$$

where it shows that only the scout bee state contained in $u(t)$ is transferred to that contained in $u(u+1)$ at the same time, so it is called the transfer of an evolutionary generation from $u(t)$ to $u(u+1)$.

Theorem 2 *The state sequence* $\{u(t), t \geq 0\}$ *of a scout bee in the evolution process constitutes a finite homogeneous Markov chain in BA. The proof is as follows:*

(1) Due to the state composition of n scout bees that make up the evolutionary generation state of scout bees, the state space X of scout bees is limited. Therefore, the evolutionary generation state space of scout bees is an n-dimensional state space of scout bees. U $\leftarrow X^n$ is finite, and the number of states including evolutionary generations is $Nu = (Nx)^n < \infty$, which satisfies the finiteness;

(2) $\forall u_i$ U and the positive integer $i_0, i_1, \ldots, i_{t+1} \leq Nu$. The equations are as follows:

$$\begin{aligned} & p\big(u(t+1) = u_{i_{t+1}} | u(t) = u_{i_t}, u(t-1) = u_{i_{t-1}}, \ldots, u(0) = u_{i_0}\big) \\ & = p\big(\big(x^1(t+1), \ldots, x^n(t+1) | x^1(t), \ldots, x^n(t)\big), \ldots, \big(x^1(0), \ldots, x^n(0)\big)\big) \\ & = p\big(x^1(t+1), \ldots, x^n(t+1) | x^1(t), \ldots, x^n(t)\big) \\ & = p\big(u(t+1) = u_{i_{t+1}} | u(t) = u_{i_t}\big) \end{aligned} \tag{15}$$

where indicate the no-aftereffect of one of the properties.

(3) $\forall$ $u_i U$, $1 \leq r$ and the positive integer $i_t, i_{t+1}, \ldots, i_{t+r} \leq Nu$. The equations are as follows:

$$\begin{aligned} & p\big(u(t+r) = u_{i_{t+r}} | u(t) = u_{i_t}\big) \\ & = \sum\nolimits_{u_{i_{t+1}} \in U} \cdots \sum\nolimits_{u_{i_{t+r-1}} \in U} p(u(t+1) = u_{i_{t+1}} | u(t) = u_{i_t}) \cdots \\ & p\big(u(t+r) = u_{i_{t+r}} | u(t+r-1) = u_{i_{t+r-1}}\big) \\ & = p^{(r)}\big(u_{i_{t+r}} | u_{i_t}\big) \end{aligned} \tag{16}$$

where it shows that the r-step state transition probability of the scout bee evolutionary generation is only related to r but not to the initial evolutionary generation t, which meets the homogeneity.

It can be seen from the above formula that the state sequence $\{u(t),\ t \geq 0\}$ of the evolutionary generation of scout bees constitutes a finite homogeneous Markov chain. Therefore, the one-step state transition probability of reconnaissance bee evolutionary generation $p\big(u(t+1) = u_{i_{t+1}} | u(t) = u_{i_t}\big)$ can be simplified as $p(u_{i_{t+1}} | u_{i_t})$.

From the perspective of the optimal state of scout bees, the scout bees reach the optimal state x^* X if $|fitness(x^*) - fitness(\xi)| \leq \varepsilon$ for the global optimal solution of the objective function.

From the perspective of the optimal state of evolutionary generation, the evolutionary generation of scout bees reaches the optimal $u^* \in U$ if $x^* \in u^*$. The evolutionary generation of scout bees reaches the optimal state when at least one scout bee in an evolutionary generation of scout bees reaches the optimal state.

Theorem 3 The evolutionary optimal state u^* is a closed set on the evolutionary state-space U of scout bees. The proof is as follows:

$$x_i^k = x^*(\forall u_i \in u^*,\ u_j \notin u^*, u_i, u_j \in S,\ u_i \in u^* \Rightarrow \exists\, x_i^k \in u_i) \tag{17}$$

$$\left|fitness\left(x_j^k\right)\right| < ||fitness(\xi)| - \varepsilon|\left(u_j \notin u^* \Rightarrow \forall\, x_j^k \in u_j\right) \tag{18}$$

According to Eqs. (9) and (11), $p(x_j^k x_i^k) = 0$, and the following formula is obtained from Eq. (14):

$$p(u_j|u_i) = \prod_{r=1}^{n} p\left(x_j^r|x_i^r\right) = 0 \tag{19}$$

which shows that once the evolutionary generation of BA reaches the optimal state, it will not be able to leave the optimal state, so u^* means the closed set on U.

From the perspective of evolution generation fitness, the fitness of a scout bee evolutionary generation is defined as the optimal fitness of the scout bee contained in the evolutionary generation, which is expressed by the following formula:

$$fitness(u) = \max_{x^i u} fitness(x^i) \tag{20}$$

Therefore, the fitness of an evolutionary generation is only determined by the scout bee with the best fitness.

From the perspective of the evolution generation group, the evolutionary generations of reconnaissance bees with the same fitness are grouped, which is expressed by G. The fitness of the evolutionary generation group is equal to the fitness of the evolutionary generation in the group, that is, $fitness(g) = fitness(U), u \in g$.

The sequence G $= \{g_1, g_2, \ldots, g_{Ng}\}$ is obtained by arranging the evolutionary generation groups in ascending order of fitness, where $fitness(g_i) < fitness(g_j), \forall 1 \le i < j \le Ng$, where Ng is the number of evolutionary generation state groups. According to the definition, the evolutionary generation group has a unique fitness, and g_{Ng} is the last group in the group sequence, which has the best fitness, so $g_{Ng} = u^*$.

From the perspective of evolution generation group transition, $\forall g_i, g_j (1 \le i, j \le Ng) g_i g_j$, the transition probability of the evolutionary generation state from g_i to g_j is recorded as $p_g(g_j|g_i)$ in the BA, which is calculated by the following formula:

$$p_g(g_j|g_i) = p(u_j \in g_j|u_i \in g_i) = \begin{cases} \sum_{u_i \in g_i} \sum_{u_j \in g_j} p\left(u_j|u_i\right), & i \le j \\ 0, & i > j \end{cases} \tag{21}$$

3.3 Convergence Analysis of BA

According to formulas (7) and (8), the probability that the optimal scout bees and other better scout bees return to x_{i_t} for the first time after k-times state transitions from state x_{i_t} can be calculated and expressed as:

$$f_{(x_{i_t})}^{(k)} = \begin{cases} 1 - p_e\left(x_{i_{t+1}}|x_{i_t}\right), & k = 1(Optimal\ scout\ bees) \\ 1 - p_b\left(x_{i_{t+1}}|x_{i_t}\right), & k = 1(Better\ scout\ bees) \\ 0, & k > 1 \end{cases} \tag{22}$$

When $fitness(x_{i_{t+1}}) > fitness(x_{i_t}), p_e\left(x_{i_{t+1}}|x_{i_t}\right) > 0$ and $p_b\left(x_{i_{t+1}}|x_{i_t}\right) > 0$, there is an equation:

$$\sum\nolimits_{k=1}^{\infty} f_{(x_{i_t})}^{(k)} = f_{(x_{i_t})}^{(1)} < 1 \tag{23}$$

This equation shows that the optimal reconnaissance bee and the better reconnaissance bee will never return to this state again with a positive probability from a certain state in the state space (determined by the greedy selection mechanism adopted in the neighbourhood search process of the BA).

The greedy selection mechanism is embodied by the decision variables represented by formula (9), which determines that the state transition of the optimal scout bees and the better scout bees are irreversible. Similarly, from Eqs. (10) and (11), the probability of random scout bees starting from state x_{i_t} and returning to x_{i_t}

$$f_{(x_{i_t})}^{(k)} = \left(\frac{Nx - 1}{Nx}\right)^{k-1} \cdot \frac{1}{Nx}, \quad k \geq 1 \tag{24}$$

An equation can be obtained according to formula (24) as follows:

$$\sum\nolimits_{k=1}^{\infty} f_{(x_{i_t})}^{(k)} = \lim_{k} \frac{1}{Nx} \cdot \frac{1 - \left(\frac{Nx-1}{Nx}\right)^{k}}{\frac{Nx-1}{Nx}} = 1 \tag{25}$$

where it is indicated that the random scout bee starts from a certain state in the state space and will return to this state again with probability 1; that is, the random scout bee always returns to the state space Ω. The random scout bees are a part of the evolutionary generation of scout bees, which determines that the individual state transition process in the evolutionary generation of the scout bee does not have convergence.

On the other hand, formula (21) can also be expressed as:

$$p_g(g_j|g_i) = \left[p_{ij}\right] = \begin{bmatrix} p_{11}\ p_{12} \cdots p_{1Ng} \\ 0\ \ p_{22} \ldots p_{2Ng} \\ \vdots\ \ \vdots\ \ \ddots\ \ \vdots \\ 0\ \ 0\ \ 0\ \ 1 \end{bmatrix} \tag{26}$$

where $\sum_{j=1}^{Ng} p_{ij} = 1, (1 \leq i \leq Ng)$. The transition probability matrix of the evolutionary generation group of BA is an upper triangular probability random matrix:

$$\begin{cases} j = \sum\limits_{i=1}^{Ng} \pi_i p_{ij} \\ \sum\limits_{j=1}^{Ng} \pi_j = 1 \end{cases} \tag{27}$$

where the static distribution $\pi = (\pi_1, \pi_2, \ldots \pi_{Ng})$ of the evolutionary generation group is equal to π = (0, 0, …, 0, 1) through the calculation of formula (27),

The static distribution represents the final distribution state of the evolutionary generation group of scout bees. The calculated static distribution shows that the evolutionary generation group will eventually converge to the end group g_{Ng} of the evolutionary generation group sequence, that is, the optimal state of the evolutionary generation.

The neighbourhood contraction strategy fails to obtain the improvement of the solution (the solution with higher fitness) in a neighbourhood search. The purpose is to make the algorithm search for the optimal solution with higher accuracy and speed, which is based on a hypothesis of the BA. This means that the scout bee has been close to the global or local optimal solution when the improvement cannot be obtained in the neighbourhood search. Due to this contraction strategy, the state transition probability of the optimal scout bees and the better scout bees can be reformulated as the following equation:

$$p_e(x(t+1) = x_{i_{t+1}}|x(t) = x_{i_t}) = \frac{nre}{\|\alpha \cdot ngh\|} \cdot q_1(x_{i_{t+1}}, x_{i_t}) \tag{28}$$

$$p_b(x(t+1) = x_{i_{t+1}}|x(t) = x_{i_t}) = \frac{nrb}{\|\alpha \cdot ngh\|} \cdot q_1(x_{i_{t+1}}, x_{i_t}) \tag{29}$$

where α is the neighbourhood contraction coefficient, satisfying $0 < \alpha < 1$. This shows that the reduced neighbourhood search range can improve the neighbourhood search process by transferring to a state with higher fitness from greater probability based on the assumption of the algorithm to improve the optimisation speed.

The strategy of site abandonment makes the algorithm discard food sources that cannot be continuously improved so that the scout bees escape the local optimal solution. The introduction of this strategy can transform Eq. (25) into the equation shown below.

$$p(g_j|g_i) = [p_{ij}] = \begin{bmatrix} p_{11} & p_{12} & \cdots & p_{1Ng} \\ p_{21} & p_{22} & \cdots & p_{2Ng} \\ \vdots & \vdots & \ddots & \vdots \\ p_{Ng1} & p_{Ng2} & \cdots & p_{NgNg} \end{bmatrix} \tag{30}$$

where $0 < p_{ij} < 1 \quad (i, j = 1,\ 2, \ldots, Ng);\ \sum_{j=1}^{Ng} p_{ij} = 1.$

In this case, the optimal state u^* is no longer a closed set in the state space Ω. The evolutionary states in the evolutionary group g_{Ng} of scout bees can be transferred to lower fitness with positive probability p_{Ngj} (j = 1, 2..., $Ng - 1$). The evolutionary sequence $\{u(t),\, t \geq 0\}$ of scout bees becomes the finite homogeneous irreducible Markov chain. The static distribution of evolutionary groups $\pi = (\pi_1, \pi_2, \ldots \pi_{Ng})$ can also be solved by Eq. (27) to $\pi_{Ng} < 1$, indicating that the probability of the evolutionary generation of the scout bees using the site abandonment strategy converging to the optimal state u^* of the evolutionary generation is less than 1.

Using the site abandonment strategy can help the algorithm escape the local optimal solution. To maintain the convergence performance of the algorithm, the strategy should be adjusted to discarding food sources that cannot obtain continuous improvement unless the food source has the current evolutionary generation of scout bees. If the best fitness is sampled, it should be stored in the current food source sequence rather than discarded. The adjusted site abandonment strategy not only helps the algorithm escape the local optimal solution but also ensures that the evolutionary generation of the scout bees converges to the optimal state u^* with probability 1.

4 Optimisation Method

In the previous section, the convergence performance was proven for BA using Markov chain theory. However, without the site abandonment strategy, BA is prone to being trapped in local optimal results. The dynamic particle swarm optimisation algorithm (DPSOA) is a dynamic intelligent optimisation algorithm simulating the migration and clustering behaviour of birds in the process of foraging [41]. In this work, a combination model of the BA and DPSOA was established to improve the local search capability of the BA by integrating the advantages of bird and bee colony foraging behaviour.

4.1 The Combination Model of the BA and DPSOA

The optimisation goal is to maximise the probability of scout bees selecting tasks, which combines the advantages of birds and bees. After the BA global search, the DPSOA is used for the local search of the BA to solve the local optimisation problem

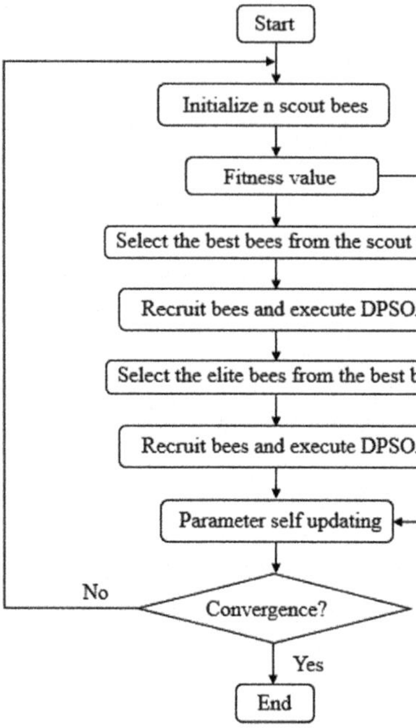

Fig. 2 Flowchart of the BA-DPSOA

of the BA. The optimisation effect is improved by updating the speed and position of the best bees, elite bees, and scout bees and iterating continuously until convergence. The flow chart for the combination model of the BA and DPSOA (BA-DPSOA) is shown in Fig. 2 based on Fig. 1.

According to previous studies [42–44], the fitness value of this method can be calculated in the following formulas.

$$P_{ik} = \frac{F_i^{\alpha}\left(\frac{1}{\|D_{ik}\|_2}\right)^{\beta}}{\sum_{j=1}^{M} F_j^{\alpha}\left(\frac{1}{\|D_{ik}\|_2}\right)^{\beta}} \tag{31}$$

$$F_i = \frac{f_i}{\sum_{j=1}^{M} f_i} \tag{32}$$

where α and β are the control parameters, α, $\beta = 1$ in this study; F represents the priority of the task, and f_i is the priority of task i; and D_{ik} represents the distance between the k-th scout bee and the target. The greater the value of P_{ik} is, the better the task allocation effect. The maximum value is the optimal position of the task allocation of scout bees.

4.2 Optimisation of Local Search Based on DPSOA

The DPSOA was used in this study based on sensitive particles, which is a typical centralised dynamic particle swarm optimisation algorithm [45–47]. The feature means "dynamic" by randomly generated sensitive particles and starting with each subsearch space divided evenly. The fitness value of sensitive particles is measured by the response threshold after the fitness value is calculated.

$$\Delta f_i = f_i(k+1) - f_i(k) \tag{33}$$

$$F = \sum_{i=1}^{n} |\Delta f| \tag{34}$$

where f_i represents the fitness value corresponding to the sensitive particle calculated at each iteration; f_i represents the fitness value difference of two adjacent iterations of sensitive particles; and F is the sum of absolute values of all differences.

This indicates that the search environment has changed and needs to be dynamically adjusted to address environmental change when the value of f is not equal to 0, as shown in the following equation:

$$\begin{cases} V_{(i)} = rand(M) \cdot V_{max} \\ X_{(i)} = rand(M) \cdot X_{max} \end{cases} \tag{35}$$

where $rand\,(M)$ is an m-dimensional vector, in which any element is a random number in [0,1]; $V_{(i)} = [V_{i1}, V_{i2}, ..., V_{im}]$, $X_{(i)} = [X_{i1}, X_{i2}, ..., X_{im}]$ represents the i-th particle selected from the population for reinitialisation, and m represents the dimension of the particle; and $V_{\max}$ represents the maximum particle velocity, generally $V_{\max} = X_{\max}$.

The algorithm will update the position and speed of particles in a certain proportion when the value of F exceeds the preset F [48]. The update formula is shown as follows:

$$\begin{cases} v_x(t+1) = w \cdot v(t) + c_1 rand(M) \cdot (pbest_x - x(t)) \\ \qquad + c_2 \cdot rand(M) \cdot (gbest_x - x(t)) \\ v_y(t+1) = w \cdot v(t) + c_1 rand(M) \cdot \left(pbest_y - x(t)\right) \\ \qquad + c_2 \cdot rand(M) \cdot \left(gbest_y - x(t)\right) \end{cases} \tag{36}$$

$$\begin{cases} x(t+1) = x(t) + v_x(t+1) \\ y(t+1) = y(t) + v_y(t+1) \end{cases} \tag{37}$$

where t represents the number of generations; w represents the inertia weight factor; c_1 and c_2 represent acceleration constants; and $pbest$ and $gbest$ represent the local optimal solution and the global optimal solution, respectively.

4.3 Evaluating Indicators

The purpose of the combination model is to allocate the scout bee to the target location to improve the accuracy of the local search. The closer to the target, the better the result of task allocation. An ordered pair [*x*,*y*] represents the actual coordinates assigned to the scout bees where $x = (x_1...x_m)^T, y = (y_1...y_m)^T, m \geq 1$; $[x^*, y^*]$ represents the target coordinates where $x^* = \left(x_1^* \ldots x_m^*\right)^T, y^* = \left(y_1^* \ldots y_m^*\right)^T$, and $m \geq 1$. The equation of the evaluation index is as follows, where the smaller the values of a and b in the equation are, the better the effect of task allocation of this method.

$$\begin{cases} x_i^* - x_i = a \\ y_i^* - y_i = b \end{cases}, 1 \leq i \leq m \tag{38}$$

where a and b are rational numbers that are as small as possible.

4.4 Experiment and Results

To verify the effectiveness of the method for the task allocation of scout bees based on the combination model of the BA and DPSOA proposed in this study. The simulation experiment was designed in MATLAB. The experimental parameters shown in Table 3 are based on Table 1.

The initial state is shown in Fig. 3. The Rosenbrock function was used to test the proposed BA-DPSOA performance, as shown in Fig. 4. As a comparison, BA performance is shown in Fig. 5.

As shown in Fig. 6, all blue points are gathered to the red target point. By comparison, there are a few blue points that are not gathered to the target point because they easily fall into local optima, as shown in Fig. 7. This shows that it works well for the effect of task allocation of scout bees for target location. The study can conclude that the probability of the combination model of the BA and DPSOA falling into local optima is much lower than that of the conventional BA from the simulation experiment.

Table 3 Experimental parameters

Parameter name	Value	Note
n	23	Marked as blue
nsc	0.8	/
w	0.7	/
v	$x_{min}/10 - x_{max}/10$	/
$pbest$	13	Marked as green
$gbest$	1	Marked as red
$stlim$	10	/

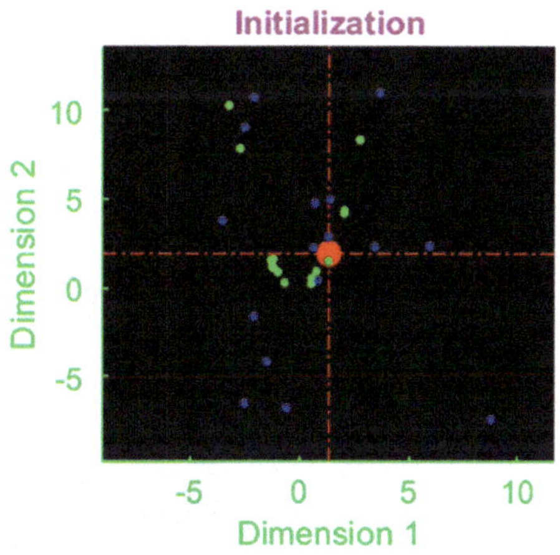

Fig. 3 Initial state

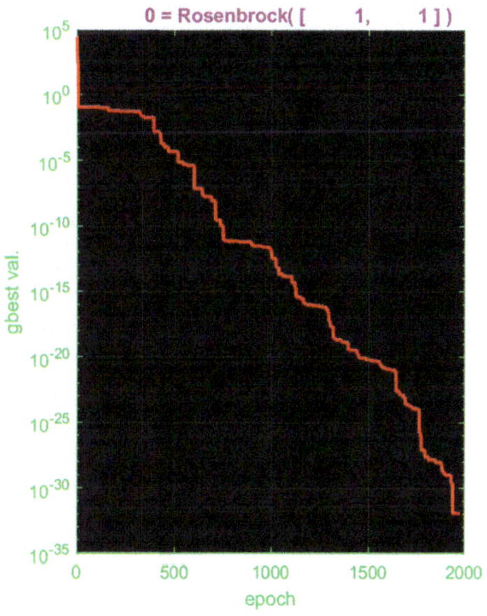

Fig. 4 BA-DPSOA convergence performance

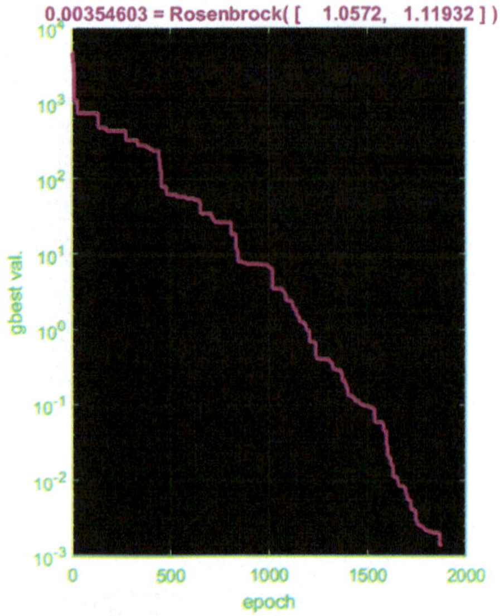

Fig. 5 BA convergence performance

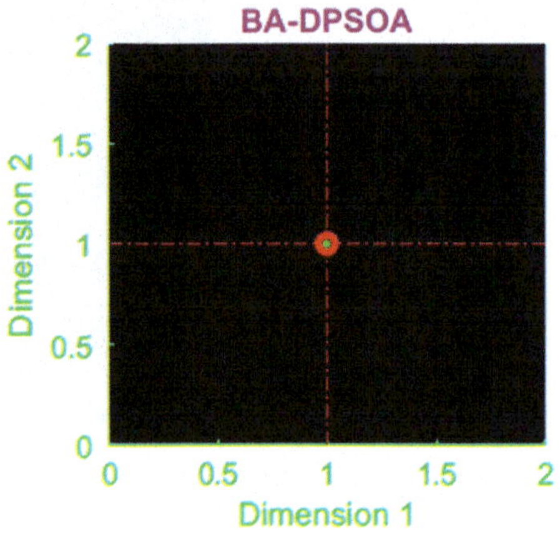

Fig. 6 Optimisation result of local search using BA-DPSOA

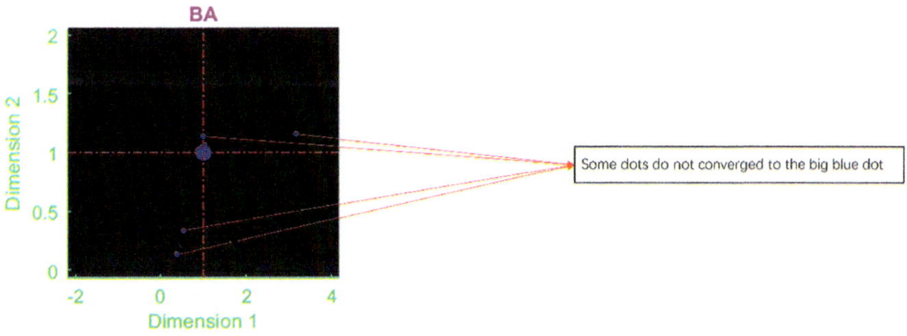

Fig. 7 Local search result using BA

5 Conclusion

As a branch of mathematics, optimisation theory includes both quantitative research on the optimal solution and finding the optimal solution. The wide application of optimisation technology has enormous economic benefits and social value. The local search process of BA lacks efficient exploitation due to the random assignment of search agents throughout neighbourhoods, which lowers algorithm accuracy and results in delayed convergence. Therefore, the BA confronted this problem of whether it can finally converge in solving many optimisation problems. The convergence performance of the BA is analysed based on the mathematical theory of the Markov chain. The calculated stationary distribution proves that the evolutionary generation of BA can converge to the optimal solution of the objective function with probability 1, which proves the effectiveness of the calculation method obtained from the objective natural phenomenon of bee colony foraging. To solve the problem that the BA easily falls into local optima, this study proposed a method that combined the BA and DPSOA (BA-DPSOA) to improve the convergence performance. The simulation experiment was conducted in MATLAB. The result shows that the BA-DPSOA can be an excellent method to help the conventional BA avoid falling into local optima. However, while the DPSOA was employed to optimise the BA, it has been insufficiently studied compared to other optimisation models. For future research, more intelligent algorithms can be used to compare the convergence performance of BA based on our proposed analysis methods to further verify the contributions of our study.

References

1. Seeley TD (1996) The wisdom of the hive: the social physiology of honey bee colony. Harvard University Press
2. Passino KM, Seeley TD, Visscher PK (2008) Swarm cognition in honey bees. Behav Ecol Sociobiol 62(3):401–414

3. Sato T, Hagiwara M (1998) Bee system: finding solution by a concentrated search. IEEJ Trans Electron Inf Syst 118(5):721–726
4. Karaboga D (2005) An idea based on honey bee swarm for numerical optimization. Technical Report TR06, Erciyes University
5. Pham D, Ghanbarzadeh A, Koc E, Otri S, Rahim S, Zaidi M (2005) The bees algorithm. Technical Note, Manufacturing Engineering Centre, Cardiff University, UK, pp 44–48
6. Tolabi HB, Ali MH, Ayob SBM, Rizwan (2014) Novel hybrid fuzzy-Bees algorithm for optimal feeder multi-objective reconfiguration by considering multiple-distributed generation. Energy 71:507–515
7. Akpinar Ş, Baykasoğlu A (2014) Modeling and solving mixed-model assembly line balancing problem with setups. Part II: a multiple colony hybrid bees algorithm. J Manuf Syst 33(4):445–461
8. Sarailoo M, Rahmani Z, Rezaie B (2015) A novel model predictive control scheme based on bees algorithm in a class of nonlinear systems: application to a three tank system. Neurocomputing 152:294–304
9. Zarei K, Atabati M, Kor K (2014) Bee algorithm and adaptive neuro-fuzzy inference system as tools for QSAR study toxicity of substituted benzenes to Tetrahymena Pyriformis. Bull Environ Contam Toxicol 92(6):642–649
10. Deghbouch H, Debbat F (2021) A hybrid bees algorithm with grasshopper optimization algorithm for optimal deployment of wireless sensor networks. Intel Artif 24(67):18–35
11. Rudolph G (1994) Convergence analysis of canonical genetic algorithms. IEEE Trans Neural Netw 5(1):96–101
12. Sun X, Yang L, Gao L, Zhang B, Li S, Li J (2015) Hyperspectral image clustering method based on artificial bee colony algorithm and Markov random fields. J Appl Remote Sens 9(1):095047
13. Sagayam KM, Hemanth DJ (2018) ABC algorithm based optimization of 1-D hidden Markov model for hand gesture recognition applications. Comput Ind 99:313–323
14. Kıran MS, Gündüz M (2013) A recombination-based hybridization of particle swarm optimization and artificial bee colony algorithm for continuous optimization problems. Appl Soft Comput 13(4):2188–2203
15. Zhang Y, Wang S, Ji G (2015) A comprehensive survey on particle swarm optimization algorithm and its applications. Math Probl Eng 2015:931256
16. Alaidi AH, Der CS, Leong YW (2021) Systematic review of enhancement of artificial bee colony algorithm using ant colony pheromone. Int J Interact Mob Technol 15(16):173
17. Ang MC, Ng KW, Pham DT (2013) Combining the Bees Algorithm and shape grammar to generate branded product concepts. Proc Inst Mech Eng, Part B: J Eng Manuf 227(12):1860–1873
18. Pham DT, Ang M, Ng K, Otri S, Darwish AH (2008) Generating branded product concepts: comparing the bees algorithm and an evolutionary algorithm, pp 398–403
19. Pham DT, Otri S, Darwish AH (2007) Application of the Bees algorithm to PCB assembly optimisation, pp 511–516
20. Pham D, Castellani M, Fahmy A (2008) Learning the inverse kinematics of a robot manipulator using the bees algorithm. IEEE, pp 493–498
21. Pham DT, Castellani M (2009) The bees algorithm: modelling foraging behaviour to solve continuous optimization problems. Proc Inst Mech Eng C J Mech Eng Sci 223(12):2919–2938
22. Darwish AH (2009) Enhanced Bees algorithm with fuzzy logic and Kalman filtering. Cardiff University, UK
23. Packianather M, Landy M, Pham D (2009) Enhancing the speed of the Bees algorithm using pheromone-based recruitment. IEEE, pp 789–794
24. Pham QT, Pham DT, Castellani M (2012) A modified bees algorithm and a statistics-based method for tuning its parameters. Proc Inst Mech Eng, Part I: J Syst Control Eng 226(3):287–301
25. Ebubekir K (2010) The bees algorithm theory, improvements and applications. In: Manufacturing Engineering Centre School of Engineering University of Wales. Cardiff United Kingdom

26. Pham D, Ghanbarzadeh A, Otri S, Koç E (2009) Optimal design of mechanical components using the bees algorithm. Proc Inst Mech Eng C J Mech Eng Sci 223(5):1051–1056
27. Ang MC, Ng KW, Pham DT, Soroka A (2013) Simulations of PCB assembly optimisation based on the Bees algorithm with TRIZ-inspired operators. Springer, pp 335–346
28. Pham DT, Haj Darwish A, Eldukhri EE (2009) Optimisation of a fuzzy logic controller using the bees algorithm. Int J Comput Aided Eng Technol 1(2):250–264
29. Packianather MS, Yuce B, Mastrocinque E, Fruggiero F, Pham DT, Lambiase A (2014) Novel genetic bees algorithm applied to single machine scheduling problem. IEEE, pp 906–911
30. Pham D, Soroka AJ, Ghanbarzadeh A, Koc E, Otri S, Packianather M (2006) Optimising neural networks for identification of wood defects using the bees algorithm. IEEE, pp 1346–1351
31. Jevtic A, Gutiérrez A, Andina D, Jamshidi M (2011) Distributed bees algorithm for task allocation in swarm of robots. IEEE Syst J 6(2):296–304
32. Pham D, Pham Q, Ghanbarzadeh A, Castellani M (2008) Dynamic optimisation of chemical engineering processes using the bees algorithm. IFAC Proc Vol 41(2):6100–6105
33. Abdullah S, Alzaqebah M (2013) A hybrid self-adaptive bees algorithm for examination timetabling problems. Appl Soft Comput 13(8):3608–3620
34. Pham DT, Al-Jabbouli H, Mahmuddin M, Otri S, Darwish AH (2008) Application of the Bees algorithm to fuzzy clustering
35. Pham DT, Afify A, Koc E (2007) Manufacturing cell formation using the Bees algorithm
36. Pham D, Lee J, Haj Darwish A, Soroka A (2008) Multi-objective environmental/economic power dispatch using the bees algorithm with pareto optimality, pp 422–430
37. Ruz GA, Goles E (2013) Learning gene regulatory networks using the bees algorithm. Neural Comput Appl 22(1):63–70
38. Xu S, Yu F, Luo Z, Ji Z, Pham DT, Qiu R (2011) Adaptive bees algorithm—bioinspiration from honeybee foraging to optimize fuel economy of a semi-track air-cushion vehicle. Comput J 54(9):1416–1426
39. Xu W, Zhou Z, Pham D, Liu Q, Ji C, Meng W (2012) Quality of service in manufacturing networks: a service framework and its implementation. Int J Adv Manuf Technol 63(9):1227–1237
40. Douc R, Moulines E, Priouret P, Soulier P (2018) Markov chains. Springer
41. Wang D, Tan D, Liu L (2018) Particle swarm optimization algorithm: an overview. Soft Comput 22(2):387–408
42. Shi Y, Eberhart RC (1999) Empirical study of particle swarm optimization. IEEE, pp 1945–1950
43. Liu Y, Qin Z, Shi Z, Lu J (2007) Center particle swarm optimization. Neurocomputing 70(4–6):672–679
44. Tkach I, Edan Y, Jevtic A, Nof SY (2013) Automatic multi-sensor task allocation using modified distributed bees algorithm. IEEE, pp 1401–1406
45. Trelea IC (2003) The particle swarm optimization algorithm: convergence analysis and parameter selection. Inf Process Lett 85(6):317–325
46. Jiao B, Lian Z, Gu X (2008) A dynamic inertia weight particle swarm optimization algorithm. Chaos Solitons Fractals 37(3):698–705
47. Nickabadi A, Ebadzadeh MM, Safabakhsh R (2011) A novel particle swarm optimization algorithm with adaptive inertia weight. Appl Soft Comput 11(4):3658–3670
48. Wang L, Singh C (2009) Multicriteria design of hybrid power generation systems based on a modified particle swarm optimization algorithm. IEEE Trans Energy Convers 24(1):163–172

Development of the Bees Algorithm Toolkit for Optimisation in LabVIEW

Murat Sahin and D. T. Pham

Abstract This chapter presents a Bees Algorithm (BA) Optimisation Toolkit developed in LabVIEW. The BA is an effective optimisation algorithm that mimics the nectar-foraging behaviour of honey bees. LabVIEW is a powerful program for data acquisition and control applications that is very popular in industry. There are tools within the scope of optimisation within LabVIEW, but there is no toolkit for the BA. In this chapter, the preparation of the BA in LabVIEW is explained step by step. The toolkit has two parts. Optimisation can be performed on standard continuous test functions with the first part, and new functions and problems can be defined and solved. It has been observed that the minimum values of complex continuous functions can be reached quickly in experimental studies. With the second part, combinatorial optimisation can be conducted on travelling salesman problems (TSPs), and new TSPs can be defined and solved. It has been shown experimentally that minimum-cost solutions can be found in a small number of iterations using this toolkit.

Keywords Bees algorithm · LabVIEW · Continuous optimisation · Combinatorial optimisation · Travelling salesman problem · Nearest neighbour method · Multi-insert · Shrinking · Test functions · Graphical programming

M. Sahin (✉)
School of Electrical and Electronic Engineering, University College Dublin, Belfield, Dublin 4, Ireland
e-mail: murat.sahin1@ucd.ie

D. T. Pham
Department of Mechanical Engineering, University of Birmingham, Birmingham, UK
e-mail: d.t.pham@bham.ac.uk

D. T. Pham and N. Hartono (eds.), *Intelligent Engineering Optimisation with the Bees Algorithm*, Springer Series in Advanced Manufacturing,
https://doi.org/10.1007/978-3-031-64936-3_4

1 Introduction

Metaheuristic optimisation methods, inspired by nature, are effectively used to solve complex problems [1]. Most of these methods are based on specific properties and behaviours of biological, molecular, insect swarm, and neurobiological systems. They are known as population-based algorithms. The aim is that the majority of the population approaches the optimum value for the solution, as in nature [2].

The Bees Algorithm (BA), developed by Pham et al., is inspired by how honey bees find food. The method starts with a random search cluster near good results in further iterations, and thanks to the global search section, it continues to search without being stuck at local minima [3]. Since the basic algorithm does not contain any equations, it has a more straightforward structure than other algorithms [4]. When the literature is examined, it is seen that it is used in mechanical systems [5], controller coefficient optimisation [6], robot designs [7], the medical field [8], and route optimisations [9–11].

It is known that the BA has been coded and used in applications such as C++ [12], MATLAB [13], and Python [14]. Another popular program is LabVIEW (Laboratory Virtual Instrument Engineering Workbench), developed by National Instruments. It is a powerful and flexible program development tool that uses the G (graphical) programming language [15]. LabVIEW is employed extensively in signal processing, energy systems, and industrial applications [16, 17]. Within the scope of optimisations carried out in these areas, the "Optimisation Toolkit" is used. This toolkit contains blocks for solving first- and second-order problems. There are also some blocks for finding the minimum values of continuous functions [18]. There is the AMA (Advanced Metaheuristics Algorithms) toolkit developed by ENIT (a community that develops a toolkit for LabVIEW) for complex optimisation. Within this toolkit, there are blocks of Particle Swarm Optimisation, Ant Colony Algorithm, Gravity Search Algorithm, Teaching–Learning Based Optimisation, and Gray Wolf Algorithm [19]. Apart from these, the researchers, within the scope of optimisation studies, Particle Swarm Optimisation [20], Bat Algorithm [21], Ant-Lion Optimisation [22], and Gray Wolf Algorithm [23] methods are coded and used in LabVIEW. When the literature was searched, it was seen that BA was not coded in LabVIEW, and it was decided to conduct this study. This chapter first gives information about the BA, especially the local search part (Sect. 2), and then details the BA toolkit for function optimisation (Sect. 3) and combinatorial optimisation (Sect. 4). The results for benchmark problems are presented to show the toolkit's effectiveness.

2 Bees Algorithm

To be able to gather food efficiently, honey bees go to more productive areas with more bees. The basic philosophy of BA is also based on this method. More searches are performed near the promising results found in random searches (Elite site), while

Fig. 1 Pseudocode of Elite site search

```
Start
for i = 1 : e
    best bee = infinitive
    for j = 1 : nep
        Get current bee n(i)
        Neighborhood search
        if 'new bee' < 'best bee'
            best bee = new bee
    if 'best bee' < 'current bee'
        current bee = best bee
End
```

fewer searches are performed in other locales (Best site). Random searches are also continued, and global optimums are tried to be reached [24, 25]. In the algorithm, the population number is named by the number of scout bees (n). The local search area (Best site) is expressed by m, the elite area by e, the number of bees searching in the elite area by nep, the number of bees searching in the other local area by nsp, the neighbourhood by ngh, and the number of iterations by I [26].

The Elite site search consists of two nested loops. In the outer loop, data are taken from the current population; in the inner loop, searches are made in the neighbourhood of these data. The best value found after the searches is compared with the current value, and if it is better than the current value, it is included in the population. The pseudocode of the Elite Site Search is given in Fig. 1.

3 Development of the LabVIEW Toolkit for Continuous Problems

The front panel of the LabVIEW toolkit is given in Fig. 2. It consists of four parts. Algorithm parameters are entered in the 'Parameters' section. There are ten different continuous optimisation functions in the 'Test Functions' section. (Equations of these functions are given in Appendix [27].) In the 'New Function' section, the user can define new functions. In the other section, optimisation of combinatorial problems can be carried out.

3.1 Parameters

The BA starts with determining the parameters. The parameter section is shown in Fig. 3.

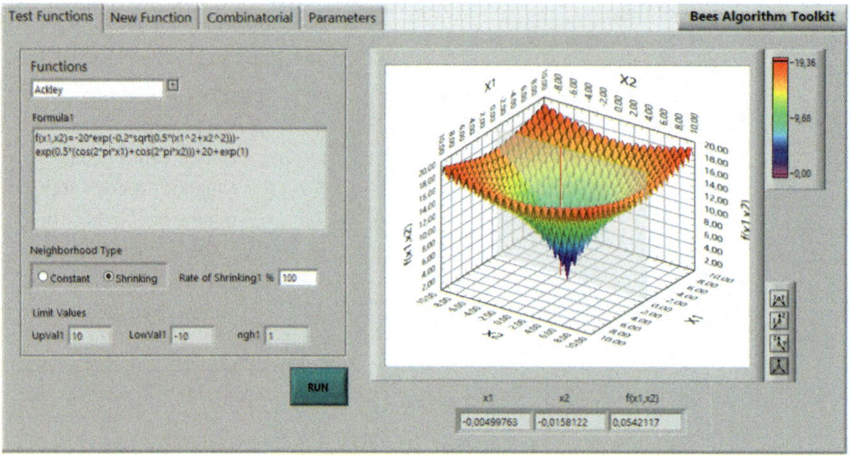

Fig. 2 Front panel of the optimisation tool (test function)

Fig. 3 BA parameter section

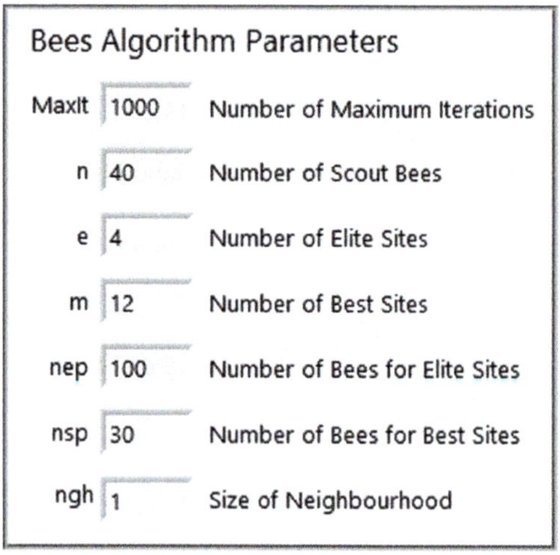

3.2 Main Algorithm of Continuous Optimisation

Before the main algorithm, the problem must be defined. In this section, the Ackley function is shown as an example. The equation of the function for two variables is given in (1) [27], and the LabVIEW code is given in Fig. 4. All functions were prepared similarly and created as subprograms.

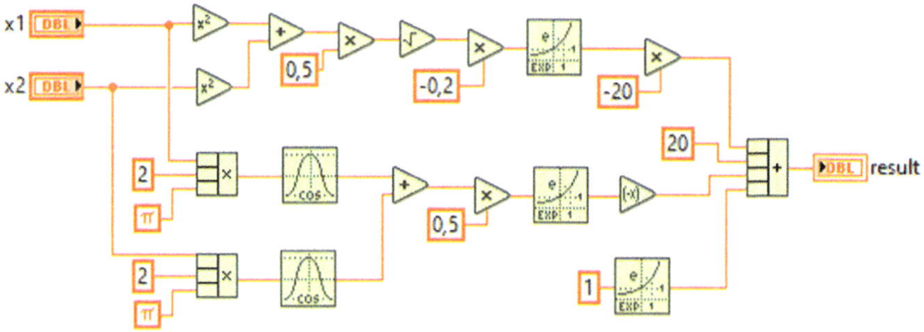

Fig. 4 LabVIEW code of the Ackley function

$$f(x) = -20\exp\left(-0.2\sqrt{\frac{1}{2}\sum_{i=1}^{2}x_i^2}\right) - \exp\left(\frac{1}{2}\sum_{i=1}^{2}\cos(cx_i)\right) + 20 + \exp(1) \quad (1)$$

The second step of the algorithm is to generate the first population. The first population is randomly generated within the limit values. It is shown in Fig. 5. The × **1** and × **2** variables are generated with the "Random Number" block. These numbers are sent to the function shown in Fig. 6, where the f(x) value is calculated. In this way, members of the first population and the cost of the members are established. There are ten different subprograms in the Case Structure. The values of UpVal1 and LowVal1 are the limits of the variables.

After the generation of the first population, the local search section begins. The first part of the local search is the elite site search. It is performed to find a better site in the neighbourhood of members from the elite site. In this context, a random value is created within the boundaries of **ngh** and added to the current member's value. The generation of a new neighbour is given in Fig. 6.

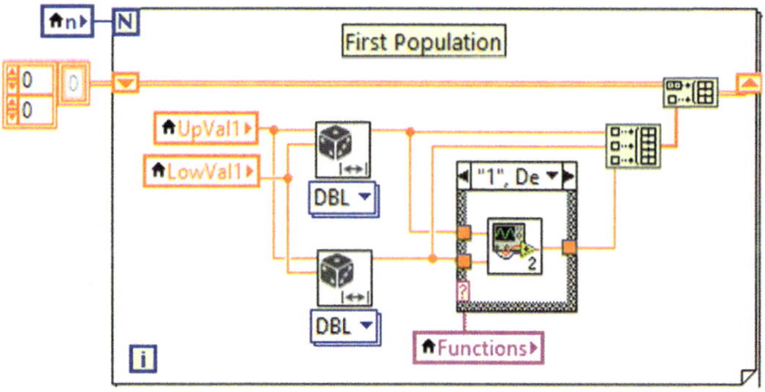

Fig. 5 Generation of the first population

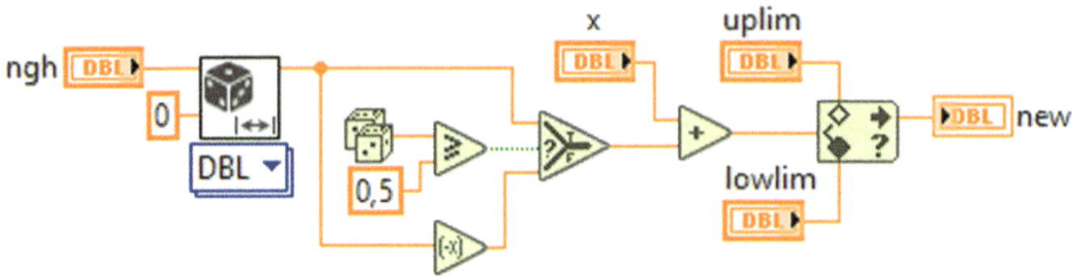

Fig. 6 Generation of a new neighbour

Fig. 7 The shrinking neighbourhood

Two different neighbourhood methods were prepared for the program. The first is a fixed neighbourhood, and the second is the shrinking neighbourhood method [28]. The code for the shrinking neighbourhood is given in Fig. 7.

The elite site search code is given in Fig. 8. It consists of two nested '**for**' loops. In the inner **for** loop, the best neighbour is chosen. In the outer **for** loop, the best neighbour is compared with the current member of the population. As a result of this process, the algorithm continues with the best member. In the Best sites search section, only the limits for loops change.

The last step is global search. The global search section is similar to the generation of the first population. For sites excluded from the local search, random new members are generated and included in the population to replace old members. The global search code is given in Fig. 9.

3.3 Creating a New Function

In the second part, the user can prepare new functions (Fig. 10). In this context, the following user interface was prepared. From this interface, variable names, equations for the function, limit values, neighbourhood size, and neighbourhood type can be filled. Many mathematical functions used in C and MATLAB can also be used here. (For example, abs, sqrt, sin, log, and sign.) The block diagram of this section is given in Fig. 11. "Eval Formula Node" is used in this section.

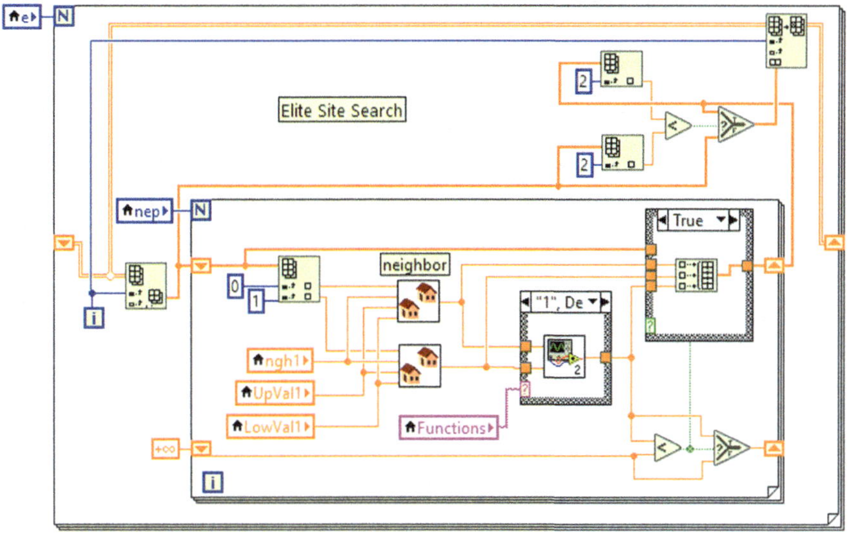

Fig. 8 Elite site search

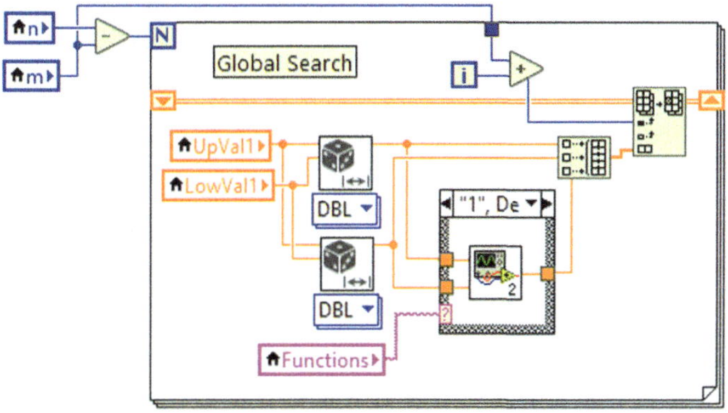

Fig. 9 Global search

3.4 Experiments and Results

In the experimental studies, the repeatability of the algorithm was checked by performing ten tests for each function. Table 1 shows information about the functions and the test results.

As seen from the table, with the Bees Algorithm Toolkit, the minimum values of the functions or values that are very close to the minimum can be found. The average values of the tests are also very close to the minimum. Thus, the repeatability of the toolkit has also been seen. The algorithm reaches optimum values quickly

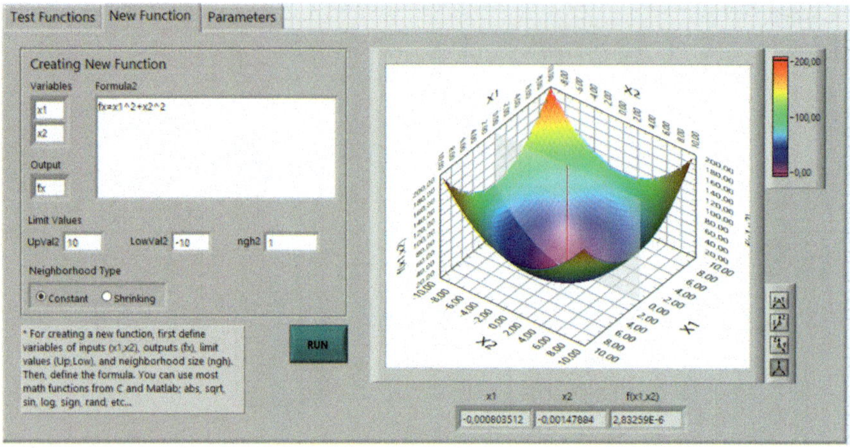

Fig. 10 Creating new functions

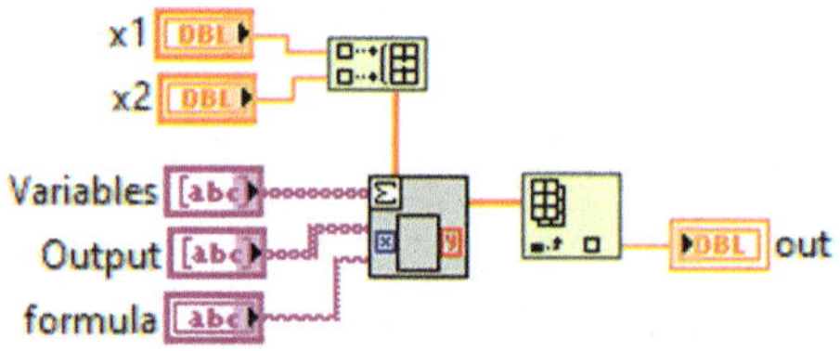

Fig. 11 Block diagram of creating new functions

Table 1 Functional data and test results

Functions			Test Results	
Name	Limit values	Best cost	Avg. cost	Min. cost
Ackley	[− 10, 10]	0	3.94E-11	6.61E-12
Cross in tray	[− 10, 10]	− 2.06261	− 2.06261	− 2.06261
Drop wave	[− 5.12, 5.12]	− 1	− 1	− 1
Egg-holder	[− 512, 512]	− 959.6407	− 959.641	− 959.641
Holder table	[− 10, 10]	− 19.2085	− 19.2085	− 19.2085
Levy	[− 10, 10]	0	1.41E-21	1.97E-21
Rastrigin	[− 5.12, 5.12]	0	0	0
Schaffer	[− 100, 100]	0	0	0
Schwefel	[− 500, 500]	0	2.55E-05	2.55E-05
Shubert	[− 10, 10]	− 186.7309	− 186.731	− 186.731

Table 2 Comparison of results

Functions		BA	NAIWO [29]	SaVOA [30]	HA [31]	EBAT [32]
Name	Best					
Ackley	0	3.94E-11	1.70E-6	2.84E-15	0.00590	**6.03E-16**
Cross in Tray	– 2.06261	**– 2.06261**	**– 2.0626**	–	**– 2.06261**	–
Drop Wave	– 1	**– 1**	**– 1**	6.87E-02	– 0.99972	**– 1**
Egg-holder	– 959.6407	**– 959.641**	– 62.4327	–	–	–
Holder Table	– 19.2085	**– 19.2085**	–	–	**– 19.2085**	–
Levy	0	1.41E-21	2.1E-12	**0**	1.46E-06	1.808471
Rastrigin	0	**0**	–	**0**	2.47E-05	3.41E-13
Schaffer	0	**0**	6.6E-17	2.55E-10	9.07E-08	3.02E-15
Schwefel	0	**2.55E-05**	7.8959	–	–	–
Shubert	– 186.7309	**– 186.731**	–	–	–	**– 186.731**

The outputs of the algorithm that obtains the best results for the problems are shown in bold

with the first iterations. Algorithm parameters have a direct relationship in finding minimum values. The parameters shown in the previous section are successful for these functions. If the results are not satisfied, the parameters can be changed. The best values found for some problems are shown in the Appendix.

To see the competitiveness of the Bees Algorithm (BA), it has been compared with different algorithms tested on the same functions. These algorithms are niche-based adaptive invasive weed optimisation (NAIWO) [29], self-adaptive virus optimisation algorithm (SaVOA) [30], hydrozoan algorithm (HA) [31], and enhanced bat algorithm (EBAT) [32]. The comparison results are given in Table 2. As seen from the table, the BA toolkit can compete with other algorithms. It achieved the best result in 8 functions. It also reached the Ackley function's third-best value and the Levy function's second-best value.

4 Development of the LabVIEW Toolkit for Combinatorial Problems

In a TSP, a salesperson has to tour different cities, starting from a city and returning to it after visiting each of the other cities only once. The optimisation aims to complete the trip with the minimum route length. The distance between two cities is found by the Euclidean equation given below [33]:

$$d(T(i), T(i+1)) = \sqrt{(x_i - x_{i+1})^2 + (y_i - y_{i+1})^2} \tag{2}$$

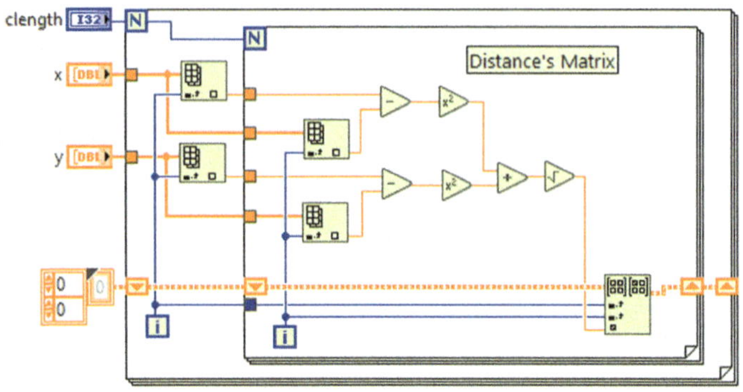

Fig. 12 Matrix of distance

x and y indicate city coordinates. The total cost is given by the following equation [33]:

$$\sum_{i=1}^{n-1} d(T(i), T(i+1)) + d(T(n), T(1)) \quad (3)$$

A matrix was created to be used in distance calculations. The distance of each city to any other city is calculated using this matrix. It is shown in Fig. 12.

4.1 Generation of Population

The nearest neighbour method (NNM) was used in the population generation section. The first city is chosen randomly. The next city is closest to the first city. This process continues in the same way [34].

In the first step, a neighbourhood matrix (Fig. 13) was created for each city within the scope of the NNM method. For each city, a close-by-distance ranking is made according to the distance to other cities.

A random city is determined in the second step, and NNM is applied, starting from that city (Fig. 14). With the first **for** loop, the randomly selected city and the second-ranked city are taken using the neighbourhood matrix. The closest to this city is found with the second for loop and added to the route. The same city should only be added to the route once. For this reason, control is made within the **while** loop.

In the last step, the length of the route created with NNM is calculated. Route and length values are written in the same array. The last value of the array is the length. It is shown in Fig. 15. The NNM shown in Fig. 14 has been added to this section as a subprogram.

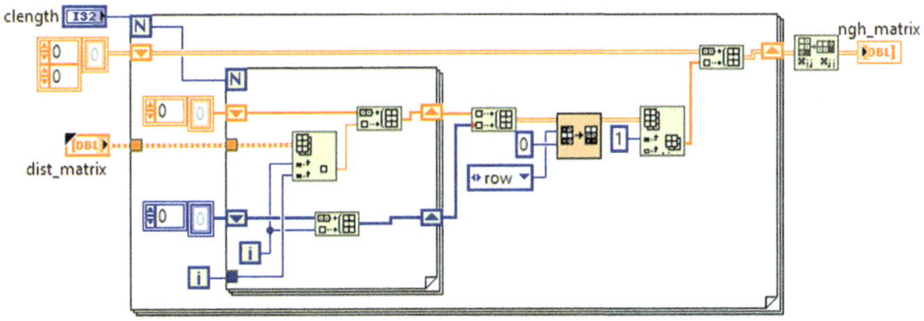

Fig. 13 Neighbourhood matrix

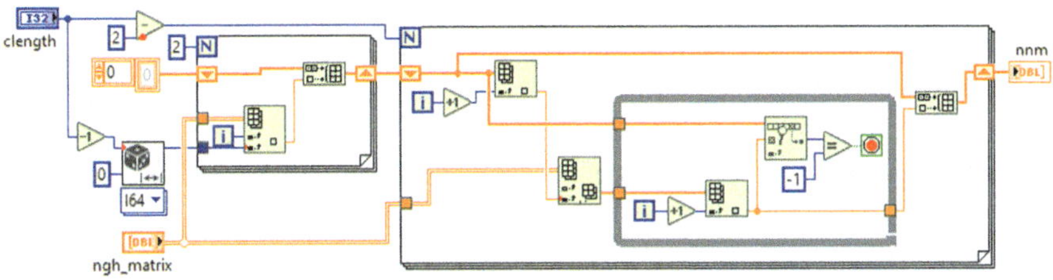

Fig. 14 Nearest neighbour method

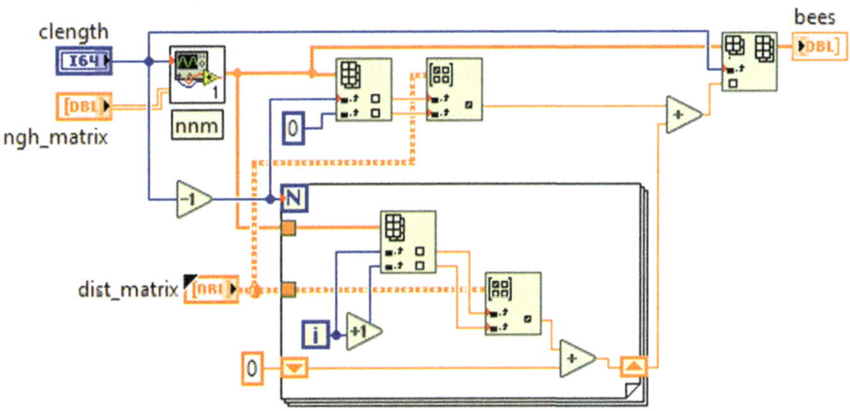

Fig. 15 Generation of the population

With NNM, a much lower cost route is produced compared to the randomly generated route. In this way, optimum routes can be reached more quickly. In Fig. 16, a randomly generated route and a route created with NNM are given for KroA100.

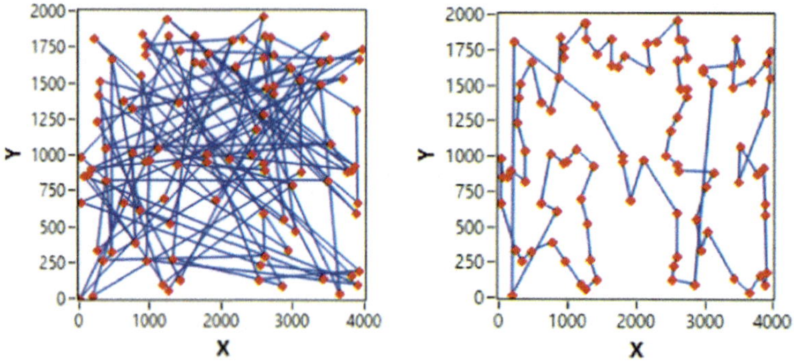

Fig. 16 Random and NNM route

4.2 *Local Search*

Local search is different from continuous problems in combinatorial problems. For TSPs, the locations of the cities on the route are changed with some functions, trying to find new routes with less cost. In one of the recent studies for the Combinatorial Bees Algorithm (CBA), Ismail et al. [11] used three functions in the local search section: swap, insert and reverse.

The order of two randomly selected cities is changed with the swap operator. Cities are chosen from the neighbourhood. The first city is randomly selected, and the other is randomly selected among the nearest neighbours (ngh). Here, too, the **while** loop is used to avoid selecting the same city. The swap function is shown in Fig. 17.

The insert function adds a randomly selected city to a tour. The insertion site is also chosen randomly. After the insert operation, the city duplicated by the inserted city is removed from the tour. The insert function is shown in Fig. 18.

The order between two randomly selected cities is reversed in the inversion function. The inversion function is shown in Fig. 19.

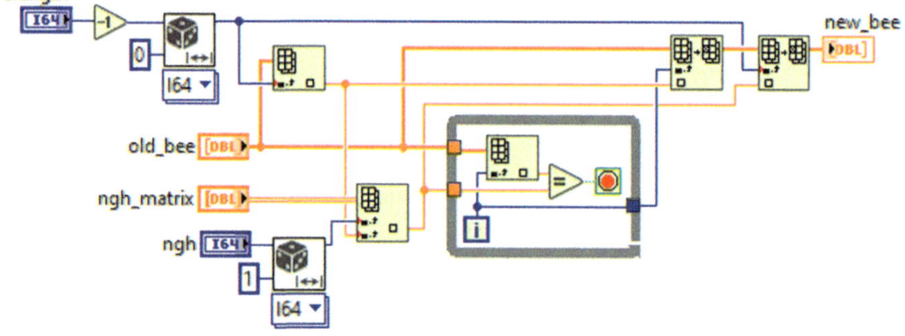

Fig. 17 Swap function

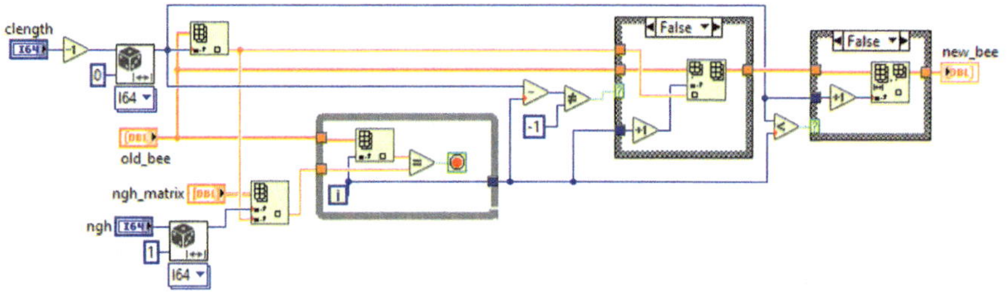

Fig. 18 Insert function

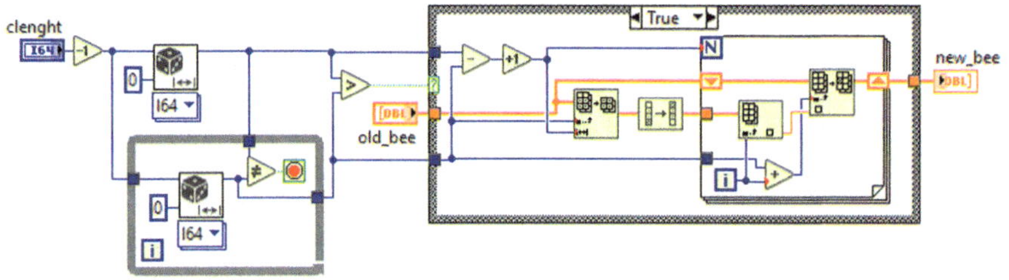

Fig. 19 Inversion function

In this study, multi-insert was used as the fourth function. Multiple cities are taken from a randomly selected region and placed in a new area in the neighbourhood. This function is more complex than others. When performing multiple selections, detecting between close neighbours and moving to a close neighbour en masse is necessary. The function is shown in Fig. 20.

These operators are used randomly in each loop, with more numbers on elite sites and fewer in the other local search section. The elite site search is shown in Fig. 21. The main structure here is the same as that given in Fig. 8. In the continuous problem, random numbers were selected in neighbourhoods, but the random function is selected in this section.

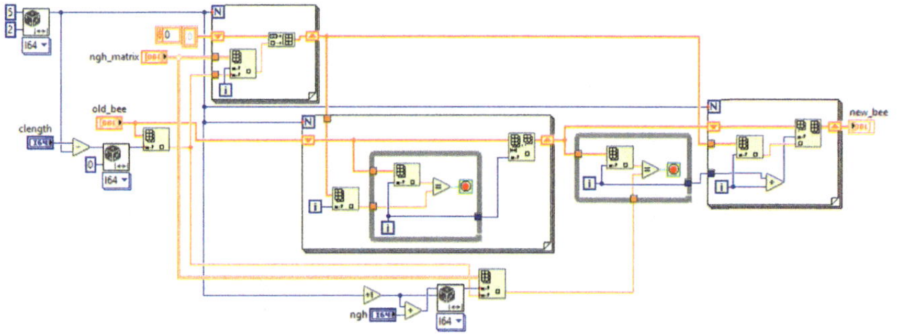

Fig. 20 Multi-insert function

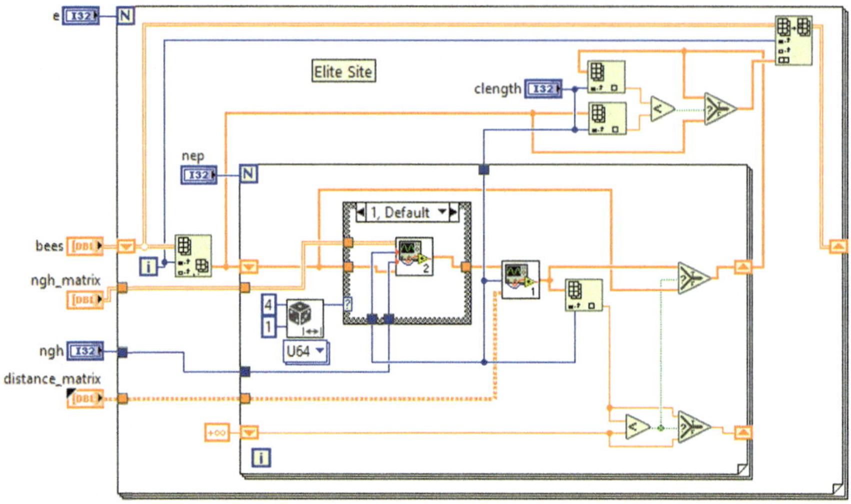

Fig. 21 Elite site search for combinatorial problems

4.3 Experiments and Results

The front panel of the Combinatorial Optimisation is given in Fig. 22. Some TSPs are available in this section. Additionally, by clicking "New", the text file containing the X–Y coordinates can be selected. TSP data are from MP-TESTDATA [35].

Previous studies were used when determining the parameters [34, 36, 37]. To avoid increasing the processing load, (n) was set to 20, which is also the local area's

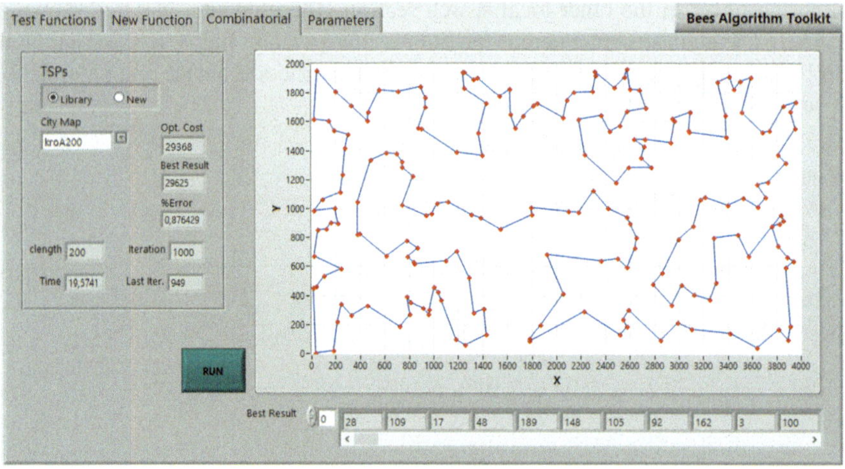

Fig. 22 Combinatorial optimisation

Table 3 Parameters of the algorithm

Parameter	Value
Runs	10
Iterations (I)	1000
Scout bees (n)	20
Elite bees (nep)	300
Best bees (nsp)	100
Elite sites (e)	5
Best sites (m)	20

value. The elite region was determined as 25% of the local region. The algorithm parameters are given in Table 3.

The experimental results are given in Table 4. The first column of Table 4 gives the name of the TSP and the best solution (BKS). Other columns show the results. In the results, the best value found (Min.), the mean of all values (Avg.), the deviation of the mean value (Davg), and the deviation of the best value found (Dbst) are given. "Avg. Iter." displays the average number of iterations for the minimum value. The last column shows the average time taken to find the minimum value. The Davg and Dbest equations are shown in (4) and (5) [11].

$$Davg = \frac{(Avg - BKS)}{(BKS)} * 100\% \tag{4}$$

$$Dbest = \frac{(Best - BKS)}{(BKS)} * 100\% \tag{5}$$

The best-known or very close values for problems up to 144 cities can be obtained. For the problems of 150–299 cities, the results were obtained with deviations of 0.34–1.86% from the best-known values. Deviations in the mean values were below 3.3%. The best values found for some problems are shown in the Appendix.

The comparison study was made with other metaheuristic algorithms. These are the Discrete Symbiotic Organisms Search (DSOS) Algorithm [38] and Heterogenous Adaptive Ant Colony Optimisation (HAACO) [39]. All Davg and Dbest data are shown in Table 5.

As given in Table 5, the results obtained with the BA Tool are at a level that can compete head-to-head with other algorithms. While better results are obtained in some TSPs, others are very close to other algorithms. In addition, it can perform these solutions at relatively low CPU times.

Table 4 Experimental results

TSP (BKS)	Min	Avg.	Dbst	Davg.	Avg. Iter	Avg. Time (s)
Berlin52 (7542)	7544	7544	0.03	0.03	58	0.32
St70 (675)	677	677	0.31	0.31	324	1.39
KroA100 (21,282)	21,285	21,290	0.01	0.04	328	2.60
KroB100 (22,141)	22,139	22,277	0	0.30	546	4.14
KroC100 (20,749)	20,750	20,834	0	0.29	520	4.9
KroD100 (21,389)	21,341	21,375	0	0	744	5.97
KroE100 (22,068)	22,068	22,142	0	0.28	470	4.21
Lin105 (14,379)	14,383	14,387	0.03	0.06	457	3.76
Pr124 (59,030)	59,030	59,229	0	0.34	667	5.83
Pr144 (58,537)	58,587	58,683	0.08	0.25	832	7.52
Ch150 (6528)	6552	6559	0.37	0.47	740	7.11
KroA150 (26,524)	26,606	26,774	0.31	0.94	812	7.69
KroB150 (26,130)	26,232	26,384	0.34	0.97	813	7.82
KroA200 (29,368)	29,452	29,609	0.28	0.82	840	10.23
KroB200 (29,437)	29,855	30,115	1.42	2.30	867	10.80
Pr226 (80,369)	80,775	81,124	0.50	0.94	813	10.45
Pr299 (48,191)	48,863	49,583	1.39	2.89	947	15.91

Table 5 Comparison of other algorithms

Dataset names (BKS)	CBA		HAACO [38]		DSOS [39]	
	Davg	Dbst	Davg	Dbst	Davg	Dbst
Berlin52 (7542)	0.03	0.03	0	0	0.01	0
St70 (675)	0.31	0.31	0.22	0	0.62	0
KroA100 (21,282)	0.04	0.01	0.39	0	0.60	0
KroB100 (22,141)	0.30	0	–	–	0.90	0
KroC100 (20,749)	0.29	0	–	–	0.64	0
KroD100 (21,389)	0	0	–	–	0.94	0
KroE100 (22,068)	0.28	0	–	–	0.74	0
Lin105 (14,379)	0.06	0.03	0.23	0	–	–
Pr124 (59,030)	0.34	0	–	–	0.68	0
Pr144 (58,535)	0.25	0.08	–	–	0.48	0.05
Ch150 (6528)	0.47	0.37	1.13	0.58	–	–
KroA200 (29,368)	0.82	0.28	0.90	0.39	–	–
Pr299 (48,191)	2.89	1.39	–	–	4.45	2.02

5 Discussion and Conclusion

This chapter has presented the development of the Bees Algorithm Toolkit for solving optimisation problems in LabVIEW. The Toolkit enables users to develop their applications. With the "New Function" module, different functions that are not included in the Toolkit can be easily defined by the user for continuous problems. In experimental studies with the Toolkit for continuous functions, it has been observed that test functions with a large number of local minima can be quickly optimised to yield their global minimum values. Due to the simple structure of the BA, it can be easily applied to combinatorial problems. In the tests, it was seen that the CBA toolkit could perform fast optimisation against large TSPs. In addition, the Toolkit can be easily adapted to many continuous problems in its current form and can be incorporated into industrial applications using LabVIEW.

Appendix 1—Equations of Test Functions

$$\boldsymbol{Ackley}, x_{1,2} \in [-10, 10],$$

$$f(x) = -20\exp\left(-0.2\sqrt{\frac{1}{2}\sum_{i=1}^{2} x_i^2}\right) - \exp\left(\frac{1}{2}\sum_{i=1}^{2}\cos(cx_i)\right) + 20 + \exp(1) \quad (6)$$

$$\boldsymbol{CrossinTray}, x_{1,2} \in [-10, 10],$$

$$f(x) = -0.0001\left(\left|\sin(x_1)\sin(x_2)\exp\left(\left|100 - \frac{\sqrt{x_1^2 + x_2^2}}{\pi}\right|\right)\right| + 1\right)^{0.1} \quad (7)$$

$$\boldsymbol{DropWave}, x_{1,2} \in [-5.12, 5.12],$$

$$f(x) = -\frac{1 + \cos\left(12\sqrt{x_1^2 + x_2^2}\right)}{0.5\left(x_1^2 + x_2^2\right) + 2} \quad (8)$$

$$\boldsymbol{Egg - holder}, x_{1,2} \in [-512, 512],$$

$$f(x) = -(x_2 + 47)\sin\left(\sqrt{\left|x_2 + \frac{x_1}{2} + 47\right|}\right) - x_1\sin\left(\sqrt{|x_1 - (x_2 + 47)|}\right) \quad (9)$$

$$\boldsymbol{HolderTable}, x_{1,2} \in [-10, 10],$$

$$f(x) = -\left|\sin(x_1)\cos(x_2)\exp\left(\left|1 - \frac{\sqrt{x_1^2 + x_2^2}}{\pi}\right|\right)\right| \tag{10}$$

Levy, $x_{1,2} \in [-10, 10]$,

$$f(x) = \sin^2(3\pi x_1) + (x_1 - 1)^2\left[1 + \sin^2(3\pi x_2)\right] + (x_2 - 1)^2\left[1 + \sin^2(2\pi x_2)\right] \tag{11}$$

Rastgirin, $x_{1,2} \in [-5.12, 5.12]$,

$$f(x) = 20 + \sum_{i=1}^{2}\left[x_i^2 - 10\cos(2\pi i)\right] \tag{12}$$

Schaffer, $x_{1,2} \in [-100, 100]$,

$$f(x) = 0.5 + \frac{\sin^2\left(x_1^2 - x_2^2\right) - 0.5}{\left[1 + 0.001\left(x_1^2 + x_2^2\right)\right]^2} \tag{13}$$

Schwefel, $x_i \in [-500, 500]$, i = 1, ..., d,

$$f(x) = 418.9829d - \sum_{i=1}^{d} x_i \sin\left(\sqrt{|x_i|}\right) \tag{14}$$

Shubert, $x_{1,2} \in [-5.12, 5.12]$,

$$f(x) = \left(\sum_{i=1}^{5} i\cos((i + 1)x_1 + i)\right)\left(\sum_{i=1}^{5} i\cos((i + 1)x_2 + i)\right) \tag{15}$$

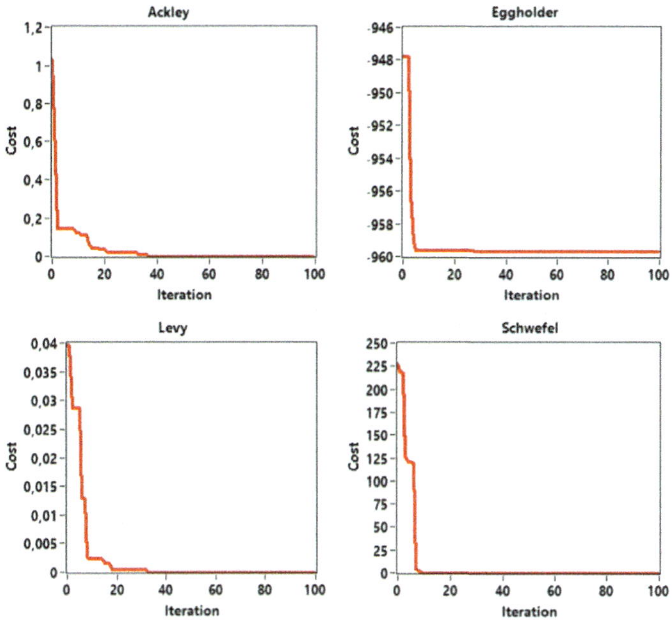

Fig. 23 Test results of some functions

Appendix 2—Test Results of Some Functions

Some results of continuous optimisation are given in Fig. 23.

Appendix 3—Best Solution of Some Combinatorial Problems

Some results of combinatorial optimisation are given in Fig. 24.

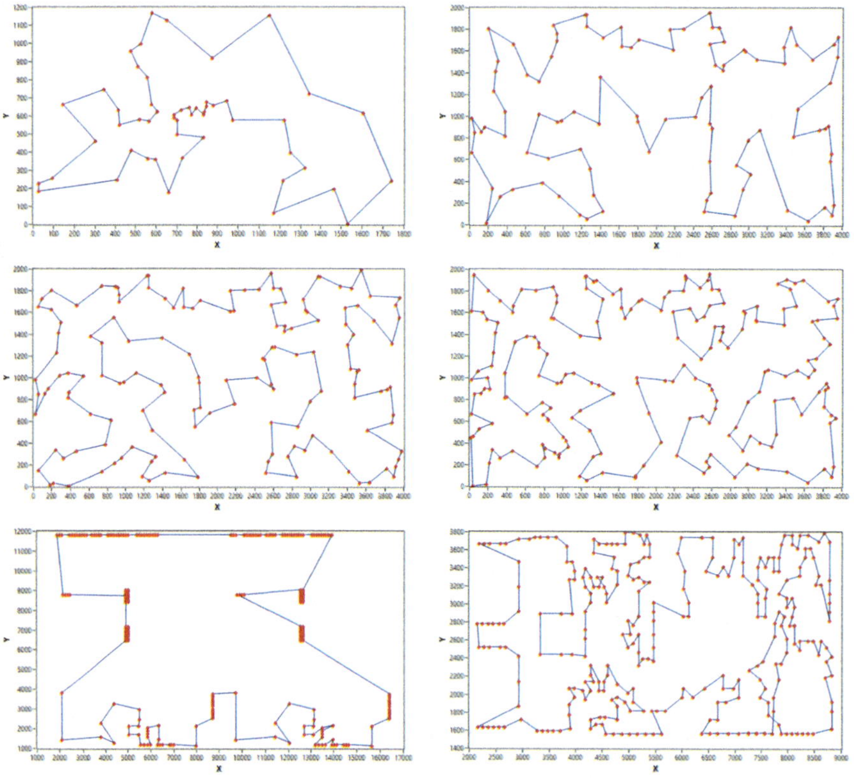

Fig. 24 The best solution of Berlin52, KroA100, Kroa150, KroA200, Pr226, and Pr299

References

1. Rao SS (2009) Engineering optimization: theory and practice. John Wiley & Sons
2. Manea D, Titan E, Serban RR, Mihai M (2019) Statistical applications of optimization methods and mathematical programming. Proc Int Conf Appl Stat 1(1):312–328
3. Pham DT, Ghanbarzadeh A, Koc E, Otri S, Rahim S, Zaidi M (2005) Bee algorithm—a novel approach to function optimisation. Cardiff University, The Manufacturing Engineering Centre
4. Zarchi M, Attaran B (2017) Performance improvement of an active vibration absorber subsystem for an aircraft model using a Bees algorithm based on multi-objective intelligent optimization. Eng Optim 49(11):1905–1921
5. Nafchi AM, Moradi A, Ghanbarzadeh A, Yaghoubi S, Moradi M (2012) An improved bees algorithm for solving optimization mechanical problems. In: 20th Annual international conference on mechanical engineering-ISME, School of Mechanical Engineering, Shiraz University, Shiraz, Iran
6. Shouran M, Anayi F, Packianather M (2021) Design of sliding mode control optimised by the Bees algorithm for LFC in the Great Britain power system. Mater Today
7. Acar O, Kalyoncu M, Hassan A (2018) The Bees algorithm for design optimization of a gripper mechanism. J Selcuk-Technic Spec. Issue 69–86

8. Onder A, Incebay O, Sen MA, Yapici R, Kalyoncu M (2021) Heuristic optimization of impeller sidewall gaps based on the Bees algorithm for a centrifugal blood pump by CFD. Int J Artif Organs 44(10):765–772
9. Alzaqebah M, Jawarneh S, Sarim HM, Abdullah S (2018) Bees algorithm for vehicle routing problems with time windows. Int J Mach Learn Comput 8(3):236–240
10. Braiwish NY, Anayi FJ, Fahmy AA, Eldukhri EE (2014) Design optimisation of permanent magnet synchronous motor for electric vehicles traction using the Bees algorithm. In: 49th International universities power engineering conference (UPEC), Cluj-Napoca, Romania. (2014)
11. Ismail AH, Hartono N, Zeybek S, Pham DT (2020) Using the Bees algorithm to solve combinatorial optimisation problems for TSPLIB. IOP Conf Ser: Mater Sci Eng 847:1–9
12. Koc E (2010) Bees algorithm: theory, improvements and applications. PhD Thesis, Cardiff University, UK
13. Heris MK (2015) Bees algorithm (BeA) in MATLAB. https://yarpiz.com/315/ypea115-bees-algorithm, Yarpiz
14. Baronti L (2020) Bees algorithm—a python implementation. https://pypi.org/project/bees-algorithm/
15. Colak I, Bulbul HI, Sahin M (2013) System identification and control of a wound rotor AC induction machine for wind turbine. In: International conference on renewable energy research and applications (ICRERA), Madrid, Spain, pp 1053–1057
16. Colak I, Bulbul HI, Sagiroglu S, Sahin M (2012) Modeling a permanent magnet synchronous generator used in wind turbine and the realization of voltage control on the model with artificial neural networks. In: International conference on renewable energy research and applications (ICRERA), Nagasaki, Japan, pp 1–6
17. Sheoran Y, Kumar V, Rana KPS, Mishra P, Kumar J, Nair SS (2015) Development of backtracking search optimization algorithm toolkit in LabVIEW™. Procedia Comput Sci 57:241–248
18. LabVIEW optimization Vis—https://www.ni.com/docs/en-US/bundle/LabVIEW-2021/page/gmath/optimization_vis.html. 25 Apr 2022
19. AMA toolkit for LabVIEW—https://www.ni.com/en-gb/support/downloads/tools-network/download.ama-toolkit-for-LabVIEW.html. 25 Apr 2022
20. Aria M (2013) Educational simulator for teaching of particle swarm optimization in LabVIEW. TELEKONTRAN 1(1):1–15
21. Thakur KS, Kumar V, Rana KPS, Mishra P, Kumar J, Nair SS (2015) Development of Bat algorithm toolkit in LabVIEW. In: International conference on computing, communication and automation (ICCCA2015), pp 5–10
22. Gupta S, Kumar V, Rana KPS, Mishra P, Kumar J (2016) Development of Ant lion optimizer toolkit in LabVIEW. In: 1st International conference on innovation and challenges in cyber security (ICICCS 2016), pp 251–256
23. Gupta S, Rana KPS, Kumar V, Mishra P, Kumar J, Nair SS (2015) Development of a grey wolf optimizer toolkit in LabVIEW. In: 1st International conference on futuristic trend in computational analysis and knowledge management (ABLAZE 2015), pp 107–113
24. Pham DT, Ghanbarzadeh A, Koc E, Otri S, Rahim S, Zaidi M (2006) The Bees algorithm—a novel tool for complex optimisation problems. In: Pham DT, Eldukhri EE, Soroka AJ (eds) Proceedings of the 2nd Virtual international conference on intelligent production machines and systems, Elsevier (Oxford), pp 454–460
25. Pham DT, Castellani M (2009) The Bees algorithm: modelling foraging behaviour to solve continuous optimization problems. Proc Inst Mech Eng C J Mech Eng Sci 223(12):2919–2938
26. Baronti L, Castellani M, Pham DT (2020) An analysis of the search mechanisms of the bees algorithm. Swarm Evol Comput 59:100746
27. Web: virtual library of simulation experiments, optimization test problems. https://www.sfu.ca/~ssurjano/optimization.html. 18 Mar 2022
28. Wasim AH, Shahnorbanun S, Siti NH (2017) The variants of the Bees algorithm (BA): a survey. Artif Intell Rev 47:67–121

29. Du X, Ma Y, Wei X, Jegatheesan V (2020) Optimal parameter estimation in activated sludge process based wastewater treatment practice. Water 12(9):2604
30. Liang YC, Cuevas Juarez JRA (2020) Self-adaptive virus optimization algorithm for continuous optimization problems. Soft Comput 24(17):13147–13166
31. Tansui D, Thammano, A (2017) Nature-inspired optimization method: hydrozoan algorithm for solving continuous problems. In: 2017 18th IEEE/ACIS International conference on software engineering, artificial intelligence, networking and parallel/distributed computing (SNPD), pp 23–28
32. Tansui D, Thammano A (2019) An enhanced Bat algorithm with random walk for solving continuous optimization problems. In: 2019 20th IEEE/ACIS International conference on software engineering, artificial intelligence, networking and parallel/distributed computing (SNPD), pp 39–44
33. Karaboga D, Gorkemli B (2019) Solving traveling salesman problem by using combinatorial artificial bee colony algorithms. Int J Artif Intell Tools 28(1):1950004
34. Sahin M (2022) Solving TSP by using combinatorial Bees algorithm with nearest neighbor method. Neural Comput Appl 1–17
35. MP-TESTDATA. The TSPLIB symmetric travelling salesman problem instances. Retrieved from http://elib.zib.de/pub/mp-testdata/tsp/tsplib/tsp/index.html. Accessed 5 March 2022
36. Lambiase A, Iannone R, Miranda S, Lambiase A, Pham DT (2016) Bees algorithm for effective supply chains configuration. Int J Eng Bus Manage 8:1–9
37. Xu W, Tang Q, Liu J, Liu Z, Zhou Z, Pham DT (2020) Disassembly sequence planning using discrete Bees algorithm for human-robot collaboration in remanufacturing. Rob Comput-Integr Manuf 62:101860
38. Ezugwu AE, Adewumi AO (2017) Discrete symbiotic organisms search algorithm for travelling salesman problem. Expert Syst Appl 87:70–78
39. Tuani AF, Keedwell E, Collett M (2020) Heterogenous adaptive ant colony optimization with 3-opt local search for the travelling salesman problem. Appl Soft Comput J 97:106720

Engineering Applications of the Bees Algorithm

Geometrical Optimisation of Smart Sandwich Plates Using the Bees Algorithm

Nguyen Dinh Duc and Tran Quoc Quan

Abstract The geometrical optimisation of smart sandwich plates on Pasternak-type elastic foundations is presented in this chapter. The basic equations are derived based on Reddy's higher-order shear deformation plate theory and Hamilton's principle, taking into account the effect of von Kármán kinematic nonlinearity. The forms of possible solutions and electric and magnetic potentials are chosen as double trigonometric functions based on boundary conditions. The expressions of the natural frequency and the critical buckling load are determined by using the Galerkin method. The Bees Algorithm is applied to obtain the maximum values of the natural frequency and the critical buckling load of the sandwich plate, which depends on geometrical and material parameters.

Keywords Geometrical optimization · Bees algorithm · Smart sandwich plates · Auxetic honeycomb core · Magneto–electro–elastic face sheets · Reddy's higher order shear deformation plate theory · Buckling · Vibration · Dynamic response · Galerkin method

1 Introduction

Smart materials are promising materials that can change their properties to respond to external environments such as stress, temperature, moisture, and electric or magnetic fields. Smart materials are mainly classified into four types: piezoelectric materials, shape memory alloys, optical fibres and magnetostrictive materials. Two smart materials that we study in this chapter are piezoelectric materials and magneto–electro–elastic materials. First, piezoelectric materials can create electricity when subjected to mechanical stress. They will also work in reverse, generating a strain

N. D. Duc (✉) · T. Q. Quan
VNU Hanoi, University of Engineering and Technology, 144—Xuan Thuy, Cau Giay, Hanoi, Vietnam
e-mail: ducnd@vnu.edu.vn

D. T. Pham and N. Hartono (eds.), *Intelligent Engineering Optimisation with the Bees Algorithm*, Springer Series in Advanced Manufacturing,
https://doi.org/10.1007/978-3-031-64936-3_5

by applying an electric field. Piezoelectric materials are found in useful applications such as the production and detection of sound, generation of high voltages, electronic frequency generation, microbalances to drive an ultrasonic nozzle, and ultrafine focusing of optical assemblies. In addition, magneto–electro–elastic materials consisting of piezoelectric and piezomagnetic phases are a typical class of smart materials. Magneto–electro–elastic material has piezoelectric and piezomagnetic properties and expresses the coupling effect between mechanical, electric, and magnetic states. Because of the ability to convert between different forms of energy, magneto–electro–elastic material has been widely used in aerospace, sensors, civil engineering, medical equipment and automotive applications. Zhang et al. [1] developed a new model for transversely isotropic magneto–electro–elastic Timoshenko beams by using a variational formulation based on Hamilton's principle. In addition, Vinyas and Harursampath [2] studied the geometrically nonlinear free vibration behaviour of higher order shear deformable carbon nanotube reinforced magneto-electro-elastic doubly curved shells. Dat et al. [3] presented an analytical approach to the nonlinear magneto–electro–elastic vibration of smart sandwich plates. The sandwich plate consists of a carbon nanotube-reinforced nanocomposite core integrated with two magneto-electroelastic face sheets. Li et al. [4, 5] introduced the coupled multiphysical cell-based smoothed finite element method to predict the temperature environment by combining the mechanical load effect on the dynamic characteristics of functionally graded magneto–electroelastic thin-walled intelligent structures.

Geometrical optimisation plays an important role in determining optimal geometrical dimensions for the maximum performance and capacity of structures. The objective function in these problems can be the critical buckling load or the natural frequency. Scientists around the world have used many methods and algorithms to solve geometrical optimisation problems. Rehman et al. [6] presented an optical-energy model for determining the optimal geometrical layouts of photovoltaic arrays by considering different tilt angles and row spacings for every row, given a field constraint. Ling et al. [7] investigated the structure and geometric dimension optimisation of interlaced microchannels for heat transfer performance enhancement using the full-size conjugate heat transfer model. Considering the duplex helical face-milling characteristics, Chen et al. [8] developed an innovative adaptive data-driven collaborative optimisation of both tooth flank geometric accuracy and loaded contact mechanical performance evaluations for nonorthogonal spiral bevel and hypoid gears. Liu et al. [9] developed an optimisation method considering thermal radiation and heat conduction to design the geometric structures of aerogel composites with minimised thermal conductivity. This work provided a guideline for the geometric design of aerogel composites for high-temperature thermal insulation applications. Bahiraei et al. [10] introduced a combined multicriterion optimisation to determine optimum geometrical parameters for the flow of an ecofriendly graphene-based nanofluid inside a tube enhanced with twisted conical strip inserts.

Duc's research group presented the first publications on the geometrical optimisation of composite structures using the Bees Algorithm [11–14]. Quang et al. [11] investigated the geometrical optimisation of a magneto-electroelastic sandwich plate on Pasternak-type elastic foundations subjected to mechanical, thermal, electric and magnetic loadings in which the Bees Algorithm was applied to obtain the maximum value of the critical buckling load. Dat et al. [12] presented global optimisation of imperfect three-dimensional pentagraphene plates subjected to uniformly distributed external pressure with simply supported and immovable edges in thermal environments. The Bees Algorithm was used to determine the optimum value of the natural frequency, which depends on five variables, including the thickness, length, and width of the pentagraphene plates and two stiffness coefficients of the elastic foundations. Anh et al. [13] used four optimisation algorithms (Bees Algorithm, Basic differential evolution algorithm, enhanced colliding bodies optimisation algorithm, and social group optimisation algorithm) to determine the maximum value of the natural frequency of the nanocomposite multilayer organic solar cell subjected to the combination of wind load and uniform temperature change. Based on classical plate theory, Dat et al. [14] applied the Bees Algorithm to maximise the value of critical buckling load with nine variables, including the thickness of five layers, the length and width of nanocomposite multilayer organic solar cells, and two stiffness coefficients of elastic foundations. The model of an organic solar cell was assumed to consist of five layers: Al, P3HT:PCBM, PEDOT:PSS, IOT and glass. The obtained results from the optimisation problems will provide suitable geometrical and material parameters for engineers and designers during the production and use of materials.

In this chapter, we use the standard Bees Algorithm to solve the two following optimisation problems:

Problem 1 Geometrical optimisation of the magneto–electro–elastic sandwich plate with auxetic honeycomb core.

Problem 2 Geometrical optimisation of piezoelectric sandwich plates with porous homogeneous cores and carbon nanotube-reinforced nanocomposite layers.

The novelty and significant contributions of this chapter may be expressed as follows:

- This is the first time that the mathematical modelling of smart sandwich plates with five layers, including porous core, CNTRC layers, and piezoelectric face sheets and three layers with auxetic honeycomb core and magneto–electro–elastic face sheets, has been introduced.
- The analytical expressions of the natural frequency and the critical buckling load of smart sandwich plates are obtained, which are the input basis for optimisation problems.
- The optimum values of the geometrical and material parameters of smart sandwich plates corresponding to the maximum values of the natural frequency and the critical buckling load are obtained by using the Bees Algorithm, which provides a scientific basis for designers and engineering in the specific objectives.

2 Geometrical Optimisation of the Magneto–Electro–Elastic Sandwich Plate with Auxetic Honeycomb Core

The global optimisation of magneto–electro–elastic sandwich plates on Pasternak-type elastic foundations subjected to mechanical, thermal, electric and magnetic loadings is presented in this section. The sandwich plate is composed of an auxetic honeycomb core with a negative Poisson's ratio and two face sheets made of magneto–electro–elastic material. The system's basic equations are derived based on Reddy's higher-order shear deformation plate theory taking into account the effect of von Kármán kinematic nonlinearity and initial imperfection. The form of possible solutions and electric and magnetic potentials are chosen as trigonometric functions. The relationship between axial compressive loading and dimensionless deflection amplitude is determined by using the Galerkin method. The Bees Algorithm is applied to obtain the maximum value of the sandwich plate's critical buckling load, which depends on five geometrical and material parameters.

2.1 Problem Description

2.1.1 Geometry and Coordinate System

The suggested sandwich plate comprising an auxetic honeycomb core and two magneto–electro–elastic face sheets is plotted in Fig. 1. The thicknesses of the auxetic core and magneto–electro–elastic face sheet are denoted as h_c and h_f, respectively. The geometrical parameters of the sandwich plate are a in the $x-$ direction, b in the $y-$ direction and the total thickness h. The Castesian coordinate system xyz is established in which the xy plane coincides with the mid-plane of the sandwich plate. The magneto–electro–elastic sandwich plate is assumed to be under axial compressive load F_x uniformly distributed at two edges $x = 0, a$, uniform temperature change ΔT, uniform electric and magnetic potentials ϕ_0 and ψ_0. u, v and w represent the displacement components in the x, y and z directions, respectively.

2.1.2 Material Properties

The core of the sandwich plate is made of auxetic material with a negative Poisson's ratio. Figure 2 shows the unit cell of the hexagonal honeycomb core layer where the length of the inclined cell rib L, the length of the vertical cell rib H, the thickness of the cell rib T and the inclined angle θ are illustrated.

The elastic modulus of the auxetic honeycomb core, such as Young's and shear modulus and Poisson's ratio, depends on the material properties of the original material and geometrical parameters of the unit cell as [15, 16].

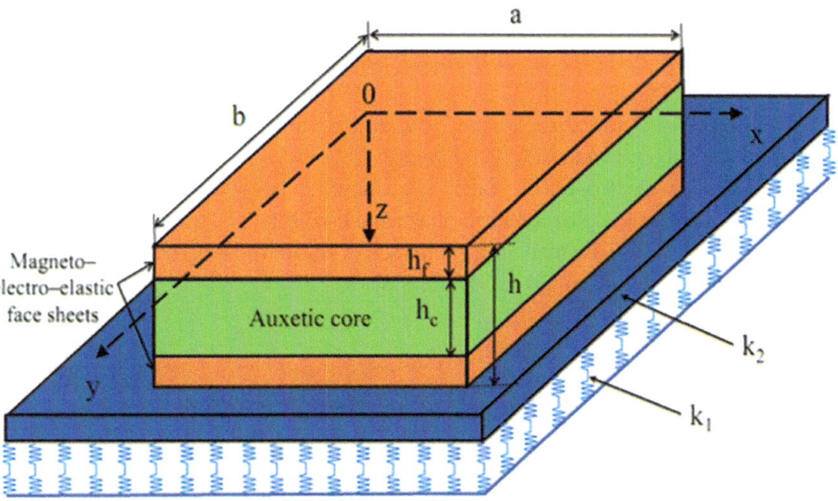

Fig. 1 Geometry of magneto–electro–elastic sandwich plate with auxetic honeycomb core in Castesian coordinate system

Fig. 2 The hexagonal honeycomb core cell

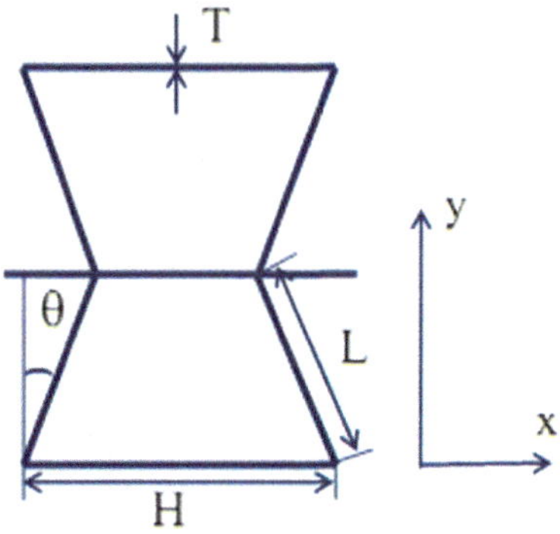

$$E_{11}^{au} = E\left(\frac{T}{L}\right)^3 \frac{\cos\theta}{\left(\frac{H}{L} + \sin\theta\right)\sin^2\theta},$$

$$E_{22}^{au} = E\left(\frac{T}{L}\right)^3 \frac{\left(\frac{H}{L} + \sin\theta\right)}{\cos^3\theta},$$

$$\nu_{12}^{au} = \frac{\cos^2\theta}{\left(\frac{H}{L} + \sin\theta\right)\sin\theta},$$

$$G_{12}^{au} = E\left(\frac{T}{L}\right)^3 \frac{\left(\frac{H}{L} + \sin\theta\right)}{\left(\frac{H}{L}\right)^2\left(2\frac{H}{L} + 1\right)\cos\theta},$$

$$G_{13}^{au} = G\frac{T}{L}\frac{\cos\theta}{\frac{H}{L} + \sin\theta},$$

$$G_{23}^{au} = G\frac{T}{L}\frac{1+2\sin^2\theta}{2\cos\theta\left(\frac{H}{L}+\sin\theta\right)}, \quad (1)$$

where *au* represents the auxetic core layer and *E* and *G* are the Young's and shear moduli of the original material, respectively.

The magneto–electro–elastic face sheet includes two phases, in which barium titanate ($BaTiO_3$) is the piezoelectric phase and cobalt ferric oxide ($CoFe_2O_4$) is the piezomagnetic phase. The mechanical characteristics of the magneto–electro–elastic face sheet are expressed in Table 1 [2, 3, 17].

2.2 Determination of the Critical Buckling Load

In this section, Reddy's higher order shear deformation plate theory [18] is used to derive a system of basic equations of sandwich plates. The strain components across the plate thickness at a distance z from the middle surface, taking into account the effect of large deflection and Von—Karman nonlinear kinematic terms, are given as

$$\begin{Bmatrix} \varepsilon_x \\ \varepsilon_y \\ \gamma_{xy} \end{Bmatrix} = \begin{Bmatrix} \varepsilon_x^0 \\ \varepsilon_y^0 \\ \gamma_{xy}^0 \end{Bmatrix} + z\begin{Bmatrix} k_x \\ k_y \\ k_{xy} \end{Bmatrix} + z^2\begin{Bmatrix} k_x^1 \\ k_y^1 \\ k_{xy}^1 \end{Bmatrix} + z^3\begin{Bmatrix} k_x^3 \\ k_y^3 \\ k_{xy}^3 \end{Bmatrix},$$
$$\begin{Bmatrix} \gamma_{xz} \\ \gamma_{yz} \end{Bmatrix} = \begin{Bmatrix} \gamma_{xz}^0 \\ \gamma_{yz}^0 \end{Bmatrix} + z^2\begin{Bmatrix} k_{xz}^2 \\ k_{yz}^2 \end{Bmatrix}, \quad (2)$$

in which the curvature and torsion components are expressed as

$$\begin{bmatrix} k_x^1 \\ k_y^1 \\ k_{xy}^1 \end{bmatrix} = \begin{bmatrix} \frac{\partial \phi_x}{\partial x} \\ \frac{\partial \phi_y}{\partial y} \\ \frac{\partial \phi_x}{\partial y} + \frac{\partial \phi_y}{\partial y} \end{bmatrix}, \begin{bmatrix} k_x^3 \\ k_y^3 \\ k_{xy}^3 \end{bmatrix} = -c_1\begin{bmatrix} \frac{\partial \phi_x}{\partial x} + \frac{\partial^2 w}{\partial x^2} \\ \frac{\partial \phi_y}{\partial y} + \frac{\partial^2 w}{\partial y^2} \\ \frac{\partial \phi_x}{\partial y} + \frac{\partial \phi_y}{\partial y} + 2\frac{\partial^2 w}{\partial x \partial y} \end{bmatrix},$$
$$\begin{bmatrix} k_{xz}^2 \\ k_{yz}^2 \end{bmatrix} = -3c_1\begin{bmatrix} \frac{\partial w}{\partial x} + \phi_x \\ \frac{\partial w}{\partial y} + \phi_y \end{bmatrix}. \quad (3)$$

where ϕ_x, ϕ_y are the rotations of the transverse normal about the y and x axes at $z = 0$, respectively, and $c_1 = 4/\left(3h^2\right)$.

The strain components in the middle surface of the plate ε_x^0, ε_y^0 and γ_{xy}^0 depend on displacement components u, v and w as

$$\begin{bmatrix} \varepsilon_x^0 \\ \varepsilon_y^0 \\ \gamma_{xy}^0 \end{bmatrix} = \begin{bmatrix} \frac{\partial u}{\partial x} + \frac{1}{2}\left(\frac{\partial w}{\partial x}\right)^2 \\ \frac{\partial v}{\partial y} + \frac{1}{2}\left(\frac{\partial w}{\partial y}\right)^2 \\ \frac{\partial u}{\partial y} + \frac{\partial v}{\partial x} + \frac{\partial w}{\partial x}\frac{\partial w}{\partial y} \end{bmatrix}, \begin{bmatrix} \gamma_{xz}^0 \\ \gamma_{yz}^0 \end{bmatrix} = \begin{bmatrix} \frac{\partial w}{\partial x} + \phi_x \\ \frac{\partial w}{\partial y} + \phi_y \end{bmatrix} \quad (4)$$

Table 1 Mechanical characteristics of magneto–electro–elastic face sheet

Material characteristics	Symbol	Values
Elastic constants (GPa)	$c_{11}^f = c_{22}^f$	220
	$c_{12}^f = c_{13}^f = c_{23}^f$	120
	c_{33}^f	215
	$c_{44}^f = c_{55}^f$	45
	c_{66}^f	0
Piezoelectric constants $\left(C/m^2\right)$	e_{31}^f	−3.5
	e_{33}^f	9.0
	e_{15}^f	0
Dielectric constants $\left(C/Nm^2\right)$	$\eta_{11}^f = \eta_{22}^f$	0.85×10^{-9}
	η_{33}^f	6.3×10^{-9}
Magnetic permeability $\left(Ns^2/C^2\right)$	$\mu_{11}^f = \mu_{22}^f$	-2×10^{-4}
	μ_{33}^f	0.9×10^{-4}
Piezomagnetic constants (N/Am)	q_{31}^f	350
	q_{33}^f	320
	q_{15}^f	200
Magneto-electric constants (Ns/VC)	$m_{11}^f = m_{22}^f$	5.5×10^{-12}
	m_{33}^f	2600×10^{-12}
Pyromagnetic constant (C/m^2K)	p_3^f	7.8×10^{-7}
Pyromagnetic constant (C/m^2K)	λ_3^f	-23×10^{-5}
Thermal expansion coefficients (K^{-1})	$\alpha_1^f = \alpha_2^f$	12.3×10^{-6}
	α_3^f	8.2×10^{-6}
Moisture expansion coefficients (m^3k^{-1})	β_1^f	0
	$\beta_2^f = \beta_3^f$	1.1×10^{-4}
Density (kg/m^3)	ρ_f	5500

where f denotes magneto–electro–elastic face sheets.

The stress—strain relationships of the auxetic honeycomb core are given in the following form [15, 16]:

$$\begin{bmatrix} \sigma_{xx} \\ \sigma_{yy} \\ \sigma_{xy} \\ \sigma_{xz} \\ \sigma_{yz} \end{bmatrix}_{au} = \begin{bmatrix} Q_{11}^{au} & Q_{12}^{au} & 0 & 0 & 0 \\ Q_{12}^{au} & Q_{22}^{au} & 0 & 0 & 0 \\ 0 & 0 & Q_{66}^{au} & 0 & 0 \\ 0 & 0 & 0 & Q_{44}^{au} & 0 \\ 0 & 0 & 0 & 0 & Q_{55}^{au} \end{bmatrix} \begin{bmatrix} \varepsilon_{xx} \\ \varepsilon_{yy} \\ \varepsilon_{xy} \\ \varepsilon_{xz} \\ \varepsilon_{yz} \end{bmatrix} - \begin{bmatrix} \alpha_{11}^{au}\Delta T \\ \alpha_{22}^{au}\Delta T \\ 0 \\ 0 \\ 0 \end{bmatrix}, \quad (5)$$

where the components of the stiffness matrix are expressed as

$$Q_{11}^{au} = \frac{E_{11}^{au}}{1 - v_{12}^{au}v_{21}^{au}}, Q_{12}^{au} = \frac{v_{12}^{au}E_{22}^{au}}{1 - v_{12}^{au}v_{21}^{au}}, Q_{22}^{au} = \frac{E_{22}^{au}}{1 - v_{12}^{au}v_{21}^{au}},$$
$$Q_{66}^{au} = G_{12}^{au}, Q_{44}^{au} = G_{23}^{au}, Q_{55}^{au} = G_{13}^{au}, \quad (6)$$

where $E_{11}^{au}, E_{22}^{au}, G_{12}^{au}, v_{12}^{au}$ and v_{21}^{au} are Young's modulus, shear modulus and Poisson's ratio auxetic core, respectively.

The stress components, electric displacements and magnetic fluxes of magneto–electro–elastic face sheets are expressed as [2]

$$\begin{aligned}
\sigma_x^f &= \overline{C_{11}}\varepsilon_{xx}^f + \overline{C_{12}}\varepsilon_{yy}^f - \overline{e_{31}}E_z - \overline{q_{31}}H_z - \overline{\alpha_1}\Delta T - \overline{\beta_1}\Delta m,\\
\sigma_y^f &= \overline{C_{12}}\varepsilon_{xx}^f + \overline{C_{22}}\varepsilon_{yy}^f - \overline{e_{32}}E_z - \overline{q_{32}}H_z - \overline{\alpha_2}\Delta T - \overline{\beta_2}\Delta m,\\
\tau_{xy}^f &= 2\sigma_{xy}^f = \overline{C_{66}}\gamma_{xy}^f,\\
\tau_{yz}^f &= 2\sigma_{yz}^f = \overline{C_{44}}\gamma_{yz}^f - \overline{e_{24}}E_y - \overline{q_{24}}H_y,\\
\tau_{xz}^f &= 2\sigma_{xz}^f = \overline{C_{55}}\gamma_{xz}^f - \overline{e_{15}}E_x - \overline{q_{15}}H_x,\\
D_x &= \overline{e_{15}}\gamma_{xz}^f + \overline{\eta_{11}}E_x + \overline{m_{11}}H_x + \overline{p_1}\Delta T + \overline{\chi_1}\Delta m, \quad (7)\\
D_y &= \overline{e_{24}}\gamma_{yz}^f + \overline{\eta_{22}}E_y + \overline{m_{22}}H_y + \overline{p_2}\Delta T + \overline{\chi_2}\Delta m,\\
D_z &= \overline{e_{31}}\varepsilon_{xx}^f + \overline{e_{32}}\varepsilon_{yy}^f + \overline{\eta_{33}}E_z + \overline{m_{33}}H_z + \overline{p_3}\Delta T + \overline{\chi_3}\Delta m,\\
B_x &= \overline{q_{15}}\gamma_{xz}^f + \overline{m_{11}}E_x + \overline{\mu_{11}}H_x + \overline{\lambda_1}\Delta T + \overline{\zeta_1}\Delta m,\\
B_y &= \overline{q_{24}}\gamma_{yz}^f + \overline{m_{22}}E_y + \overline{\mu_{22}}H_y + \overline{\lambda_2}\Delta T + \overline{\zeta_2}\Delta m,\\
B_z &= \overline{q_{31}}\varepsilon_{xx}^f + \overline{q_{32}}\varepsilon_{yy}^f + \overline{m_{33}}E_z + \overline{\mu_{33}}H_z + \overline{\lambda_3}\Delta T + \overline{\zeta_3}\Delta m.
\end{aligned}$$

where

$$\begin{aligned}
&\overline{C_{11}} = C_{11} - \frac{C_{13}^2}{C_{33}}, \overline{C_{12}} = C_{12} - \frac{C_{13}C_{23}}{C_{33}}, \overline{C_{22}} = C_{22} - \frac{C_{23}^2}{C_{33}}, \overline{C_{66}} = C_{66},\\
&\overline{e_{31}} = e_{31} - \frac{C_{13}e_{33}}{C_{33}}, \overline{e_{32}} = e_{32} - \frac{C_{23}e_{33}}{C_{33}}, \overline{q_{31}} = q_{31} - \frac{C_{13}q_{33}}{C_{33}},\\
&\overline{q_{32}} = q_{32} - \frac{C_{23}q_{33}}{C_{33}}, \overline{\mu_{11}} = \mu_{11} + \frac{q_{15}^2}{C_{55}}, \overline{\mu_{22}} = \mu_{22} + \frac{q_{24}^2}{C_{44}}, \overline{\mu_{33}} = \mu_{33} + \frac{q_{33}^2}{C_{33}},\\
&\overline{\eta_{11}} = \eta_{11} + \frac{e_{15}^2}{C_{55}}, \overline{\eta_{22}} = \eta_{22} + \frac{e_{24}^2}{C_{44}}, \overline{\eta_{33}} = \eta_{33} + \frac{e_{33}^2}{C_{33}},
\end{aligned}$$

$$\overline{m_{11}} = m_{11} + \frac{e_{15}q_{15}}{C_{55}}, \overline{m_{22}} = m_{22} + \frac{e_{24}q_{24}}{C_{44}}, \overline{m_{33}} = m_{33} + \frac{e_{33}q_{33}}{C_{33}},$$
$$\overline{\alpha_1} = \alpha_1 - \frac{C_{13}\alpha_3}{C_{33}}, \overline{\alpha_2} = \alpha_2 - \frac{C_{23}\alpha_3}{C_{33}}, \overline{\beta_1} = \beta_1 - \frac{C_{13}\beta_3}{C_{33}}, \overline{\beta_2} = \beta_2 - \frac{C_{23}\beta_3}{C_{33}},$$
$$\overline{p_1} = p_1, \overline{p_2} = p_2, \overline{p_3} = p_3 + \frac{e_{33}\alpha_{33}}{C_{33}}, \overline{\lambda_1} = \lambda_1, \overline{\lambda_2} = \lambda_2, \overline{\lambda_3} = \lambda_3 + \frac{q_{33}\alpha_{33}}{C_{33}},$$
$$\overline{\chi_1} = \chi_1, \overline{\chi_2} = \chi_2, \overline{\chi_3} = \chi_3 + \frac{e_{33}\beta_{33}}{C_{33}}, \overline{\zeta_1} = \zeta_1, \overline{\zeta_2} = \zeta_2, \overline{\zeta_3} = \zeta_3 + \frac{q_{33}\beta_{33}}{C_{33}}. \quad (8)$$

The electric and magnetic intensity can be determined from gradients of the scalar electric and magnetic potentials as [2]

$$\begin{Bmatrix} E_x \\ E_y \\ E_z \end{Bmatrix} = \begin{Bmatrix} -\frac{\partial\tilde{\Phi}}{\partial x} \\ -\frac{\partial\tilde{\Phi}}{\partial y} \\ -\frac{\partial\tilde{\Phi}}{\partial z} \end{Bmatrix} = \begin{Bmatrix} \cos(\beta z)\frac{\partial\Phi}{\partial x} \\ \cos(\beta z)\frac{\partial\Phi}{\partial y} \\ -\beta\sin(\beta z)\Phi - \frac{2\phi_0}{h} \end{Bmatrix},$$
$$\begin{Bmatrix} H_x \\ H_y \\ H_z \end{Bmatrix} = \begin{Bmatrix} -\frac{\partial\tilde{\Psi}}{\partial x} \\ -\frac{\partial\tilde{\Psi}}{\partial y} \\ -\frac{\partial\tilde{\Psi}}{\partial z} \end{Bmatrix} = \begin{Bmatrix} \cos(\beta z)\frac{\partial\Psi}{\partial x} \\ \cos(\beta z)\frac{\partial\Psi}{\partial y} \\ -\beta\sin(\beta z)\Psi - \frac{2\psi_0}{h} \end{Bmatrix}. \quad (9)$$

where $\beta = \frac{\pi}{h}$ and $\Phi(x, y)$, $\Psi(x, y)$ are the spatial variations in the electric and magnetic potentials, respectively, and ϕ_0 and ψ_0 are the initial electric and magnetic potentials, respectively.

The force and moment resultants of the magneto–electro–elastic sandwich plate are found from the following equation:

$$\begin{aligned}
N_x &= A_{11}\varepsilon_x^0 + A_{12}\varepsilon_y^0 + B_{11}k_x^1 + B_{12}k_y^1 + E_{11}k_x^3 + E_{12}k_y^3 - G_1E_z \\
&\quad - H_1H_z - \alpha_1\Delta T - \beta_1\Delta m, \\
N_y &= A_{12}\varepsilon_x^0 + A_{22}\varepsilon_y^0 + B_{12}k_x^1 + B_{22}k_y^1 + E_{12}k_x^3 + E_{22}k_y^3 - G_2E_z \\
&\quad - H_2H_z - \alpha_2\Delta T - \beta_2\Delta m, \\
N_{xy} &= A_{66}\gamma_{xy}^0 + B_{66}k_{xy}^1 + E_{66}k_{xy}^3, \\
M_x &= B_{11}\varepsilon_x^0 + B_{12}\varepsilon_y^0 + D_{11}k_x^1 + D_{12}k_y^1 + F_{11}k_x^3 + F_{12}k_y^3 - G_3E_z \\
&\quad - H_3H_z - \alpha_3\Delta T - \beta_3\Delta m, \\
M_y &= B_{12}\varepsilon_x^0 + B_{22}\varepsilon_y^0 + D_{12}k_x^1 + D_{22}k_y^1 + F_{11}k_x^3 + F_{12}k_y^3 - G_4E_z \\
&\quad - H_4H_z - \alpha_4\Delta T - \beta_4\Delta m, \\
M_{xy} &= B_{66}\gamma_{xy}^0 + D_{66}k_{xy}^1 + F_{66}k_{xy}^1, \\
P_x &= E_{11}\varepsilon_x^0 + E_{12}\varepsilon_y^0 + F_{11}k_x^1 + F_{12}k_y^1 + H_{11}k_x^3 + H_{12}k_y^3 - G_3E_z \\
&\quad - H_3H_z - \alpha_3\Delta T - \beta_3\Delta m, \\
P_y &= E_{12}\varepsilon_x^0 + E_{22}\varepsilon_y^0 + F_{12}k_x^1 + F_{22}k_y^1 + H_{11}k_x^3 + H_{12}k_y^3 - G_4E_z
\end{aligned}$$

$$
\begin{aligned}
&- H_4H_z - \alpha_4\Delta T - \beta_4\Delta m,\\
P_{xy} &= E_{66}\gamma_{xy}^0 + F_{66}k_{xy}^1 + H_{66}k_{xy}^3,\\
Q_x &= A_{44}\gamma_{xz}^0 + D_{44}k_{xz}^2 - G_7E_x - H_7H_z,\\
Q_y &= A_{55}\gamma_{yz}^0 + D_{55}k_{yz}^2 - G_8E_x - H_8H_z,\\
R_x &= D_{44}\gamma_{xz}^0 + F_{44}k_{xz}^2 - G_9E_x - H_9H_z,\\
R_y &= D_{55}\gamma_{yz}^0 + F_{55}k_{yz}^2 - G_{10}E_x - H_{10}H_z,
\end{aligned} \tag{10}
$$

in which the coefficients $A_{ij}, B_{ij}, E_{ij}, F_{ij}, H_{ij}(ij = 11, 12, 22, 44, 55, 66)$, $G_k, H_k\left(k = \overline{1, 10}\right)$; and $\alpha_l, \beta_l\left(l = \overline{1, 6}\right)$ are given in Appendix 1.

Airy's stress function, $f(x, y)$, is defined as

$$
N_x = \frac{\partial^2 f}{\partial y^2}, N_y = \frac{\partial^2 f}{\partial x^2}, N_{xy} = -\frac{\partial^2 f}{\partial x \partial y}. \tag{11}
$$

The nonlinear equilibrium equations of the magneto–electro–elastic sandwich plate are [18]

$$
\frac{\partial N_x}{\partial x} + \frac{\partial N_{xy}}{\partial y} = 0, \tag{12a}
$$

$$
\frac{\partial N_{xy}}{\partial x} + \frac{\partial N_y}{\partial y} = 0, \tag{12b}
$$

$$
\begin{aligned}
&\frac{\partial Q_x}{\partial x} + \frac{\partial Q_y}{\partial y} - 3c_1\left(\frac{\partial R_x}{\partial x} + \frac{\partial R_y}{\partial y}\right) + \left(\frac{\partial^2 P_x}{\partial x^2} + 2\frac{\partial^2 P_y}{\partial x \partial y} + \frac{\partial^2 P_y}{\partial y^2}\right)\\
&+ N_x\frac{\partial^2 w}{\partial x^2} + 2N_{xy}\frac{\partial^2 w}{\partial x \partial y} + N_y\frac{\partial^2 w}{\partial y^2} - k_1 w + k_2\left(\frac{\partial^2 w}{\partial x^2} + \frac{\partial^2 w}{\partial y^2}\right) = 0.
\end{aligned} \tag{12c}
$$

$$
\frac{\partial M_x}{\partial x} + \frac{\partial M_{xy}}{\partial y} - Q_x + 3c_1R_x - c_1\left(\frac{\partial P_x}{\partial x} + \frac{\partial P_{xy}}{\partial y}\right) = 0, \tag{12d}
$$

$$
\frac{\partial M_{xy}}{\partial x} + \frac{\partial M_y}{\partial y} - Q_y + 3c_1R_y - c_1\left(\frac{\partial P_{xy}}{\partial x} + \frac{\partial P_y}{\partial y}\right) = 0, \tag{12e}
$$

$$
\begin{aligned}
&\int\limits_{-h/2}^{h_3-h/2}\left(\frac{\partial D_x}{\partial x}\cos(\beta z) + \frac{\partial D_y}{\partial y}\cos(\beta z) + D_z\beta\sin(\beta z)\right)dz\\
&+ \int\limits_{h/2-h_1}^{h/2}\left(\frac{\partial D_x}{\partial x}\cos(\beta z) + \frac{\partial D_y}{\partial y}\cos(\beta z) + D_z\beta\sin(\beta z)\right)dz = 0,
\end{aligned} \tag{12f}
$$

$$\int_{-h/2}^{h_3-h/2}\left(\frac{\partial B_x}{\partial x}\cos(\beta z)+\frac{\partial B_y}{\partial y}\cos(\beta z)+B_z\beta\sin(\beta z)\right)dz$$
$$+\int_{h/2-h_1}^{h/2}\left(\frac{\partial B_x}{\partial x}\cos(\beta z)+\frac{\partial B_y}{\partial y}\cos(\beta z)+B_z\beta\sin(\beta z)\right)dz=0. \tag{12g}$$

where k_1 is the modulus of the Winkler independent spring system and k_2 is the stiffness of the Pasternak shear layer.

The geometrical compatibility equation of the sandwich plate is defined as [18]

$$\frac{\partial^2\varepsilon_x^0}{\partial y^2}+\frac{\partial^2\varepsilon_y^0}{\partial x^2}-\frac{\partial^2\gamma_{xy}^0}{\partial x\partial y}=\left(\frac{\partial^2 w}{\partial x\partial y}\right)^2-\frac{\partial^2 w}{\partial x^2}\frac{\partial^2 w}{\partial y^2}. \tag{13}$$

Substitution of Eqs. (10) and (11) into Eq. (12a–g), the nonlinear equilibrium equations of the sandwich plate can be determined to depend on deflection, rotations, stress function, electric and magnetic potentials as

$$L_{11}(w)+L_{12}(\phi_x)+L_{13}(\phi_y)+L_{14}(f)+L_{15}(\Phi)+L_{16}(\Psi)+S(w,f)=0,$$
$$L_{21}(w)+L_{22}(\phi_x)+L_{23}(\phi_y)+L_{24}(f)+L_{25}(\Phi)+L_{26}(\Psi)=0,$$
$$L_{31}(w)+L_{32}(\phi_x)+L_{33}(\phi_y)+L_{34}(f)+L_{35}(\Phi)+L_{36}(\Psi)=0,$$
$$\int_{-\frac{h}{2}}^{h_3-\frac{h}{2}}\left(\frac{\partial D_x}{\partial x}\cos(\beta z)+\frac{\partial D_y}{\partial y}\cos(\beta z)+D_z\beta\sin(\beta z)\right)dz$$
$$+\int_{\frac{h}{2}-h_1}^{\frac{h}{2}}\left(\frac{\partial D_x}{\partial x}\cos(\beta z)+\frac{\partial D_y}{\partial y}\cos(\beta z)+D_z\beta\sin(\beta z)\right)dz=0,$$
$$\int_{-h/2}^{-h/2+h_3}\left(\frac{\partial B_x}{\partial x}\cos(\beta z)+\frac{\partial B_y}{\partial y}\cos(\beta z)+B_z\beta\sin(\beta z)\right)dz$$
$$+\int_{h/2-h_1}^{h/2}\left(\frac{\partial B_x}{\partial x}\cos(\beta z)+\frac{\partial B_y}{\partial y}\cos(\beta z)+B_z\beta\sin(\beta z)\right)dz=0, \tag{14}$$

in which the details of coefficients $L_{1i}, L_{2i}, L_{3i}\left(i=\overline{1,6}\right)$ and S are expressed in Appendix 2.

Substituting Eqs. (11) and (12a–g) into Eq. (13), the geometrical compatibility equation of the sandwich plate has the form as follows:

$$I_{21}\frac{\partial^4 f}{\partial x^4}+I_{11}\frac{\partial^4 f}{\partial y^4}+J_1\frac{\partial^4 f}{\partial x^2\partial y^2}+J_2\frac{\partial^3\phi_x}{\partial x^3}+J_3\frac{\partial^3\phi_x}{\partial x\partial y^2}+J_4\frac{\partial^3\phi_y}{\partial y^3}+J_5\frac{\partial^3\phi_y}{\partial x^2\partial y}$$
$$-c_1I_{25}\frac{\partial^4 w}{\partial x^4}-c_1I_{16}\frac{\partial^4 w}{\partial y^4}+J_6\frac{\partial^4 w}{\partial x^2\partial y^2}-\left(\left(\frac{\partial^2 w}{\partial x\partial y}\right)^2-\frac{\partial^2 w}{\partial x^2}\frac{\partial^2 w}{\partial y^2}\right)=0, \quad (15)$$

in which

$$\begin{aligned}
&J_1=I_{31}-2I_{12}, J_2=I_{23}-c_1I_{25},\\
&J_3=I_{13}-c_1I_{15}-I_{32}+c_1I_{33},\\
&J_4=I_{14}-c_1I_{16}, J_5=I_{24}-c_1I_{26}-I_{32}+c_1I_{33},\\
&J_6=-c_1I_{15}-c_1I_{26}+2c_1I_{33}.
\end{aligned} \quad (16)$$

Four edges of the magneto–electro–elastic sandwich plate are assumed to be simply supported and freely movable. The boundary conditions in this case are

$$\begin{aligned}
&w=N_{xy}=\varphi_y=M_x=P_x=0, N_x=N_{x0}\text{at}x=0,a\\
&w=N_{xy}=\varphi_x=M_y=P_y=0, N_y=N_{y0}\text{at}y=0,b
\end{aligned} \quad (17)$$

where N_{x0}, N_{y0} are in-plane compressive loads at four edges.

The displacement, rotations, electric and magnetic potentials and stress functions are chosen as the combination of undetermined coefficients and known trigonometric functions as follows:

$$\begin{aligned}
w(x,y)&=W\sin\lambda_m x\sin\delta_n y,\\
\phi_x(x,y)&=\Phi_x\cos\lambda_m x\sin\delta_n y,\\
\phi_y(x,y)&=\Phi_y\sin\lambda_m x\cos\delta_n y,\\
\Phi(x,y)&=\phi\sin\lambda_m x\sin\delta_n y,\\
\Psi(x,y)&=\psi\sin\lambda_m x\sin\delta_n y,\\
f(x,y)&=A_1\cos 2\lambda_m x+A_2\cos 2\delta_n y+A_3\sin\lambda_m x\sin\delta_n y\\
&\quad+\frac{1}{2}N_{x0}y^2+\frac{1}{2}N_{y0}x^2,
\end{aligned} \quad (18)$$

where $\lambda_m=m\pi/a, \delta_n=n\pi/b$ and m, n are odd natural numbers representing the number of half waves and W, Φ_x, Φ_y, ϕ and ψ are the amplitudes of the deflection, rotation angles, and electric and magnetic potentials, respectively.

Three coefficients $A_i (i=1\div 3)$ of the stress function are determined by substituting Eq. (18) into Eq. (15) and balancing the coefficients on both sides of Eq. (17) as

$$A_1=M_1W^2, A_2=M_2W^2, A_3=M_3W+M_4\Phi_x+M_5\Phi_y, \quad (19)$$

where

$$M_1 = \frac{\delta_n^2}{32 I_{21} \lambda_m^2}, M_2 = \frac{\lambda_m^2}{32 I_{11} \delta_n^2},$$
$$M_3 = \frac{c_1 I_{25} \lambda_m^4 + c_1 I_{16} \delta_n^4 - J_6 \lambda_m^2 \delta_n^2}{I_{21} \lambda_m^4 + J_1 \lambda_m^2 \delta_n^2 + I_{11} \delta_n^4},$$
$$M_4 = \frac{-\left(J_2 \lambda_m^3 + J_3 \lambda_m \delta_n^2\right)}{I_{21} \lambda_m^4 + J_1 \lambda_m^2 \delta_n^2 + I_{11} \delta_n^4}, M_5 = \frac{-\left(J_4 \delta_n^3 + J_5 \lambda_m^2 \delta_n\right)}{I_{21} \lambda_m^4 + J_1 \lambda_m^2 \delta_n^2 + I_{11} \delta_n^4}. \quad (20)$$

Substituting Eqs. (18) and (19) into Eq. (14) and applying the Galerkin method for obtained results (multiplying both sides of the equations in the system Eqs. (14) by $\sin \lambda_m x \sin \delta_n y$, $\cos \lambda_m x \sin \delta_n y$ and $\sin \lambda_m x \cos \delta_n y$ then integrating results in the domain $0 \le x \le a, 0 \le y \le b$, the compatibility equation becomes

$$\left(l_{11} + n_1 - N_{x0}\lambda_m^2 - N_{y0}\delta_n^2\right)W + l_{12}\Phi_x + l_{13}\Phi_y + l_{14}W\Phi_x$$
$$+ l_{15}W\Phi_y + l_{16}\phi + l_{17}\psi + (n_2 + n_3)W^2 + n_4 W^3 = 0, \quad (21a)$$

$$(l_{21} + n_6)W + l_{22}\Phi_x + l_{23}\Phi_y + l_{24}\phi + l_{25}\psi + n_7 W^2 = 0, \quad (21b)$$

$$(l_{31} + n_8)W + l_{32}\Phi_x + l_{33}\Phi_y + l_{34}\phi + l_{35}\psi + n_9 W^2 = 0, \quad (21c)$$

$$l_{41}W + l_{42}\Phi_x + l_{43}\Phi_y + l_{44}\phi + l_{45}\psi = 0, \quad (21d)$$

$$l_{51}W + l_{52}\Phi_x + l_{53}\Phi_y + l_{54}\phi + l_{55}\psi = 0, \quad (21e)$$

in which the coefficients $l_{1i}\left(i = \overline{1,7}\right)$, $l_{jk}\left(j = \overline{2,5}; k = \overline{1,5}\right)$ and $n_m\left(m = \overline{1,9}\right)$ are expressed in Appendix 3.

Substituting the amplitude of electric and magnetic potentials φ and ψ from Eqs. (21d) and (21e) into three Eqs. (21a–c), we have the following equations.

$$d_1 W + d_2\Phi_x + d_3\Phi_y + l_{14}(W + \mu h)\Phi_x + l_{15}(W + \mu h)\Phi_y$$
$$+ \left[n_1 - N_{x0}\lambda_m^2 - N_{y0}\delta_n^2\right](W + \mu h) + n_2 W(W + \mu h)$$
$$+ n_3 W(W + 2\mu h) + n_4 W(W + \mu h)(W + 2\mu h) = 0, \quad (22a)$$

$$d_4 W + d_5\Phi_x + d_6\Phi_y + n_6(W + \mu h) + n_7 W(W + 2\mu h) = 0, \quad (22b)$$

$$d_7 W + d_8\Phi_x + d_9\Phi_y + n_8(W + \mu h) + n_9 W(W + 2\mu h) = 0, \quad (22c)$$

in which the coefficients $d_i\left(i = \overline{1,9}\right)$ are given in Appendix 4.

Two amplitudes of the rotation angles are obtained from Eqs. (22b) and (22c) as follows:

$$\Phi_x = s_1\overline{W} + s_2\left(\overline{W} + \mu\right) + s_3\overline{W}\left(\overline{W} + 2\mu\right),$$
$$\Phi_y = s_4\overline{W} + s_5\left(\overline{W} + \mu\right) + s_6\overline{W}\left(\overline{W} + 2\mu\right), \quad (23)$$

in which $\overline{W} = \frac{W}{h}$ is the dimensionless deflection and the coefficients $s_i\left(i = \overline{1,6}\right)$ are defined as

$$s_1 = \frac{h(d_4d_9 - d_6d_7)}{d_8d_6 - d_9d_5}, s_2 = \frac{h(d_9n_6 - n_8d_6)}{d_8d_6 - d_9d_5},$$
$$s_3 = \frac{h^2(n_7d_9 - n_9d_6)}{d_8d_6 - d_9d_5}, s_4 = \frac{h(d_7d_5 - d_8d_4)}{d_8d_6 - d_9d_5},$$
$$s_5 = \frac{h(n_8d_5 - d_8n_6)}{d_8d_6 - d_9d_5}, s_6 = \frac{h^2(-d_8n_7 + n_9d_5)}{d_8d_6 - d_9d_5}. \quad (24)$$

The magneto–electro–elastic sandwich plate is subjected to axial compressive loads F_x (Pascals) uniformly distributed at two edges $x = 0, a$ in the absence of thermal loading. In this case, we have

$$N_{x0} = -F_xh, N_{y0} = 0. \quad (25)$$

Substituting Eqs. (23) and (25) into Eqs. (22a–c), the relationship between the axial compressive load and dimensionless deflection of the magneto–electro–elastic sandwich plate is given as

$$F_x = (a_1 + a_2 + a_5)\overline{W} + +a_3\overline{W}^2 + a_4 + a_6, \quad (26)$$

where

$$a_1 = -\frac{hl_{14}s_1 + hl_{15}s_4 + h^2n_2}{h^2\lambda_m^2}, a_2 = -\frac{hl_{14}s_2 + hl_{15}s_5}{h^2\lambda_m^2},$$
$$a_3 = -\frac{hl_{14}s_3 + hl_{15}s_6 + h^3n_4}{h^2\lambda_m^2}, a_4 = -\frac{hd_1 + d_2s_1 + d_3s_4}{h^2\lambda_m^2},$$
$$a_5 = -\frac{d_2s_3 + d_3s_6 + h^2n_3}{h^2\lambda_m^2}, a_6 = -\frac{d_2s_2 + d_3s_5 + hn_1}{h^2\lambda_m^2}. \quad (27)$$

Specifically, the critical buckling load of the perfect magneto–electro–elastic sandwich plate is obtained by taking the limit of the function F_x when $\overline{W} \to 0$ as

$$F_{xCritical} = a_4 + a_6. \quad (28)$$

2.3 Optimisation Problem

From the expressions in the previous sections, it is easy to see that a change in the geometrical dimensions and stiffness of elastic foundations impacts the post–buckling behaviors of the magneto–electro–elastic sandwich plate. In this section, we determine the optimum values of these parameters, which correspond to the maximum value of the critical buckling load of the sandwich plate subjected to axial compression loading. The standard Bees Algorithm [19] is used to solve the optimisation problem. Figure 3 indicates the process of a standard Bees Algorithm.

The Bees Algorithm is performed 20 times, and the optimum value of the objective function is the average value of 20 outputs. Because we do not know the exact optimum value of the objective function, the Bees Algorithm stops when the number of evaluation objective functions reaches 50,000. The objective function is the critical buckling load, which is obtained from Eq. (28). It depends on five variables, including the plate length a, the plate width b, the total thickness h and two elastic foundation

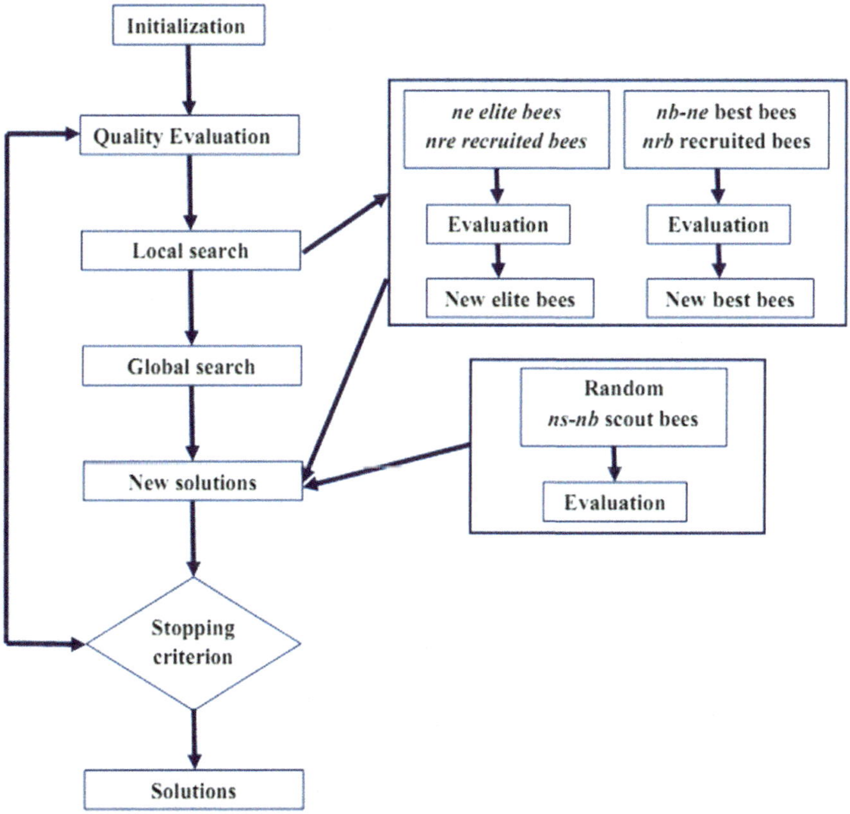

Fig. 3 Diagram of the standard Bees Algorithm

Table 2 Variables and domain of each variable

Variable	Domain
$a(m)$	$[1, 5]$
$b(m)$	$[1, 5]$
$h(m)$	$[0.01, 0.05]$
$k_1\left(\frac{GPa}{m}\right)$	$[0, 1]$
$k_2(GPa.m)$	$[0, 0.01]$

Table 3 Values of seven parameters in the Bees Algorithm

Symbol	Value
ns	26
ne	2
nb	6
nre	20
nrb	10
ngh	0.8
stlim	10

coefficients k_1, k_2. The values of each variable belong to a given range, which is shown in Table 2.

In the Bees Algorithm, there are seven flexible parameters that we can choose based on the class of problem. They are the number of scout bees *ns*, number of elite bees *ne*, number of best bees *nb*, number of recruited bees for each elite site *nre*, number of recruited bees for each best site *nrb*, initial size of neighborhood *ngh* and number of cycles of local stagnation *stlim*. The values of these parameters are chosen as shown in Table 3.

2.4 Numerical Results

Based on the Bees Algorithm, the obtained maximum value of the critical buckling load of the sandwich plate is $43.45399(GPa)$ when the optimum values of the corresponding variables are presented in Table 4. The standard deviation of 20 obtained values of each variable is also shown in this table. It is easy to see that the standard deviation is small compared to the optimum value of each variable, which indicates the low dispersion of the results by using the Bees Algorithm.

Figure 4 presents the convergence of the maximum value of the critical buckling load of the sandwich plate according to the number of objective function evaluations in three loops of the Bees Algorithm. The maximum value of the critical buckling load rises rapidly at approximately 7000 times the objective function evaluation and then gradually reaches the convergent value. Furthermore, the difference in the critical

Table 4 The optimum values of variables corresponding to the maximum value of the critical buckling load of the sandwich plate by using the Bees Algorithm

Variable	Optimum value	Standard deviation
$a(m)$	3.45079	0.10766
$b(m)$	1.00366	0.02589
$h(m)$	0.01002	0.00058
$k_1\left(\frac{GPa}{m}\right)$	0.84888	0.03819
$k_2(GPa.m)$	0.05853	0.00380

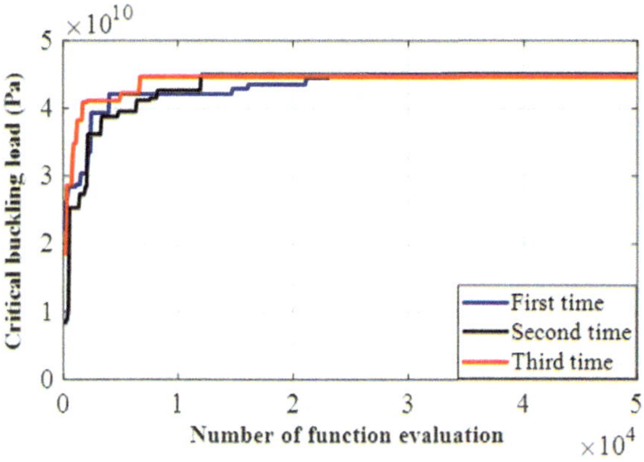

Fig. 4 Convergence of the maximum value of the critical buckling load of the sandwich plate by using the Bees Algorithm

buckling load between the three loops is not large, which shows the high consistency of the Bees Algorithm.

2.5 Conclusions

This section investigates the geometrical optimisation of a magneto–electro–elastic sandwich plate with an auxetic honeycomb core using the Bees Algorithm. The nonlinear equilibrium system equations are established based on Reddy's higher order shear deformation plate theory. The expression of the critical buckling load is determined by using the Galerkin method. The Bees Algorithm is used to maximise the value of the critical buckling load of the sandwich plate subjected to an axial compressive load. The obtained maximum value of the critical buckling load of the sandwich plate with the given range of variables based on the Bees Algorithm is 43.45399(GPa).

3 Geometrical Optimisation of Piezoelectric Sandwich Plates with Porous Homogeneous Cores and Carbon Nanotube-Reinforced Nanocomposite Layers

The global optimisation of smart sandwich plates subjected to mechanical, thermal and electric loadings is presented. The sandwich plate comprises a porous homogeneous core, two carbon nanotube reinforced composite (CNTRC) layers, and two piezoelectric face sheets. Basic equations are derived based on Reddy's higher order shear deformation plate theory and Hamilton's principle, in which the initial imperfection and Pasternak-type elastic foundations are included. The Bees Algorithm is used to determine the maximum value of the natural frequency of the smart sandwich plate and the corresponding optimum geometrical and material parameter values.

3.1 Problem Description

3.1.1 Geometry and Coordinate System

As shown in Fig. 5, a smart sandwich plate with total thickness h, length a and width b is considered. The Cartesian coordinates (x, y, z) are located on the middle surface of the plate in which z is in the thickness direction. The smart sandwich plate consists of five layers, which have a functionally graded porous homogeneous core, two carbon nanotube reinforced composite layers and two piezoelectric face sheets. The thickness of the core layer, each carbon nanotube reinforced composite layer and each piezoelectric face sheet are h_c, h_f and h_p, respectively.

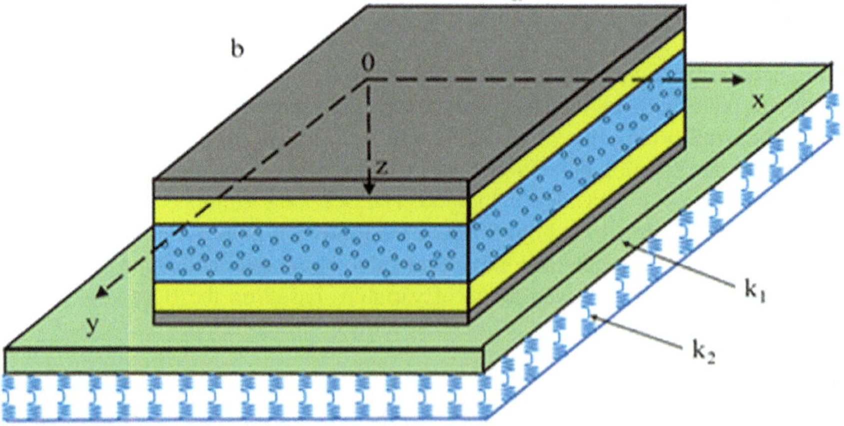

Fig. 5 Schematic diagram of a smart sandwich plate on elastic foundations

The displacement components in the middle surface to the coordinates (x, y, z) are u, v and w, respectively, and ϕ_x, ϕ_y denote the rotations of the transverse normal about the y and x axes, respectively, at $z = 0$. The smart sandwich plate is assumed to rest on Pasternak-type elastic foundations. The interaction between the sandwich plate and elastic foundations is modelled by a Winkler independent spring system with modulus k_1 and a Pasternak shear layer with stiffness k_2.

3.1.2 Material Properties

The Young's modulus E^c, mass density ρ^c and thermal expansion coefficients α^c of the porous core are summed to be location-dependent as [20, 21]

$$\begin{aligned} E^c &= E_1[1 - e_0\lambda(z)], \\ \rho^c &= \rho_1[1 - e_m\lambda(z)], \\ \alpha^c &= \alpha_1[1 - e_m\lambda(z)], \end{aligned} \tag{29}$$

where E_1, ρ_1, α_1 are the material properties of the nonporous original material of the core layer and e_0 and e_m are the coefficient of porosity and the mass density coefficient, respectively. The relationship between e_0 and e_m can be given as

$$e_m = 1 - \sqrt{1 - e_0}(0 \le e_0 \le 1). \tag{30}$$

Three different types of porosity distributions, namely, nonuniform symmetric distribution, nonuniform asymmetric distribution and uniform distribution, are considered. The function $\lambda(z)$ is expressed in three different cases as [20]

$$\lambda(z) = \begin{cases} \cos\left(\frac{\pi z}{h_c}\right) & Non - uniformsymmetricdistribution \\ \cos\left(\frac{\pi z}{2h_c} + \frac{\pi}{4}\right) & Non - uniformasymmetricdistribution \\ \lambda_0 & Uniformdistribution \end{cases} \tag{31}$$

in which

$$\lambda_0 = \frac{1}{e_0} - \frac{1}{e_0}\left(\frac{2}{\pi}\sqrt{1 - e_0} - \frac{2}{\pi} + 1\right)^2. \tag{32}$$

The titanium alloy (Ti-6Al-4 V) is chosen for the original material without porosity of the core layer. Except for Poisson's ratio and mass density, the elastic modulus of this homogeneous material is assumed to be nonlinearly temperature-dependent as [22]

$$\nu_1 = 0.29,$$
$$\rho_1 = 4429\frac{kg}{m^3},$$
$$\alpha_1 = 7.5788\left(1 + 6.638 \times 10^{-4}T + 3.147 \times 10^{-6}T^2\right) \times 10^{-6}K^{-1},$$
$$E_1 = 122.56\left(1 - 4.586 \times 10^{-4}T\right)GPa \tag{33}$$

in which $T = T_0 + \Delta T$, T_0 is the room temperature and ΔT is the temperature increment in the environment containing the material.

The FG-CNTRC material is composed of a homogeneous matrix reinforced by (10,10) single-walled carbon nanotubes (SWCNTs). The effective Young's modulus, shear modulus, Poisson's ratio, mass density and longitudinal and transverse thermal expansion coefficients of the FG-CNTRC material are expressed as [22]

$$E_{11} = \eta_1 V_{CNT} E_{11}^{CNT} + V_m E_m,$$
$$\frac{\eta_2}{E_{22}} = \frac{V_{CNT}}{E_{22}^{CNT}} + \frac{V_m}{E_m},$$
$$\frac{\eta_3}{G_{12}} = \frac{V_{CNT}}{G_{12}^{CNT}} + \frac{V_m}{G_m},$$
$$G_{13} = G_{12},$$
$$G_{23} = 1.2G_{12},$$
$$\nu_{12} = V_{CNT}^{*}\nu_{12}^{CNT} + V_m\nu_m,$$
$$\rho = V_{CNT}\rho_{CNT} + V_m\rho_m,$$
$$\alpha_{11} = \frac{V_{CNT}E_{11}^{CNT}\alpha_{11}^{CNT} + V_m E_m \alpha_m}{V_{CNT}E_{11}^{CNT} + V_m E_m},$$
$$\alpha_{22} = \left(1 + \nu_{12}^{CNT}\right)V_{CNT}\alpha_{22}^{CNT} + (1 + \nu_m)V_m\alpha_m - \nu_{12}\alpha_{11}, \tag{34}$$

where the subscript and superscript CNT and m denote carbon nanotube and homogeneous matrix components, respectively. V_{CNT} and V_{CNT} are the volume fractions of the carbon nanotube and the matrix, respectively, and $\eta_i\left(i = \overline{1,3}\right)$ are the CNT efficiency parameters, which are determined by matching the effective material properties of FG-CNTRC obtained from molecular simulation results with those from the rule of mixtures [22].

In this study, it is assumed that the volume fractions of the CNTs and homogeneous matrix are distributed according to linear functions throughout the thickness direction for the top and bottom layers as

$$V_{CNT}(z) = \begin{cases} 2\frac{\left(z - \frac{h_c}{2}\right)V_{CNT}^{*}}{h_f} & Topface \\ 2\frac{\left(-z - \frac{h_c}{2}\right)V_{CNT}^{*}}{h_f} & Bottomface \end{cases},$$
$$V_m(z) = 1 - V_{CNT}(z), \tag{35}$$

where

$$V_{CNT}^* = \frac{w_{CNT}}{w_{CNT} + \frac{\rho_{CNT}}{\rho_m} - w_{CNT}\frac{\rho_{CNT}}{\rho_m}}, \tag{36}$$

in which w_{CNT} is the mass fraction of CNTs.

Except for Poisson's ratio, the material properties of the CNT and homogeneous matrix are assumed to strongly depend on temperature as [22]

$$\begin{aligned}
\nu_m &= 0.34, \\
E_m &= (3.52 - 0.0034T)GPa, \\
\alpha_m &= 45(1 + 0.0005\Delta T) \times 10^{-6}K^{-1}, \\
\nu_{12}^{CNT} &= 0.175, \\
E_{11}^{CNT} &= \left(6.18387 - 0.00286T + 4.22867 \times 10^{-6}T^2 - 2.2724 \times 10^{-9}T^3\right)TPa, \\
E_{22}^{CNT} &= \left(7.75348 - 0.00358T + 5.30057 \times 10^{-6}T^2 - 2.84868 \times 10^{-9}T^3\right)TPa, \\
G_{12}^{CNT} &= \left(1.80126 - 7.7845 \times 10^{-4}T - 1.1279 \times 10^{-6}T^2 + 4.93484 \times 10^{-9}T^3\right) \\
&\quad \times TPa, \\
\alpha_{11}^{CNT} &= \left(-1.12148 + 0.02289T - 2.88155 \times 10^{-5}T^2 + 1.13253 \times 10^{-8}T^3\right) \\
&\quad \times 10^{-6}K^{-1}, \\
\alpha_{22}^{CNT} &= (5.43874 - 9.95498 \times 10^{-4}T + 3.13525 \times 10^{-7}T^2 \\
&\quad - 3.36332 \times 10^{-12}T^3) \times 10^{-6}K^{-1}
\end{aligned} \tag{37}$$

3.2 Determination of the Natural Frequency

Reddy's higher order shear deformation plate theory is used to establish basic equations and investigate the nonlinear vibration and dynamic buckling analysis of the smart sandwich plate.

For small strains and moderate deflection, the strain field of the smart sandwich plate is defined as [18]

$$\begin{bmatrix} \varepsilon_x \\ \varepsilon_y \\ \gamma_{xy} \end{bmatrix} = \begin{bmatrix} \varepsilon_x^0 \\ \varepsilon_y^0 \\ \gamma_{xy}^0 \end{bmatrix} + z\begin{bmatrix} k_x^1 \\ k_y^1 \\ k_{xy}^1 \end{bmatrix} + z^3\begin{bmatrix} k_x^3 \\ k_y^3 \\ k_{xy}^3 \end{bmatrix},$$
$$\begin{bmatrix} \gamma_{xz} \\ \gamma_{yz} \end{bmatrix} = \begin{bmatrix} \gamma_{xz}^0 \\ \gamma_{yz}^0 \end{bmatrix} + z^2\begin{bmatrix} k_{xz}^2 \\ k_{yz}^2 \end{bmatrix}, \tag{38}$$

where

$$\begin{bmatrix} \varepsilon_x^0 \\ \varepsilon_y^0 \\ \gamma_{xy}^0 \end{bmatrix} = \begin{bmatrix} \frac{\partial u}{\partial x} + \frac{1}{2}\left(\frac{\partial w}{\partial x}\right)^2 \\ \frac{\partial v}{\partial y} + \frac{1}{2}\left(\frac{\partial w}{\partial y}\right)^2 \\ \frac{\partial u}{\partial y} + \frac{\partial v}{\partial x} + \frac{\partial w}{\partial x}\frac{\partial w}{\partial y} \end{bmatrix}, \begin{bmatrix} \gamma_{xz}^0 \\ \gamma_{yz}^0 \end{bmatrix} = -3c_1 \begin{bmatrix} \frac{\partial w}{\partial x} + \phi_x \\ \frac{\partial w}{\partial y} + \phi_y \end{bmatrix},$$

$$\begin{bmatrix} k_x^1 \\ k_y^1 \\ k_{xy}^1 \end{bmatrix} = \begin{bmatrix} \frac{\partial \phi_x}{\partial x} \\ \frac{\partial \phi_y}{\partial y} \\ \frac{\partial \phi_x}{\partial y} + \frac{\partial \phi_y}{\partial y} \end{bmatrix}, \begin{bmatrix} k_x^3 \\ k_y^3 \\ k_{xy}^3 \end{bmatrix} = -c_1 \begin{bmatrix} \frac{\partial \phi_x}{\partial x} + \frac{\partial^2 w}{\partial x^2} \\ \frac{\partial \phi_y}{\partial y} + \frac{\partial^2 w}{\partial y^2} \\ \frac{\partial \phi_x}{\partial y} + \frac{\partial \phi_y}{\partial y} + 2\frac{\partial^2 w}{\partial x \partial y} \end{bmatrix},$$

$$\begin{bmatrix} k_{xz}^2 \\ k_{yz}^2 \end{bmatrix} = -3c_1 \begin{bmatrix} \frac{\partial w}{\partial x} + \phi_x \\ \frac{\partial w}{\partial y} + \phi_y \end{bmatrix}, \tag{39}$$

with $c_1 = 4/\left(3h^2\right)$.

The stress field of the functionally graded porous homogeneous core is given as follows [20]:

$$\begin{bmatrix} \sigma_{xx} \\ \sigma_{yy} \\ \sigma_{xy} \\ \sigma_{xz} \\ \sigma_{yz} \end{bmatrix}_C = \begin{bmatrix} Q_{11}^C & Q_{12}^C & 0 & 0 & 0 \\ Q_{12}^C & Q_{22}^C & 0 & 0 & 0 \\ 0 & 0 & Q_{66}^C & 0 & 0 \\ 0 & 0 & 0 & Q_{44}^C & 0 \\ 0 & 0 & 0 & 0 & Q_{55}^C \end{bmatrix} \begin{bmatrix} (\varepsilon_{xx})_C - \alpha^C \Delta T \\ \left(\varepsilon_{yy}\right)_C - \alpha^C \Delta T \\ \left(\varepsilon_{xy}\right)_C \\ (\varepsilon_{xz})_C \\ \left(\varepsilon_{yz}\right)_C \end{bmatrix}, \tag{40}$$

in which the elastic stiffness is

$$Q_{11}^C = Q_{22}^C = \frac{E^C}{1 - \left(v^C\right)^2}, Q_{12}^C = \frac{v^C E^C}{1 - \left(v^C\right)^2},$$
$$Q_{66}^C = Q_{44}^C = Q_{55}^C = \frac{E^C}{2\left(1 + v^C\right)}. \tag{41}$$

For carbon nanotube reinforced composite layers, the linear constitutive relation is expressed by Hooke's law as [22]

$$\begin{bmatrix} \sigma_{xx} \\ \sigma_{yy} \\ \sigma_{xy} \\ \sigma_{xz} \\ \sigma_{yz} \end{bmatrix}_f = \begin{bmatrix} Q_{11}^f & Q_{12}^f & 0 & 0 & 0 \\ Q_{12}^f & Q_{22}^f & 0 & 0 & 0 \\ 0 & 0 & Q_{66}^f & 0 & 0 \\ 0 & 0 & 0 & Q_{44}^f & 0 \\ 0 & 0 & 0 & 0 & Q_{55}^f \end{bmatrix} \begin{bmatrix} (\varepsilon_{xx})_f - \alpha_{11} \Delta T \\ \left(\varepsilon_{yy}\right)_f - \alpha_{22} \Delta T \\ \left(\varepsilon_{xy}\right)_f \\ (\varepsilon_{xz})_f \\ \left(\varepsilon_{yz}\right)_f \end{bmatrix}, \tag{42}$$

where

$$Q_{11}^f = \frac{E_{11}}{1 - \nu_{12}\nu_{21}}, \; Q_{22}^f = \frac{E_{22}}{1 - \nu_{12}\nu_{21}}, \; Q_{12}^f = \frac{\nu_{21}E_{11}}{1 - \nu_{12}\nu_{21}},$$
$$Q_{44}^f = G_{23}, \; Q_{55}^f = G_{13}, \; Q_{66}^f = G_{12}. \tag{43}$$

By taking into account the effect of temperature and electric field, the stress–strain relation of isotropic piezoelectric face sheets can be written as

$$\begin{bmatrix} \sigma_{xx} \\ \sigma_{yy} \\ \sigma_{xy} \\ \sigma_{xz} \\ \sigma_{yz} \end{bmatrix}_p = \begin{bmatrix} Q_{11}^p & Q_{12}^p & 0 & 0 & 0 \\ Q_{12}^p & Q_{22}^p & 0 & 0 & 0 \\ 0 & 0 & Q_{66}^p & 0 & 0 \\ 0 & 0 & 0 & Q_{44}^p & 0 \\ 0 & 0 & 0 & 0 & Q_{55}^p \end{bmatrix} \begin{bmatrix} (\varepsilon_{xx})_p - \alpha^p \Delta T \\ (\varepsilon_{yy})_p - \alpha^p \Delta T \\ (\varepsilon_{xy})_p \\ (\varepsilon_{xz})_p \\ (\varepsilon_{yz})_p \end{bmatrix} - \begin{bmatrix} 0 & 0 & e_{31} \\ 0 & 0 & e_{32} \\ 0 & e_{24} & 0 \\ e_{15} & 0 & 0 \\ 0 & 0 & 0 \end{bmatrix} \begin{bmatrix} E_x \\ E_y \\ E_z \end{bmatrix}, \tag{44}$$

in which

$$Q_{11}^p = Q_{22}^p = \frac{E^p}{1 - (\nu^p)^2}, \; Q_{12}^p = \frac{\nu^p E^p}{1 - (\nu^p)^2},$$
$$Q_{66}^p = Q_{44}^p = Q_{55}^p = \frac{E^p}{2(1 + \nu^p)}, \tag{45}$$

and $e_{31}, e_{32}, e_{15}, e_{24}$ are the piezoelectric stiffness of piezoelectric layers, which can be determined from dielectric constants $d_{31}, d_{32}, d_{15}, d_{24}$ and elastic stiffness as

$$\begin{aligned} e_{31} &= d_{31}Q_{11}^p + d_{32}Q_{12}^p, \\ e_{32} &= d_{31}Q_{12}^p + d_{32}Q_{22}^p, \\ e_{15} &= d_{15}Q_{55}^p, \\ e_{24} &= d_{24}Q_{44}^p. \end{aligned} \tag{46}$$

In this study, we assume that the piezoelectric polarisation is along the z direction. Therefore, the components of the electric field are expressed as

$$\left[E_x \; E_y \; E_z \right]^T = \left[0 \; 0 \; V_p/h_p \right]^T, \tag{47}$$

where V_p is the applied voltage to the piezoelectric layers in the thickness direction.

By integrating the stress through the thickness, the force and moment resultants of the smart sandwich plate can be determined as

$$(N_i, M_i, P_i) = \sum_{k=1}^{5} \int_{z_k}^{z_{k+1}} \sigma_i^k \left(1, z, z^3\right) dz, \; i = x, y, xy$$

$$(Q_i, R_i) = \sum_{k=1}^{5} \int_{z_k}^{z_{k+1}} \sigma_{iz}^k \left(1, z^2\right) dz, \; i = x, y, \tag{48}$$

in which z_k and z_{k+1} are the coordinates of the *kth* layer of the smart sandwich plate.

By substituting Eq. (38) into Eqs. (40–44) then obtained results in Eq. (48), the force and moment—strain relationship of the smart sandwich plate can be written as

$$\begin{aligned}
N_x &= A_{11}\varepsilon_x^0 + A_{12}\varepsilon_y^0 + B_{11}k_x^1 + B_{12}k_y^1 + E_{11}k_x^3 + E_{12}k_y^3 - \Phi_1 - 2V_a e_{31}, \\
N_y &= A_{12}\varepsilon_x^0 + A_{22}\varepsilon_y^0 + B_{12}k_x^1 + B_{22}k_y^1 + E_{12}k_x^3 + E_{22}k_y^3 - \Phi_2 - 2V_a e_{32}, \\
N_{xy} &= A_{66}\gamma_{xy}^0 + B_{66}k_{xy}^1 + E_{66}k_{xy}^3, \\
M_x &= B_{11}\varepsilon_x^0 + B_{12}\varepsilon_y^0 + D_{11}k_x^1 + D_{12}k_y^1 + F_{11}k_x^3 + F_{12}k_y^3 - \Phi_3, \\
M_y &= B_{12}\varepsilon_x^0 + B_{22}\varepsilon_y^0 + D_{12}k_x^1 + D_{22}k_y^1 + F_{12}k_x^3 + F_{22}k_y^3 - \Phi_4, \\
M_{xy} &= B_{66}\gamma_{xy}^0 + D_{66}k_{xy}^1 + F_{66}k_{xy}^3, P_x = E_{11}\varepsilon_x^0 + E_{12}\varepsilon_y^0 + F_{11}k_x^1 \\
&\quad + F_{12}k_y^1 + H_{11}k_x^3 + H_{12}k_y^3 - \Phi_5, \\
P_y &= E_{12}\varepsilon_x^0 + E_{22}\varepsilon_y^0 + F_{12}k_x^1 + F_{22}k_y^1 + H_{12}k_x^3 + H_{22}k_y^3 - \Phi_6, \\
P_{xy} &= E_{66}\gamma_{xy}^0 + F_{66}k_{xy}^1 + H_{66}k_{xy}^3, \\
Q_x &= A_{44}\gamma_{xz}^0 + D_{44}k_{xz}^2, Q_y = A_{55}\gamma_{yz}^0 + D_{55}k_{yz}^2, \\
R_x &= D_{44}\gamma_{xz}^0 + F_{44}k_{xz}^2, R_y = D_{55}\gamma_{yz}^0 + F_{55}k_{yz}^2,
\end{aligned} \tag{49}$$

in which the details of coefficients $A_{ij}, B_{ij}, D_{ij}, E_{ij}, F_{ij}, H_{ij}$ $(ij = 11, 12, 22,), A_{kl}, D_{kl}, F_{kl} \quad (kl = 44, 55), \Phi_m (m = 1, 2, 3, 4, 5, 6)$ may be found in Appendix 5.

Airy's stress function $f(x, y, t)$ is defined as

$$N_x = \frac{\partial^2 f}{\partial y^2}, N_y = \frac{\partial^2 f}{\partial x^2}, N_{xy} = -\frac{\partial^2 f}{\partial x \partial y}. \tag{50}$$

Based on the Hamilton principle and the basis of Reddy's higher order shear deformation plate theory, the nonlinear motion equation of the smart sandwich plate is defined as [18]

$$\frac{\partial N_x}{\partial x} + \frac{\partial N_{xy}}{\partial y} = \overline{J_1}\frac{\partial^2 u}{\partial t^2} + \overline{J_2}\frac{\partial^2 \phi_x}{\partial t^2} - \overline{J_3}\frac{\partial^3 w}{\partial t^2 \partial x}, \tag{51a}$$

$$\frac{\partial N_{xy}}{\partial x} + \frac{\partial N_y}{\partial y} = \overline{J_1^*}\frac{\partial^2 v}{\partial t^2} + \overline{J_2^*}\frac{\partial^2 \phi_y}{\partial t^2} - \overline{J_3^*}\frac{\partial^3 w}{\partial t^2 \partial y}, \tag{51b}$$

$$
\begin{aligned}
&\frac{\partial Q_x}{\partial x}+\frac{\partial Q_y}{\partial y}-3c_1\left(\frac{\partial R_x}{\partial x}+\frac{\partial R_y}{\partial y}\right)+c_1\left(\frac{\partial^2 P_x}{\partial x^2}+2\frac{\partial^2 P_y}{\partial x\partial y}+\frac{\partial^2 P_y}{\partial y^2}\right)\\
&\quad+q+N_x\frac{\partial^2 w}{\partial x^2}+2N_{xy}\frac{\partial^2 w}{\partial x\partial y}+N_y\frac{\partial^2 w}{\partial y^2}-k_1 w+k_2\nabla^2 w\\
&\quad=J_1\frac{\partial^2 w}{\partial t^2}+2\varepsilon J_1\frac{\partial w}{\partial t}+\overline{J_3}\frac{\partial^3 u}{\partial t^2\partial x}+\overline{J_5}\frac{\partial^3 \phi_x}{\partial t^2\partial x}\\
&\quad+\overline{J_3^*}\frac{\partial^3 v}{\partial t^2\partial y}+\overline{J_5^*}\frac{\partial^3 \phi_y}{\partial t^2\partial y}-c_1^2 J_7\left(\frac{\partial^4 w}{\partial t^2\partial x^2}+\frac{\partial^4 w}{\partial t^2\partial y^2}\right),
\end{aligned}
\tag{51c}
$$

$$
\begin{aligned}
\frac{\partial M_x}{\partial x}+\frac{\partial M_{xy}}{\partial y}-Q_x+3c_1R_x-c_1\left(\frac{\partial P_x}{\partial x}+\frac{\partial P_{xy}}{\partial y}\right)&=\overline{J_2}\frac{\partial^2 u}{\partial t^2}+\overline{J_4}\frac{\partial^2 \phi_x}{\partial t^2}\\
&\quad-\overline{J_5}\frac{\partial^3 w}{\partial t^2\partial x},
\end{aligned}
\tag{51d}
$$

$$
\begin{aligned}
\frac{\partial M_{xy}}{\partial x}+\frac{\partial M_y}{\partial y}-Q_y+3c_1R_y-c_1\left(\frac{\partial P_{xy}}{\partial x}+\frac{\partial P_y}{\partial y}\right)&=\overline{J_2^*}\frac{\partial^2 v}{\partial t^2}+\overline{J_4^*}\frac{\partial^2 \phi_y}{\partial t^2}\\
&\quad-\overline{J_5^*}\frac{\partial^3 w}{\partial t^2\partial y}
\end{aligned}
\tag{51e}
$$

in which ε is the viscous damping coefficient and

$$
\begin{aligned}
&\overline{J_1}=J_1,\overline{J_2}=J_2-c_1J_4,\overline{J_3}=c_1J_4,\\
&\overline{J_4}=J_3-2c_1J_5+c_1^2J_7,\overline{J_5}=c_1J_5-c_1^2J_7,\\
&\qquad(J_1,J_2,J_3,J_4,J_5,J_7)=\int_{-\frac{hc}{2}}^{\frac{hc}{2}}\rho_c(z)\left(1,z,z^2,z^3,z^4,z^6\right)dz\\
&\qquad\qquad+\int_{-hp-hf-\frac{hc}{2}}^{-hf-\frac{hc}{2}}\rho_p(z)\left(1,z,z^2,z^3,z^4,z^6\right)dz\\
&\qquad\qquad+\int_{-hf-\frac{hc}{2}}^{-\frac{hc}{2}}\rho_f(z)\left(1,z,z^2,z^3,z^4,z^6\right)dz\\
&\qquad\qquad+\int_{\frac{hc}{2}}^{\frac{hc}{2}+hf}\rho_f(z)\left(1,z,z^2,z^3,z^4,z^6\right)dz
\end{aligned}
$$

$$+\int_{hc/2+hf}^{hc/2+hf+hp} \rho_p(z)\left(1, z, z^2, z^3, z^4, z^6\right)dz. \quad (52)$$

By replacing Eq. (50) into two first equations of system Eqs. (51a and 51b), we can collect the second derivative of the displacements with respect to time as

$$\frac{\partial^2 u}{\partial t^2} = -\frac{\overline{J_2}}{\overline{J_1}}\frac{\partial^2 \phi_x}{\partial t^2} + \frac{\overline{J_3}}{\overline{J_1}}\frac{\partial^3 w}{\partial t^2 \partial x}, \quad (53a)$$

$$\frac{\partial^2 v}{\partial t^2} = -\frac{\overline{J_2^*}}{\overline{J_1^*}}\frac{\partial^2 \phi_y}{\partial t^2} + \frac{\overline{J_3^*}}{\overline{J_1^*}}\frac{\partial^3 w}{\partial t^2 \partial y}. \quad (53b)$$

Substituting Eqs. (53a) and (53b) into Eqs. (51c–e), the differential system equations can be obtained as follows

$$\begin{aligned}
&\frac{\partial Q_x}{\partial x} + \frac{\partial Q_y}{\partial y} - 3c_1\left(\frac{\partial R_x}{\partial x} + \frac{\partial R_y}{\partial y}\right) + c_1\left(\frac{\partial^2 P_x}{\partial x^2} + 2\frac{\partial^2 P_{xy}}{\partial x \partial y} + \frac{\partial^2 P_y}{\partial y^2}\right) \\
&+ \frac{\partial^2 f}{\partial y^2}\frac{\partial^2 w}{\partial x^2} - 2\frac{\partial^2 f}{\partial x \partial y}\frac{\partial^2 w}{\partial x \partial y} + \frac{\partial^2 f}{\partial x^2}\frac{\partial^2 w}{\partial y^2} \\
&+ q - k_1 w + k_2 \nabla^2 w + \frac{1}{R_x}\frac{\partial^2 f}{\partial x^2} + \frac{1}{R_y}\frac{\partial^2 f}{\partial y^2} \\
&= J_1 \frac{\partial^2 w}{\partial t^2} + 2\varepsilon J_1 \frac{\partial w}{\partial t} + \overline{J_5}\frac{\partial^3 \phi_x}{\partial t^2 \partial x} + \overline{J_5^*}\frac{\partial^3 \phi_y}{\partial t^2 \partial y} \\
&+ \overline{J_7}\frac{\partial^4 w}{\partial t^2 \partial x^2} + \overline{J_7^*}\frac{\partial^4 w}{\partial t^2 \partial y^2},
\end{aligned} \quad (54a)$$

$$\frac{\partial M_x}{\partial x} + \frac{\partial M_{xy}}{\partial y} - Q_x + 3c_1 R_x - c_1\left(\frac{\partial P_x}{\partial x} + \frac{\partial P_{xy}}{\partial y}\right) = \overline{J_3}\frac{\partial^2 \phi_x}{\partial t^2} - \overline{J_5}\frac{\partial^3 w}{\partial t^2 \partial x}, \quad (54b)$$

$$\frac{\partial M_{xy}}{\partial x} + \frac{\partial M_y}{\partial y} - Q_y + 3c_1 R_y - c_1\left(\frac{\partial P_{xy}}{\partial x} + \frac{\partial P_y}{\partial y}\right) = \overline{J_3^*}\frac{\partial^2 \phi_y}{\partial t^2} - \overline{J_5^*}\frac{\partial^3 w}{\partial t^2 \partial y}, \quad (54c)$$

in which

$$\begin{aligned}
&\overline{J_3} = \overline{J_4} - \left(\overline{J_2}\right)^2/\overline{J_1}, \overline{J_3^*} = \overline{J_4^*} - \left(\overline{J_2^*}\right)^2/\overline{J_1^*}, \\
&\overline{J_5} = \overline{J_5} - \overline{J_2 J_3}/\overline{J_1}, \overline{J_5^*} = \overline{J_5^*} - \overline{J_2^* J_3^*}/\overline{J_1^*}, \\
&\overline{J_7} = \left(\overline{J_3}\right)^2/\overline{J_1} - c_1^2 J_7, \overline{J_7^*} = \left(\overline{J_3^*}\right)^2/\overline{J_1^*} - c_1^2 J_7.
\end{aligned} \quad (55)$$

By substituting Eqs. (49) and (50) into Eq. (54a–c) leads to

$$
\begin{aligned}
L_{11}(w) + L_{12}(\phi_x) + L_{13}(\phi_y) + L_{14}(f) + S(w,f) + q = {} & J_1 \frac{\partial^2 w}{\partial t^2} + 2\varepsilon J_1 \frac{\partial w}{\partial t} \\
& + \overline{J_5} \frac{\partial^3 \phi_x}{\partial t^2 \partial x} + \overline{J_5^*} \frac{\partial^3 \phi_y}{\partial t^2 \partial y} \\
& + \overline{J_7} \frac{\partial^4 w}{\partial t^2 \partial x^2} + \overline{J_7^*} \frac{\partial^4 w}{\partial t^2 \partial y^2}, \\
L_{21}(w) + L_{22}(\phi_x) + L_{23}(\phi_y) + L_{24}(f) = {} & \overline{I_3} \frac{\partial^2 \phi_x}{\partial t^2} - \overline{I_5} \frac{\partial^3 w}{\partial t^2 \partial x}, \\
L_{31}(w) + L_{32}(\phi_x) + L_{33}(\phi_y) + L_{34}(f) = {} & \overline{J_3^*} \frac{\partial^2 \phi_y}{\partial t^2} - \overline{J_5^*} \frac{\partial^3 w}{\partial t^2 \partial y}.
\end{aligned} \tag{56}
$$

where the details of operators $L_{1i}, L_{2i}, L_{3i}\left(i = \overline{1,4}\right), S$ are given in Appendix 6.

To find the displacements of the sandwich plate, we need to use one equation that shows the relationship between the strain components. Specifically, the geometrical compatibility equation for a perfect smart sandwich plate is introduced as [18]

$$
\frac{\partial^2 \varepsilon_x^0}{\partial y^2} + \frac{\partial^2 \varepsilon_y^0}{\partial x^2} - \frac{\partial^2 \gamma_{xy}^0}{\partial x \partial y} = {\frac{\partial^2 w}{\partial x \partial y}}^2 - \frac{\partial^2 w}{\partial x^2} \frac{\partial^2 w}{\partial y^2}. \tag{57}
$$

Inserting Eqs. (49) and (50) into Eq. (57), the compatibility equation of the perfect sandwich plate becomes

$$
\begin{aligned}
& j_{21} \frac{\partial^4 f}{\partial x^4} + j_{11} \frac{\partial^4 f}{\partial y^4} + H_1 \frac{\partial^4 f}{\partial x^2 \partial y^2} + H_2 \frac{\partial^3 \phi_x}{\partial x^3} + H_3 \frac{\partial^3 \phi_x}{\partial x \partial y^2} + H_4 \frac{\partial^3 \phi_y}{\partial y^3} + H_5 \frac{\partial^3 \phi_y}{\partial y \partial x^2} \\
& - c_1 j_{25} \frac{\partial^4 w}{\partial x^4} - c_1 j_{16} \frac{\partial^4 w}{\partial y^4} + j_6 \frac{\partial^4 w}{\partial x^2 \partial y^2} - \left({\frac{\partial^2 w}{\partial x \partial y}}^2 - \frac{\partial^2 w}{\partial x^2} \frac{\partial^2 w}{\partial y^2} \right) = 0,
\end{aligned} \tag{58}
$$

in which

$$
\begin{aligned}
& H_1 = j_{31} - 2 j_{12}, H_2 = j_{23} - c_1 j_{25}, \\
& H_3 = j_{13} - c_1 j_{15} - j_{32} + c_1 j_{33}, \\
& H_4 = j_{14} - c_1 j_{16}, \\
& H_5 = j_{24} - c_1 j_{26} - j_{32} + c_1 j_{33}, \\
& H_6 = -c_1 j_{15} - c_1 j_{26} + 2 c_1 j_{33}.
\end{aligned} \tag{59}
$$

The four edges of the smart sandwich plate are assumed to be simply supported and immovable. The boundary conditions are given as

$$
\begin{aligned}
& x = 0, a : w = u = \phi_y = M_x = P_x = 0, N_x = N_{x0}, \\
& y = 0, b : w = v = \phi_x = M_y = P_y = 0, N_y = N_{y0},
\end{aligned} \tag{60}
$$

in which N_{x0} and N_{y0} are fictitious compressive edge loads of the sandwich plate.

The deflection and rotations of the smart sandwich plate are defined in terms of unknown time-dependent coefficients and one-term mode shapes based on boundary conditions as follows:

$$\begin{bmatrix} w(x,y,t) \\ \phi_x(x,y,t) \\ \phi_y(x,y,t) \end{bmatrix} = \begin{bmatrix} W(t)\sin\lambda_m x\sin\delta_n y \\ \Phi_x(t)\cos\lambda_m x\sin\delta_n y \\ \Phi_y(t)\sin\lambda_m x\cos\delta_n y \end{bmatrix}, \tag{61}$$

where $\lambda_m = m\pi/a$, $\delta_n = n\pi/b$ with m, n are the number of half sine and cosine waves and are responsible for denoting the mode shape of the sandwich plate; W, Φ_x and Φ_y are the time-dependent amplitudes of the deflection and rotation angles.

By substituting Eq. (61) into Eq. (58) and balancing coefficients on both sides, the stress function is determined as

$$\begin{aligned} f(x,y,t) = {} & C_1(t)\cos 2\lambda_m x + C_2(t)\cos 2\delta_n y \\ & + C_3(t)\sin\lambda_m x\sin\delta_n y \\ & + \frac{1}{2}N_{y0}x^2 + \frac{1}{2}N_{x0}y^2, \end{aligned} \tag{62}$$

where

$$\begin{aligned} C_1 &= \frac{\delta_n^2}{32I_{21}\lambda_m^2}W(W+2\mu h), \\ C_2 &= \frac{\lambda_m^2}{32I_{11}\delta_n^2}W(W+2\mu h), \\ C_3 &= G_1W + G_2\Phi_x + G_3\Phi_y, \end{aligned} \tag{63}$$

and

$$\begin{aligned} G_1 &= \frac{c_1 j_{25}\lambda_m^4 + c_1 j_{16}\delta_n^4 - H_6\lambda_m^2\delta_n^2}{j_{21}\lambda_m^4 + H_1\lambda_m^2\delta_n^2 + j_{11}\delta_n^4}, \\ G_2 &= \frac{-\left(H_2\lambda_m^3 + H_3\lambda_m\delta_n^2\right)}{j_{21}\lambda_m^4 + H_1\lambda_m^2\delta_n^2 + j_{11}\delta_n^4}, \\ G_3 &= \frac{-\left(H_4\delta_n^3 + H_5\lambda_m^2\delta_n\right)}{j_{21}\lambda_m^4 + H_1\lambda_m^2\delta_n^2 + j_{11}\delta_n^4}. \end{aligned} \tag{64}$$

Substituting Eqs. (61 and 62) into Eq. (56), then applying the Galerkin method for the obtained equations, we obtain the nonlinear system of differential equations to investigate the nonlinear vibration and dynamic buckling of the smart sandwich plate with a porous core, CNTRC layers, and piezoelectric face sheets as

$$\begin{aligned}
&l_{11}W + l_{12}\Phi_x + l_{13}\Phi_y + l_{14}(W + W_0)\Phi_x + l_{15}(W + W_0)\Phi_y \\
&\quad + \left[n_1 - N_{x0}\lambda_m^2 - N_{y0}\delta_n^2\right](W + W_0) + n_2 W(W + W_0) \\
&\quad + n_3 W(W + 2W_0) + n_4 W(W + W_0)(W + 2W_0) + n_5 q \\
&= J_0\frac{\partial^2 W}{\partial t^2} + 2\varepsilon J_1\frac{\partial W}{\partial t} - \lambda_m\overline{J_5}\frac{\partial^2 \Phi_x}{\partial t^2} - \delta_n\overline{J_5^*}\frac{\partial^2 \Phi_y}{\partial t^2}, \\
&l_{21}W + l_{22}\Phi_x + l_{23}\Phi_y + n_6(W + W_0) + n_7 W(W + 2W_0) \\
&= \overline{J_3}\frac{\partial^2 \Phi_x}{\partial t^2} - \lambda_m\overline{J_5}\frac{\partial^2 W}{\partial t^2}, \\
&l_{31}W + l_{32}\Phi_x + l_{33}\Phi_y + n_8(W + W_0) + n_9 W(W + 2W_0) \\
&= \overline{J_3^*}\frac{\partial^2 \Phi_y}{\partial t^2} - \delta_n\overline{J_5^*}\frac{\partial^2 W}{\partial t^2},
\end{aligned} \tag{65}$$

in which the details of coefficients $l_{1i}\left(i = \overline{1,3}\right)$, $l_{jk}\left(j = \overline{2,3}, k = \overline{1,2}\right)$, $n_m\left(m = \overline{1,9}\right)$, J_0 may be found in Appendix 7.

To ensure the existence of thermal stresses and consider the effect of temperature on the vibration of the smart sandwich plate, four edges of the plate are assumed to be immovable in the transverse plane. In other words, displacement components are eliminated in four edges, i.e., $u = 0$ on $x = 0, a$ and $v = 0$ on $y = 0, b$. These conditions are satisfied on the average sense on the whole area of the plate surface as

$$\int_0^b\int_0^a \frac{\partial u}{\partial x}dxdy = 0, \int_0^a\int_0^b \frac{\partial v}{\partial x}dydx = 0, \tag{66}$$

The derivative of the displacement components can be obtained by substituting Eq. (39) into Eq. (49) as

$$\begin{aligned}
\frac{\partial u}{\partial x} &= j_{11}\frac{\partial^2 f}{\partial y^2} - j_{12}\frac{\partial^2 f}{\partial x^2} + j_{13}\frac{\partial \phi_x}{\partial x} + j_{14}\frac{\partial \phi_y}{\partial y} \\
&\quad - c_1 j_{15}\left(\frac{\partial^2 w}{\partial x^2} + \frac{\partial \phi_x}{\partial x}\right) - c_1 j_{16}\left(\frac{\partial^2 w}{\partial y^2} + \frac{\partial \phi_y}{\partial y}\right) \\
&\quad + j_{17}(\Phi_1 + 2V_p e_{31}) + j_{18}(\Phi_2 + 2V_p e_{32}) \\
&\quad - \frac{1}{2}\left(\frac{\partial w}{\partial x}\right)^2 - \frac{\partial w}{\partial x}\frac{\partial w^*}{\partial x}, \\
\frac{\partial v}{\partial y} &= j_{21}\frac{\partial^2 f}{\partial x^2} - j_{12}\frac{\partial^2 f}{\partial y^2} + j_{23}\frac{\partial \phi_x}{\partial x} + j_{24}\frac{\partial \phi_y}{\partial y} \\
&\quad - c_1 j_{25}\left(\frac{\partial^2 w}{\partial x^2} + \frac{\partial \phi_x}{\partial x}\right) - c_1 j_{26}\left(\frac{\partial^2 w}{\partial y^2} + \frac{\partial \phi_y}{\partial y}\right) \\
&\quad + j_{27}(\Phi_1 + 2V_p e_{31}) + j_{28}(\Phi_2 + 2V_p e_{32}) \\
&\quad - \frac{1}{2}\left(\frac{\partial w}{\partial y}\right)^2 - \frac{\partial w}{\partial y}\frac{\partial w^*}{\partial y}.
\end{aligned} \tag{67}$$

Replacing Eqs. (61 and 62) into Eq. (67), then substituting the obtained results into Eq. (66), the fictitious compressive edge loads are expressed to depend on the amplitude of the deflection and rotation angles as

$$\begin{aligned} N_{x0} &= g_1 W + g_4 (W + W_0) W + g_2 \Phi_x \\ &\quad + g_3 \Phi_y + g_5 \left(\Phi_1 + 2V_p e_{31} \right) \\ &\quad + g_6 \left(\Phi_2 + 2V_p e_{32} \right), \\ N_{y0} &= f_1 W + f_4 (W + W_0) W + f_2 \Phi_x \\ &\quad + f_3 \Phi_y + f_5 \left(\Phi_1 + 2V_p e_{31} \right) \\ &\quad + f_6 \left(\Phi_2 + 2V_p e_{32} \right), \end{aligned} \tag{68}$$

in which the details of coefficients $g_i, f_i \left(i = \overline{1,6} \right)$ may be found in Appendix 8.

Substituting Eq. (68) into Eq. (65), the nonlinear system of differential equations becomes

$$\begin{aligned} & l_{11} W + l_{12} \Phi_x + l_{13} \Phi_y + l_{14}^1 (W + W_0) \Phi_x + l_{15}^1 (W + W_0) \Phi_y + n_1^1 (W + W_0) \\ & \quad + n_2^1 W (W + W_0) + n_3 W (W + 2W_0) + n_4^1 W (W + W_0)(W + 2W_0) + n_5 q \\ & = J_0 \frac{\partial^2 W}{\partial t^2} + 2\varepsilon J_1 \frac{\partial W}{\partial t} - \lambda_m \overline{J_5} \frac{\partial^2 \Phi_x}{\partial t^2} - \delta_n \overline{J_5^*} \frac{\partial^2 \Phi_y}{\partial t^2}, \\ & l_{21} W + l_{22} \Phi_x + l_{23} \Phi_y + n_6 (W + W_0) + n_7 W (W + 2W_0) \\ & = \overline{J_3} \frac{\partial^2 \Phi_x}{\partial t^2} - \lambda_m \overline{J_5} \frac{\partial^2 W}{\partial t^2}, \\ & l_{31} W + l_{32} \Phi_x + l_{33} \Phi_y + n_8 (W + W_0) + n_9 W (W + 2W_0) \\ & = \overline{J_3^*} \frac{\partial^2 \Phi_y}{\partial t^2} - \delta_n \overline{J_5^*} \frac{\partial^2 W}{\partial t^2}, \end{aligned} \tag{69}$$

where

$$\begin{aligned} l_{14}^1 &= \left(l_{14} - \lambda_m^2 g_2 - \delta_n^2 f_2 \right), \\ l_{15}^1 &= \left(l_{15} - \lambda_m^2 g_3 - \delta_n^2 f_3 \right), \\ n_1^1 &= \begin{bmatrix} n_1 - \left(\lambda_m^2 g_5 + \delta_n^2 f_5 \right) \left(\Phi_1 + 2V_p e_{31} \right) \\ - \left(\lambda_m^2 g_6 + \delta_n^2 f_6 \right) \left(\Phi_2 + 2V_p e_{32} \right) \end{bmatrix}, \\ n_2^1 &= \left(n_2 - \lambda_m^2 g_1 - \delta_n^2 f_1 \right), \\ n_4^1 &= \left(n_4 - \lambda_m^2 g_4 - \delta_n^2 f_4 \right). \end{aligned} \tag{70}$$

The natural frequencies of the perfect smart sandwich plate are determined by solving the following equation in which the minimum value of three solutions is chosen:

$$\begin{vmatrix} l_{11}+n_1^1+J_0\omega^2 & l_{12}-\lambda_m\overline{J_5}\omega^2 & l_{13}-\delta_n\overline{J_5^*}\omega^2 \\ l_{21}+n_6-\lambda_m\overline{J_5}\omega^2 & l_{22}+\overline{J_3}\omega^2 & l_{23} \\ l_{31}+n_8-\delta_n\overline{J_5^*}\omega^2 & l_{32} & l_{33}+\overline{J_3^*}\omega^2 \end{vmatrix}=0. \quad (71)$$

3.3 Optimisation Problem

The natural frequency of the smart sandwich plate is determined as Eq. (71). It is assumed to depend on nine geometrical and material parameters: the thickness of the porous core, CNTRC layer and piezoelectric face sheet h_c, h_f, h_p, the length and width of the plate a, b, two moduli of elastic foundations k_1, k_2, the temperature increment ΔT and the coefficient of porosity e_0. The standard Bees Algorithm [19] is used to calculate the maximum value of the natural frequency and optimum values of the above parameters. Nine variables are assumed to be in the constant domain, as shown in Table 5, and the basic parameters of the Bees Algorithm are chosen, as shown in Table 6.

Table 5 The domain of nine variables of global optimisation

Variable	Domain
$h_c(m)$	$[10^{-3}, 10^{-1}]$
$h_f(m)$	$[10^{-3}, 10^{-1}]$
$h_p(m)$	$[10^{-3}, 10^{-1}]$
$a(m)$	[0.05, 5]
$b(m)$	[0.05, 5]
$k_1(GPa/m)$	[0, 0.9]
$k_2(GPa.m)$	[0, 0.9]
$\Delta T(K)$	[0, 1000]
e_0	[0, 1]

Table 6 The values of the parameters of the standard Bees Algorithm

Parameter	Value
Number of scout bees	25
Number of elite bees	3
Number of best bees	8
Number of recruited bees around each elite bee	20
Number of recruited bees around each best bee	10
Initial size of neighbourhood	0.08
Number of limit loop for site abandonment	10

Table 7 The optimum values of nine geometrical and material variables of the smart sandwich plate by using the Bees Algorithm

Variable	Optimum value	Standard deviation
$h_c(m)$	0.752	0.01604
$h_f(m)$	0.069	0.01697
$h_p(m)$	0.027	0.00704
$a(m)$	1.49	0.04243
$b(m)$	1.03	0.05692
$k_1(GPa/m)$	0.253	0.00506
$k_2(GPa.m)$	4.57	0.11696
$\Delta T(K)$	157.12	1.6456
e_0	0.291	0.0206

3.4 Numerical Results

The Bees Algorithm is run 25 times, and the optimum value of the natural frequency of the smart sandwich plate is defined as the average value of the 25 obtained values. Each time, the loop stops if the number of objective function evaluations reaches 40,000. Finally, the maximum value of the natural frequency is $70118.45(\frac{1}{s})$ when the optimum values of the nine variables are shown in Table 7. The standard deviation of the 25 obtained values of each variable is also determined. The results show the low dispersion and the high consistency of the obtained results by using the Bees Algorithm.

Figure 6 illustrates the convergence of the optimum value of the natural frequency according to the number of objective function evaluations in one loop of the Bees Algorithm. As can be observed, the optimum value of the natural frequency rises sharply in the first number of objective function evaluations and then gradually reaches the final value.

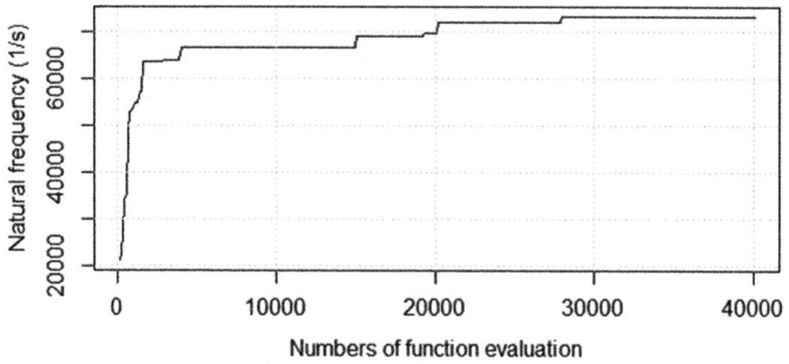

Fig. 6 Convergence of the optimum value of the natural frequency by using the Bees Algorithm

4 Conclusions

This section introduces the geometrical optimisation of the smart sandwich plate on elastic foundations based on Reddy's higher order shear deformation plate theory. The sandwich plate has five layers: one porous core, two CNTRC layers, and two piezoelectric face sheets. The Bees Algorithm is used to determine the maximum value of the natural frequency. The optimum value of the natural frequency with constant geometrical and material parameters is $70118.45\left(\frac{1}{s}\right)$.

Appendix 1

$$
\begin{aligned}
\left(A_{ij}, B_{ij}, D_{ij}E_{ij}, F_{ij}H_{ij}\right) = & \int_{-\frac{h}{2}}^{h_3-\frac{h}{2}} \widetilde{C_{ij}}\left(1, z, z^2, z^3, z^4, z^6\right)dz \\
& + \int_{h_3-\frac{h}{2}}^{\frac{h}{2}-h_1} Q_{ij}\left(1, z, z^2, z^3, z^4, z^6\right)dz \\
& + \int_{\frac{h}{2}-h_1}^{\frac{h}{2}} \widetilde{C_{ij}}\left(1, z, z^2, z^3, z^4, z^6\right)dz, \; ij = 11, 12, 22, 66,
\end{aligned}
$$

$$
\begin{aligned}
\left(A_{kl}, D_{kl}, F_{kl}\right) = & \int_{-\frac{h}{2}}^{h_3-\frac{h}{2}} \overline{C_{ij}}\left(1, z^2, z^4\right)dz + \int_{\frac{h}{2}-h_1}^{\frac{h_c}{2}} Q_{ij}\left(1, z^2, z^4\right)dz \\
& + \int_{\frac{h_c}{2}}^{\frac{h}{2}-h_1} \overline{C_{ij}}\left(1, z^2, z^4\right)dz, \; ij = 44, 55
\end{aligned}
$$

$$
G_i = \int_{-\frac{h}{2}}^{h_3-\frac{h}{2}} \overline{e_{31}}\left(1, z, z^3\right)dz + \int_{\frac{h}{2}-h_1}^{\frac{h}{2}} \overline{e_{31}}\left(1, z, z^3\right)dz,
$$

$$
H_i = \int_{-h/2}^{h_3-h/2} \overline{q_{31}}\left(1, z, z^3\right)dz + \int_{h/2-h_1}^{h/2} \overline{q_{31}}\left(1, z, z^3\right)dz, \; i = 1, 3, 5,
$$

$$G_j = \int_{-h/2}^{h_3-h/2} \overline{e_{32}}(1, z, z^3)dz + \int_{h/2-h_1}^{h/2} \overline{e_{32}}(1, z, z^3)dz,$$

$$H_j = \int_{-\frac{h}{2}}^{h_3-\frac{h}{2}} \overline{q_{32}}(1, z, z^3)dz + \int_{\frac{h}{2}-h_1}^{\frac{h}{2}} \overline{q_{32}}(1, z, z^3)dz, i = 2, 4, 6,$$

$$G_k = \int_{-\frac{h}{2}}^{h_3-\frac{h}{2}} \overline{e_{24}}(1, z^2,)dz + \int_{\frac{h}{2}-h_1}^{\frac{h}{2}} \overline{e_{24}}(1, z^2)dz,$$

$$H_k = \int_{-\frac{h}{2}}^{h_3-\frac{h}{2}} \overline{q_{24}}(1, z^2)dz + \int_{\frac{h}{2}-h_1}^{\frac{h}{2}} \overline{q_{24}}(1, z^2)dz, k = 8, 10,$$

$$G_l = \int_{-\frac{h}{2}}^{h_3-\frac{h}{2}} \overline{e_{15}}(1, z^2,)dz + \int_{\frac{h}{2}-h_1}^{\frac{h}{2}} \overline{e_{15}}(1, z^2)dz,$$

$$H_l = \int_{-h/2}^{h_3-h/2} \overline{q_{15}}(1, z^2)dz + \int_{h/2-h_1}^{h/2} \overline{q_{15}}(1, z^2)dz, l = 7, 9,$$

$$\alpha_i = \int_{-h/2}^{h_3-h/2} (\overline{C_{11}} + \overline{C_{12}})\overline{\alpha_1}(1, z, z^3)dz + \int_{h/2-h_1}^{h/2} (\overline{C_{11}} + \overline{C_{12}})\overline{\alpha_1}(1, z, z^3)dz$$
$$+ \int_{h_3-h/2}^{h/2-h_1} (Q_{11} + Q_{12})\alpha_{11}(1, z, z^3)dz, i = 1, 3, 5,$$

$$\alpha_j = \int_{-h/2}^{h_3-h/2} (\overline{C_{11}} + \overline{C_{12}})\overline{\alpha_2}(1, z, z^3)dz + \int_{h/2-h_1}^{h/2} (\overline{C_{11}} + \overline{C_{12}})\overline{\alpha_2}(1, z, z^3)dz$$
$$+ \int_{h_3-h/2}^{h/2-h_1} (Q_{12} + Q_{22})\alpha_{11}(1, z, z^3)dz, i = 1, 3, 5, j = 2, 4, 6,$$

$$\beta_i = \int_{-h/2}^{h_3-h/2} \overline{\beta_1}(1, z, z^3)dz + \int_{h/2-h_1}^{h/2} \overline{\beta_1}(1, z, z^3)dz, i = 1, 3, 5,$$

$$\beta_j = \int_{-h/2}^{-h_c/2} \overline{\beta_2}\left(1, z, z^3\right)dz + \int_{h/2-h_1}^{h/2} \overline{\beta_2}\left(1, z, z^3\right)dz, j = 2, 4, 6.$$

Appendix 2

$$L_{11}(w) = Y_{11}\frac{\partial^2 w}{\partial x^2} + Y_{12}\frac{\partial^2 w}{\partial y^2} + Y_{13}\frac{\partial^4 w}{\partial x^4} + Y_{14}\frac{\partial^4 w}{\partial x^2 \partial y^2} + Y_{15}\frac{\partial^4 w}{\partial y^4} - k_1 w + k_2\left(\frac{\partial^2 w}{\partial x^2} + \frac{\partial^2 w}{\partial y^2}\right),$$

$$L_{12}(\phi_x) = Y_{11}\frac{\partial \phi_x}{\partial x} + Y_{16}\frac{\partial^3 \phi_x}{\partial x^3} + Y_{17}\frac{\partial^3 \phi_x}{\partial x \partial y^2},$$

$$L_{13}\left(\phi_y\right) = Y_{12}\frac{\partial \phi_y}{\partial y} + Y_{16}\frac{\partial^3 \phi_y}{\partial y^3} + Y_{17}\frac{\partial^3 \phi_y}{\partial x^2 \partial y},$$

$$L_{14}(f) = Y_{110}\frac{\partial^4 f}{\partial x^4} + Y_{111}\frac{\partial^4 f}{\partial x^2 \partial y^2} + Y_{112}\frac{\partial^4 f}{\partial y^4},$$

$$L_{15}(\Phi) = (Y_{114}\cos(\beta z) - Y_{115}\beta\sin(\beta z))\frac{\partial^2 \Phi}{\partial x^2} + (Y_{113}\cos(\beta z) - Y_{116}\beta\sin(\beta z))\frac{\partial^2 \Phi}{\partial y^2},$$

$$L_{16}(\Psi) = (Y_{118}\cos(\beta z) - Y_{119}\beta\sin(\beta z))\frac{\partial^2 \Psi}{\partial x^2} + (Y_{117}\cos(\beta z) - Y_{120}\beta\sin(\beta z))\frac{\partial^2 \Psi}{\partial y^2},$$

$$S(w, f) = \frac{\partial^2 f}{\partial y^2}\frac{\partial^2 w}{\partial x^2} - 2\frac{\partial^2 f}{\partial x \partial y}\frac{\partial^2 w}{\partial y^2} + \frac{\partial^2 f}{\partial x^2}\frac{\partial^2 w}{\partial y^2},$$

$$L_{21}(w) = Y_{21}\frac{\partial w}{\partial x} + Y_{22}\frac{\partial^3 w}{\partial x^3} + Y_{23}\frac{\partial^3 w}{\partial x \partial y^2},$$

$$L_{22}(\phi_x) = Y_{21}\phi_x + Y_{24}\frac{\partial^2 \phi_x}{\partial x^2} + Y_{25}\frac{\partial^2 \phi_x}{\partial y^2},$$

$$L_{23}\left(\phi_y\right) = Y_{26}\frac{\partial^2 \phi_y}{\partial x \partial y}, L_{24}(f) = Y_{27}\frac{\partial^3 f}{\partial x^3} + Y_{28}\frac{\partial^3 f}{\partial x \partial y^2},$$

$$L_{25}(\Phi) = -(Y_{114}\cos(\beta z) + Y_{29}\beta\sin(\beta z))\frac{\partial \Phi}{\partial x},$$

$$L_{26}(\Psi) = -(Y_{118}\cos(\beta z) + Y_{210}\beta\sin(\beta z))\frac{\partial \Psi}{\partial x},$$

$$L_{31}(w) = Y_{31}\frac{\partial w}{\partial y} + Y_{32}\frac{\partial^3 w}{\partial x^2 \partial x} + Y_{33}\frac{\partial^3 w}{\partial y^3},$$

$$L_{33}(\phi_y) = Y_{31}\phi_y + Y_{35}\frac{\partial^2 \phi_y}{\partial x^2} + Y_{36}\frac{\partial^2 \phi_y}{\partial y^2},$$

$$L_{32}(\phi_x) = Y_{34}\frac{\partial^2 \phi_x}{\partial x \partial y}, L_{34}(f) = Y_{37}\frac{\partial^3 f}{\partial x^2 \partial y} + Y_{38}\frac{\partial^3 f}{\partial y^3},$$

$$L_{35}(\Phi) = -(Y_{113}\cos(\beta z) + Y_{39}\beta\sin(\beta z))\frac{\partial \Phi}{\partial y},$$

$$L_{36}(\Psi) = -(Y_{117}\cos(\beta z) + Y_{310}\beta\sin(\beta z))\frac{\partial \Psi}{\partial y},$$

$$Y_{111} = -c_1(2E_{66}I_{31} - E_{11}I_{11} + 2E_{12}I_{12} - E_{22}I_{21}),$$
$$Y_{112} = c_1(E_{12}I_{11} - E_{22}I_{12}),$$

$$Y_{113} = G_8 - 3c_1G_{10}, Y_{114} = G_7 - 3c_1G_9, Y_{115} = -c_1G_5, Y_{116} = -c_1G_6,$$

$$Y_{117} = H_8 - 3c_1H_{10},$$

$$Y_{118} = H_7 - 3c_1H_9, Y_{119} = -c_1H_5, Y_{120} = -c_1H_6,$$

$$Y_{21} = -A_{44} + 6c_1D_{44} - 9c_1^2F_{44},$$

$$Y_{22} = -c_1(F_{11} + B_{11}I_{15} + B_{12}I_{25} - c_1E_{11}I_{15} - c_1H_{11} - c_1E_{12}I_{25}),$$

$$Y_{23} = -c_1(B_{11}I_{16} + B_{12}I_{26} + F_{12} + 2B_{66}I_{33} + 2F_{66} - 2c_1E_{66}I_{33} - 2c_1H_{66} - c_1E_{12}I_{26} - c_1E_{11}I_{16} - c_1H_{12}),$$

$$\begin{aligned} Y_{24} = {} & B_{11}I_{13} - c_1B_{11}I_{15} + D_{11} - c_1F_{11} \\ & + B_{12}I_{23} - c_1B_{12}I_{25} - c_1E_{11}I_{13} \\ & + c_1^2E_{11}I_{15} - c_1F_{11} + c_1^2H_{11} \\ & - c_1E_{12}I_{23} + c_1^2E_{12}I_{25}, \end{aligned}$$

$$Y_{25} = B_{66}I_{32} - c_1B_{66}I_{33} + D_{66} - c_1F_{66} - c_1E_{66}I_{32} + c_1^2E_{66}I_{33} - c_1F_{66} + c_1^2H_{66},$$

$$Y_{26} = B_{11}I_{14} - c_1B_{11}I_{16} + B_{12}I_{24} - c_1B_{12}I_{26} + D_{12} - c_1F_{12} + B_{66}I_{32} - c_1B_{66}I_{33} + D_{66} - c_1F_{66} - c_1E_{66}I_{32} + c_1^2E_{66}I_{33} - c_1F_{66} + c_1^2H_{66} - c_1E_{11}I_{14} + c_1^2E_{11}I_{16} - c_1E_{12}I_{24} - c_1F_{12} + c_1^2E_{12}I_{26} + c_1^2H_{12},$$

$$Y_{27} = -B_{11}I_{12} + B_{12}I_{21} + c_1E_{11}I_{12} - c_1E_{12}I_{21},$$

$$Y_{28} = B_{11}I_{11} - B_{66}I_{31} - B_{12}I_{12} - c_1E_{11}I_{11} + c_1E_{12}I_{12} + c_1E_{66}I_{31},$$

$$Y_{29} = -G_3 + c_1G_5, Y_{210} = -H_3 + c_1H_5, Y_{31} = -A_{55} + 6c_1D_{55} - 9c_1^2F_{55},$$

$$Y_{32} = -c_1(2B_{66}I_{33} + 2F_{66} + B_{12}I_{15} + F_{12} + B_{22}I_{25} - 2c_1H_{66} - c_1E_{12}I_{15} - c_1E_{22}I_{25},$$

$$Y_{33} = -c_1(B_{12}I_{16} + B_{22}I_{26} + F_{22} - c_1E_{12}I_{16} - c_1E_{22}I_{26} - c_1H_{22}),$$

$$Y_{34} = B_{66}I_{32} - c_1B_{66}I_{33} + D_{66} - c_1F_{66} + B_{12}I_{13} - c_1B_{12}I_{15} + D_{12} - c_1F_{12} + B_{22}I_{23} - c_1B_{22}I_{25} - c_1E_{66}I_{32} + c_1^2E_{66}I_{33} - c_1F_{66} + c_1^2h_{66} - c_1E_{12}I_{13} + c_1^2E_{12}I_{15} - c_1F_{12} + c_1^2H_{12} - c_1E_{22}I_{23} + c_1^2E_{22}I_{25},$$

$$Y_{35} = B_{66}I_{32} - c_1B_{66}I_{33} + D_{66} - c_1F_{66} - c_1E_{66}I_{32} + c_1^2E_{66}I_{33} - c_1F_{66} + c_1^2H_{66},$$

$$Y_{36} = B_{12}I_{14} - c_1B_{12}I_{16} + B_{22}I_{24} - c_1B_{22}I_{26} + D_{22} - c_1F_{22} - c_1E_{12}I_{14} + c_1^2E_{12}I_{16} - c_1E_{22}I_{26} - c_1F_{22} + c_1^2H_{22},$$

$$Y_{37} = -B_{12}I_{12} - B_{66}I_{31} + B_{22}I_{21} + c_1E_{66}I_{31} + c_1E_{12}I_{12} - c_1E_{22}I_{21},$$

$$Y_{38} = B_{12}I_{11} - B_{22}I_{12} - c_1E_{12}I_{11} + c_1E_{22}I_{12},$$

$$Y_{39} = -G_4 + c_1G_6, Y_{310} = -H_4 + c_1H_6,$$

$$\Delta = A_{11}A_{22} - A_{12}^2, I_{11} = \frac{A_{22}}{\Delta}, I_{12} = \frac{A_{12}}{\Delta}, I_{13} = \frac{B_{12}A_{12} - B_{11}A_{22}}{\Delta},$$

$$I_{14} = \frac{B_{22}A_{12} - B_{12}A_{22}}{\Delta}, I_{15} = \frac{E_{12}A_{12} - E_{11}A_{22}}{\Delta},$$

$$I_{16} = \frac{E_{22}A_{12} - E_{12}A_{22}}{\Delta}, I_{17} = 2\frac{2e_{32}A_{12} - 2e_{31}A_{22}}{\Delta},$$

$$I_{18} = 2\frac{2q_{32}A_{12} - 2q_{31}A_{22}}{\Delta}, I_{19} = \frac{-\alpha_2 A_{12} + \alpha_1 A_{22}}{\Delta},$$

$$I_{20} = \frac{-\beta_2 A_{12} + \beta_1 A_{22}}{\Delta}, I_{21} = \frac{A_{11}}{\Delta}, I_{23} = \frac{B_{11}A_{12} - B_{12}A_{11}}{\Delta},$$

$$I_{24} = \frac{B_{12}A_{12} - B_{22}A_{11}}{\Delta}, I_{25=}\frac{E_{11}A_{12} - E_{12}A_{11}}{\Delta},$$

$$I_{26} = \frac{E_{12}A_{12} - E_{22}A_{11}}{\Delta}, I_{27} = 2\frac{2e_{31}A_{12} - 2e_{32}A_{11}}{\Delta},$$

$$I_{28} = 2\frac{2q_{31}A_{12} - 2q_{32}A_{11}}{\Delta}, I_{29} = \frac{-\alpha_1 A_{12} + \alpha_2 A_{11}}{\Delta}, I_{30} = \frac{-\beta_1 A_{12} + \beta_2 A_{11}}{\Delta},$$

$$I_{21} = \frac{A_{11}}{\Delta}, I_{23} = \frac{B_{11}A_{12} - B_{12}A_{11}}{\Delta}, I_{24} = \frac{B_{12}A_{12} - B_{22}A_{11}}{\Delta},$$
$$I_{31} = \frac{1}{A_{66}}, I_{32} = -\frac{B_{66}}{A_{66}}, I_{33} = -\frac{A_{66}}{A_{66}}.$$

Appendix 3

$$l_{11} = -k_1 - k_2\left(\lambda_m^2 + \delta_n^2\right) + T_{13}\lambda_m^4 + T_{14}\lambda_m^2\delta_n^2 + T_{15}\delta_n^4 + M_3\left(T_{110}\lambda_m^4 + T_{111}M_3\lambda_m^2\delta_n^2 + T_{112}M_3\delta_n^4\right),$$

$$l_{12} = -T_{11}\lambda_m + T_{16}\lambda_m^3 + T_{17}\lambda_m\delta_n^2 + T_{110}M_4\lambda_m^4 + T_{111}M_4\lambda_m^2\delta_n^2 + T_{112}M_4\delta_n^4,$$

$$l_{13} = -T_{12}\delta_n + T_{18}\delta_n^3 + T_{19}\lambda_m^2\delta_n + T_{110}M_5\lambda_m^4 + T_{111}M_5\lambda_m^2\delta_n^2 + T_{112}M_5\delta_n^4,$$

$$l_{14} = \frac{32M_4\lambda_m\delta_n}{3ab}, l_{15} = \frac{32M_5\lambda_m\delta_n}{3ab}, l_{16} = \beta\sin(\beta z)\left(\lambda_m^2 T_{115} + \delta_n^2 T_{116}\right),$$

$$l_{17} = \beta\sin(\beta z)\left(T_{119}\lambda_m^2 + T_{120}\delta_n^2\right),$$

$$n_1 = -T_{11}\lambda_m^2 - T_{12}\delta_n^2,\ n_2 = \frac{32M_3\lambda_m\delta_n}{3ab},$$

$$n_3 = \frac{8T_{110}\lambda_m\delta_n}{3abI_{21}} - \frac{8T_{112}\lambda_m\delta_n}{3abI_{11}},\ n_4 = -\frac{\lambda_m^4}{16I_{11}} - \frac{\delta_n^4}{16I_{21}},$$

$$l_{21} = -\lambda_m^3(T_{22} + M_3T_{27}) - \lambda_m\delta_n^2(T_{23} + M_3T_{28}),$$

$$l_{22} = T_{21} - T_{24}\lambda_m^2 - T_{25}\delta_n^2 - T_{27}M_4\lambda_m^3 - T_{28}M_4\lambda_m\delta_n^2,$$

$$l_{23} = -T_{26}\lambda_m\delta_n - T_{27}M_5\lambda_m^3 - T_{28}M_5\lambda_m\delta_n^2,$$

$$l_{24} = -T_{29}\beta\sin(\beta z)\lambda_m,\ l_{25} = -T_{210}\beta\sin(\beta z)\lambda_m,\ n_6 = T_{21}\lambda_m,\ n_7 = \frac{8T_{27}\delta_n}{3abI_{21}},$$

$$l_{31} = -\delta_n^3(T_{33} + M_3T_{38}) - \lambda_m^2\delta_n(T_{32} + M_3T_{37}),$$

$$l_{32} = -T_{34}\lambda_m\delta_n - T_{38}M_4\delta_n^3 - T_{37}M_4\lambda_m^2\delta_n,$$

$$l_{33} = T_{31} - T_{35}\lambda_m^2 - T_{36}\delta_n^2 - T_{38}M_5\delta_n^3 - T_{37}M_5\lambda_m^2\delta_n,$$

$$l_{34} = -T_{39}\beta\sin(\beta z)\delta_n,$$

$$l_{35} = -T_{310}\beta\sin(\beta z)\delta_n,\ n_8 = T_{31}\delta_n,\ n_9 = \frac{8Y_{38}\lambda_m}{3abI_{11}},$$

$$l_{41} = -\left(O_{11}\lambda_m^2 + O_{12}\delta_n^2\right) + c_1O_{14}\lambda_m^2 + c_1O_{16}\delta_n^2,$$

$$l_{42} = -(O_{11} + O_{13} - c_1O_{14})\lambda_m,\ l_{43} = -(O_{12} + O_{15} - c_1O_{16})\delta_n,$$

$$l_{44} = \left(-\lambda_m^2\eta_{11}^* - \delta_n^2\eta_{22}^*\right)\beta\cos(\beta z) - \beta\sin(\beta z)\eta_{33}^*,$$

$$l_{45} = \left(-\lambda_m^2m_{11}^* - \delta_n^2m_{22}^*\right)\beta\cos(\beta z) - \beta\sin(\beta z)m_{33}^*,$$

$$l_{51} = -\left(O_{21}\lambda_m^2 + O_{22}\delta_n^2\right) + c_1O_{24}\lambda_m^2 + c_1O_{26}\delta_n^2,$$

$$l_{52} = -(O_{21} + O_{23} - c_1O_{24})\lambda_m,\ l_{53} = -(O_{22} + O_{25} - c_1O_{26})\delta_n,$$

$$l_{54} = \left(-\lambda_m^2m_{11}^* - \delta_n^2m_{22}^*\right)\beta\cos(\beta z) - \beta\sin(\beta z)m_{33}^*,$$

$$l_{55} = \left(-\lambda_m^2\mu_{11}^* - \delta_n^2\mu_{22}^*\right)\beta\cos(\beta z) - \beta\sin(\beta z)\mu_{33}^*,$$

$$O_{11} = \int_{-h/2}^{-h/2+h_3} \overline{e_{15}}\left(1 - 3c_1z^2\right)\cos(\beta z)dz + \int_{h/2-h_1}^{h/2} \overline{e_{15}}\left(1 - 3c_1z^2\right)\cos(\beta z)dz$$

$$\eta_{11}^* = \int_{-h/2}^{-h/2+h_3} \overline{\eta_{11}}\cos(\beta z)dz + \int_{h/2-h_1}^{h/2} \overline{\eta_{11}}\cos(\beta z)dz$$

$$m_{11}^* = \int_{-h/2}^{-h/2+h_3} \overline{m_{11}}\cos(\beta z)dz + \int_{h/2-h_1}^{h/2} \overline{m_{11}}\cos(\beta z)dz$$

$$O_{12} = \int_{-h/2}^{-h/2+h_3} \overline{e_{24}}\left(1 - 3c_1z^2\right)\cos(\beta z)dz + \int_{h/2-h_1}^{h/2} \overline{e_{24}}\left(1 - 3c_1z^2\right)\cos(\beta z)dz$$

$$O_{13} = \int_{-h/2}^{-h/2+h_3} \overline{e_{31}}z\cos(\beta z)dz + \int_{h/2-h_1}^{h/2} \overline{e_{31}}z\cos(\beta z)dz$$

$$\mu_{33}^* = \int_{-h/2}^{-h/2+h_3} \overline{\mu_{33}}\cos(\beta z)dz + \int_{h/2-h_1}^{h/2} \overline{\mu_{33}}\cos(\beta z)dz.$$

Appendix 4

$$d_1 = \frac{l_{11}l_{55}l_{44} - l_{11}l_{54}l_{45} - l_{17}l_{51}l_{44} + l_{17}l_{54}l_{41} - l_{16}l_{41}l_{55} + l_{16}l_{45}l_{51}}{l_{55}l_{44} - l_{54}l_{45}},$$

$$d_2 = \frac{l_{12}l_{55}l_{44} - l_{12}l_{54}l_{45} - l_{17}l_{52}l_{44} + l_{17}l_{54}l_{42} - l_{16}l_{42}l_{55} + l_{16}l_{45}l_{52}}{l_{55}l_{44} - l_{54}l_{45}},$$

$$d_3 = \frac{l_{13}l_{55}l_{44} - l_{13}l_{54}l_{45} - l_{17}l_{53}l_{44} + l_{17}l_{54}l_{43} - l_{16}l_{43}l_{55} + l_{16}l_{45}l_{53}}{l_{55}l_{44} - l_{54}l_{45}},$$

$$d_4 = \frac{l_{21}l_{55}l_{44} - l_{21}l_{54}l_{45} - l_{25}l_{51}l_{44} + l_{25}l_{54}l_{41} - l_{24}l_{41}l_{55} + l_{24}l_{45}l_{51}}{(l_{55}l_{44} - l_{54}l_{45})},$$

$$d_5 = \frac{l_{22}l_{55}l_{44} - l_{22}l_{54}l_{45} - l_{25}l_{52}l_{44} + l_{25}l_{54}l_{42} - l_{24}l_{42}l_{55} + l_{24}l_{45}l_{52}}{l_{55}l_{44} - l_{54}l_{45}},$$

$$d_6 = \frac{l_{23}l_{55}l_{44} - l_{23}l_{54}l_{45} - l_{25}l_{53}l_{44} + l_{25}l_{54}l_{43} - l_{24}l_{43}l_{55} + l_{24}l_{45}l_{53}}{l_{55}l_{44} - l_{54}l_{45}},$$

$$d_7 = \frac{l_{31}l_{55}l_{44} - l_{31}l_{54}l_{45} - l_{35}l_{51}l_{44} + l_{35}l_{54}l_{41}}{l_{55}l_{44} - l_{54}l_{45}} + \frac{-l_{34}l_{41}l_{55} + l_{34}l_{45}l_{51}}{l_{55}l_{44} - l_{54}l_{45}},$$

$$d_8 = \frac{l_{32}l_{55}l_{44} - l_{32}l_{54}l_{45} - l_{35}l_{52}l_{44} + l_{35}l_{54}l_{42} - l_{34}l_{42}l_{55} + l_{34}l_{45}l_{52}}{l_{55}l_{44} - l_{54}l_{45}},$$

$$d_9 = \frac{l_{33}l_{55}l_{44} - l_{33}l_{54}l_{45} - l_{35}l_{53}l_{44} + l_{35}l_{54}l_{43} - l_{34}l_{43}l_{55} + l_{34}l_{45}l_{53}}{l_{55}l_{44} - l_{54}l_{45}}.$$

Appendix 5

$$\begin{aligned}\left(A_{ij}, B_{ij}, D_{ij}, E_{ij}, F_{ij}, H_{ij}\right) = & \int\limits_{-h_f-\frac{h_c}{2}-h_p}^{-\frac{h_c}{2}-h_f} Q_{ij}\left(1, z, z^2, z^3, z^4, z^6\right)dz \\ & + \int\limits_{-h_f-\frac{h_c}{2}}^{-\frac{h_c}{2}} Q_{ij}\left(1, z, z^2, z^3, z^4, z^6\right)dz \\ & + \int\limits_{-\frac{h_c}{2}}^{\frac{h_c}{2}} Q_{ij}\left(1, z, z^2, z^3, z^4, z^6\right)dz \\ & + \int\limits_{\frac{h_c}{2}}^{\frac{h_c}{2}+h_f} Q_{ij}\left(1, z, z^2, z^3, z^4, z^6\right)dz \\ & + \int\limits_{\frac{h_c}{2}+h_f}^{\frac{h_c}{2}+h_f+h_p} Q_{ij}\left(1, z, z^2, z^3, z^4, z^6\right)dz, \\ & ij = 11, 12, 22, 66,\end{aligned}$$

$$(A_{kl}, D_{kl}, F_{kl}) = \int\limits_{-h_p-h_f-\frac{h_c}{2}}^{-h_f-\frac{h_c}{2}} Q_{ij}\left(1, z^2, z^4\right)dz + \int\limits_{-h_f-h_c/2}^{-h_c/2} Q_{ij}\left(1, z^2, z^4\right)dz$$

$$+\int_{-h_c/2}^{h_c/2} Q_{ij}\left(1, z^2, z^4\right)dz + \int_{h_c/2}^{h_c/2+h_f} Q_{ij}\left(1, z^2, z^4\right)dz$$

$$+\int_{h_c/2+h_f}^{h_c/2+h_f+h_p} Q_{ij}\left(1, z^2, z^4\right)dz,\ kl = 44, 55,$$

$$(\Phi_1, \Phi_3, \Phi_5) = \int_{-h_f-\frac{h_c}{2}-h_p}^{-\frac{h_c}{2}-h_f} \left(Q_{11}^p\alpha_{11}\Delta T + Q_{12}^p\alpha_{22}\Delta T\right)\left(1, z, z^2\right)dz$$

$$+\int_{-h_f-\frac{h_c}{2}}^{-\frac{h_c}{2}} \left(Q_{11}^f\alpha_{11}\Delta T + Q_{12}^f\alpha_{22}\Delta T\right)\left(1, z, z^2\right)dz$$

$$+\int_{-\frac{h_c}{2}}^{\frac{h_c}{2}} \left(Q_{11}^c\alpha_{11}\Delta T + Q_{12}^c\alpha_{22}\Delta T\right)\left(1, z, z^2\right)dz$$

$$+\int_{\frac{h_c}{2}}^{h_f+\frac{h_c}{2}} \left(Q_{11}^f\alpha_{11}\Delta T + Q_{12}^f\alpha_{22}\Delta T\right)\left(1, z, z^2\right)dz$$

$$+\int_{\frac{h_c}{2}+h_f}^{h_f+\frac{h_c}{2}+h_p} \left(Q_{11}^p\alpha_{11}\Delta T + Q_{12}^p\alpha_{22}\Delta T\right)\left(1, z, z^2\right)dz,$$

$$(\Phi_2, \Phi_4, \Phi_6) = \int_{-h_f-\frac{h_c}{2}-h_p}^{-\frac{h_c}{2}-h_f} \left(Q_{12}^p\alpha_{11}\Delta T + Q_{22}^p\alpha_{22}\Delta T\right)\left(1, z, z^2\right)dz$$

$$+\int_{-h_f-\frac{h_c}{2}}^{-\frac{h_c}{2}} \left(Q_{12}^f\alpha_{11}\Delta T + Q_{22}^f\alpha_{22}\Delta T\right)\left(1, z, z^2\right)dz$$

$$+\int_{-\frac{h_c}{2}}^{\frac{h_c}{2}} \left(Q_{12}^c\alpha_{11}\Delta T + Q_{22}^c\alpha_{22}\Delta T\right)\left(1, z, z^2\right)dz$$

$$+\int_{\frac{h_c}{2}}^{h_f+\frac{h_c}{2}} \left(Q_{12}^f\alpha_{11}\Delta T + Q_{22}^f\alpha_{22}\Delta T\right)\left(1, z, z^2\right)dz$$

$$+\int_{h_c/2+h_f}^{h_f+h_c/2+h_p}\left(Q_{12}^{p}\alpha_{11}\Delta T+Q_{22}^{p}\alpha_{22}\Delta T\right)\left(1,z,z^2\right)dz,$$

Appendix 6

$$L_{11}(w)=O_{11}\frac{\partial^2 w}{\partial x^2}+O_{12}\frac{\partial^2 w}{\partial y^2}+O_{13}\frac{\partial^4 w}{\partial x^4}+O_{14}\frac{\partial^4 w}{\partial x^2\partial y^2}$$
$$+O_{15}\frac{\partial^4 w}{\partial y^4}-k_1 w+k_2\left(\frac{\partial^2 w}{\partial x^2}+\frac{\partial^2 w}{\partial y^2}\right),$$
$$L_{12}(\phi_x)=O_{11}\frac{\partial\phi_x}{\partial x}+O_{16}\frac{\partial^3\phi_x}{\partial x^3}+XO_{17}\frac{\partial^3\phi_x}{\partial x\partial y^2},$$

$$L_{13}(\phi_y)=O_{12}\frac{\partial\phi_y}{\partial y}+O_{18}\frac{\partial^3\phi_y}{\partial y^3}+O_{19}\frac{\partial^3\phi_y}{\partial x^2\partial y},$$
$$L_{14}(f)=O_{110}\frac{\partial^4 f}{\partial x^4}+O_{111}\frac{\partial^4 f}{\partial x^2\partial y^2}+O_{112}\frac{\partial^4 f}{\partial y^4},$$

$$S(w,f)=\frac{\partial^2 f}{\partial y^2}\frac{\partial^2 w}{\partial x^2}-2\frac{\partial^2 f}{\partial x\partial y}\frac{\partial^2 w}{\partial y^2}+\frac{\partial^2 f}{\partial x^2}\frac{\partial^2 w}{\partial y^2},$$

$$L_{21}(w)=O_{21}\frac{\partial w}{\partial x}+O_{22}\frac{\partial^3 w}{\partial x^3}+O_{23}\frac{\partial^3 w}{\partial x\partial y^2},$$
$$L_{22}(\phi_x)=O_{21}\phi_x+O_{24}\frac{\partial^2\phi_x}{\partial x^2}+O_{25}\frac{\partial^2\phi_x}{\partial y^2},$$

$$L_{31}(w)=O_{31}\frac{\partial w}{\partial y}+O_{32}\frac{\partial^3 w}{\partial x^2\partial y}+O_{33}\frac{\partial^3 w}{\partial y^3},$$
$$L_{32}(\phi_x)=O_{34}\frac{\partial^2\phi_x}{\partial x\partial y},$$

$$L_{33}(\phi_y)=O_{31}\phi_y+O_{35}\frac{\partial^2\phi_y}{\partial x^2}+O_{36}\frac{\partial^2\phi_y}{\partial y^2},$$
$$L_{34}(f)=O_{37}\frac{\partial^3 f}{\partial x^2\partial y}+O_{38}\frac{\partial^3 f}{\partial y^3},$$

$$O_{11}=A_{44}-6c_1D_{44}+9c_1^2F_{44},$$
$$O_{12}=A_{55}-6c_1D_{55}+9c_1^2F_{55},$$

$$O_{13} = -c_1^2(E_{11}I_{15} + E_{12}I_{25} + H_{11}),$$

$$O_{14} = -c_1^2(4E_{66}I_{33} + 4H_{66} + E_{11}I_{16} + E_{12}I_{26} + 2H_{12} + E_{12}I_{15} + E_{22}I_{25}),$$

$$O_{15} = -c_1^2(E_{12}I_{16} + E_{22}I_{26} + H_{22}),$$

$$O_{16} = c_1(E_{11}I_{13} - c_1E_{11}I_{15} + F_{11} - c_1H_{11} + E_{12}I_{23} - c_1E_{12}I_{25}),$$

$$\begin{aligned} O_{17} = c_1(&2E_{66}I_{32} - 2c_1E_{66}I_{33} + 2F_{66} - 2c_1H_{66} + c_1E_{12}I_{13} \\ &- c_1E_{12}I_{15} + F_{12} - c_1H_{12} + E_{22}I_{23} - c_1E_{22}I_{25}), \end{aligned}$$

$$O_{18} = c_1(E_{12}I_{14} - c_1E_{12}I_{16} + E_{22}I_{24} - c_1E_{22}I_{26} + F_{22} - c_1H_{22}),$$

$$\begin{aligned} O_{19} = c_1(&2E_{66}I_{32} - 2c_1E_{66}I_{33} + 2F_{66} - 2c_1H_{66} + E_{11}I_{14} \\ &- c_1E_{11}I_{16} + E_{12}I_{24} - c_1E_{12}I_{26} + F_{12} - c_1H_{12}), \end{aligned}$$

$$O_{110} = -c_1(E_{11}I_{12} - E_{12}I_{21}),$$

$$O_{111} = -c_1(2E_{66}I_{31} - E_{11}I_{11} + 2E_{12}I_{12} - E_{22}I_{21}),$$

$$O_{112} = c_1(E_{12}I_{11} - E_{22}I_{12}),$$

$$O_{21} = -A_{44} + 6c_1D_{44} - 9c_1^2F_{44},$$

$$O_{22} = -c_1(B_{11}I_{15} + F_{11} + B_{12}I_{25} - c_1E_{11}I_{15} - c_1H_{11} - c_1E_{12}I_{25}),$$

$$\begin{aligned} O_{23} = -c_1(&B_{11}I_{16} + B_{12}I_{26} + F_{12} + 2B_{66}I_{33} + 2F_{66} - 2c_1E_{66}I_{33} \\ &- 2c_1H_{66} - c_1E_{11}I_{16} - c_1E_{12}I_{26} - c_1H_{12}), \end{aligned}$$

$$\begin{aligned} O_{24} = &B_{11}I_{13} - c_1B_{11}I_{15} + D_{11} - c_1F_{11} \\ &+ B_{12}I_{23} - c_1B_{12}I_{25} - c_1E_{11}I_{13} \\ &+ c_1^2E_{11}I_{15} - c_1F_{11} + c_1^2H_{11} \\ &- c_1E_{12}I_{23} + c_1^2E_{12}I_{25}, \end{aligned}$$

$$O_{25} = B_{66}I_{32} - c_1B_{66}I_{33} + D_{66} - c_1F_{66}$$

$$- c_1E_{66}I_{32} + c_1^2E_{66}I_{33} - c_1F_{66} + c_1^2H_{66},$$

$$\begin{aligned} O_{26} = B_{11}I_{14} - c_1B_{11}I_{16} + B_{12}I_{24} - c_1B_{12}I_{26} \\ + D_{12} - c_1F_{12} + B_{66}I_{32} - c_1B_{66}I_{33} \\ + D_{66} - c_1F_{66} - c_1E_{66}I_{32} \\ + c_1^2E_{66}I_{33} - c_1F_{66} + c_1^2H_{66} \\ - c_1E_{11}I_{14} + c_1^2E_{11}I_{16} - c_1E_{12}I_{24} \\ + c_1^2E_{12}I_{26} - c_1F_{12} + c_1^2H_{12}, \end{aligned}$$

$$O_{27} = -B_{11}I_{12} + B_{12}I_{21} + c_1E_{11}I_{12} - c_1E_{12}I_{21},$$

$$O_{28} = B_{11}I_{11} - B_{12}I_{12} - B_{66}I_{31} - c_1E_{11}I_{11} + c_1E_{12}I_{12} + c_1E_{66}I_{31},$$

$$O_{31} = -A_{55} + 6c_1D_{55} - 9c_1^2F_{55},$$

$$\begin{aligned} O_{32} = -c_1(2B_{66}I_{33} + 2F_{66} + B_{12}I_{15} + F_{12} + B_{22}I_{25} - 2c_1E_{66}I_{33} \\ - 2c_1H_{66} - c_1E_{12}I_{15} - c_1H_{12} - c_1E_{22}I_{25}), \end{aligned}$$

$$O_{33} = -c_1(B_{12}I_{16} + B_{22}I_{26} + F_{22} - c_1E_{12}I_{16} - c_1E_{22}I_{26} - c_1H_{22}),$$

$$\begin{aligned} O_{34} = B_{66}I_{32} - c_1B_{66}I_{33} + D_{66} - c_1F_{66} + B_{12}I_{13} - c_1B_{12}I_{15} \\ + D_{12} - c_1F_{12} + B_{22}I_{23} - c_1B_{22}I_{25} - c_1E_{66}I_{32} \\ + c_1^2E_{66}I_{33} - c_1F_{66} + c_1^2H_{66} - c_1E_{12}I_{13} + c_1^2E_{12}I_{15} \\ - c_1F_{12} + c_1^2H_{12} - c_1E_{22}I_{23} + c_1^2E_{22}I_{25}, \end{aligned}$$

$$\begin{aligned} O_{35} = B_{66}I_{32} - c_1B_{66}I_{33} + D_{66} - c_1F_{66} - c_1E_{66}I_{32} \\ + c_1^2E_{66}I_{33} - c_1F_{66} + c_1^2H_{66}, \end{aligned}$$

$$\begin{aligned} O_{36} = B_{12}I_{14} - c_1B_{12}I_{16} + B_{22}I_{24} - c_1B_{22}I_{26} \\ + D_{22} - c_1F_{22} - c_1E_{12}I_{14} + c_1^2E_{12}I_{16} \\ - c_1E_{22}I_{24} + c_1^2E_{22}I_{26} - c_1F_{22} + c_1^2H_{22}, \end{aligned}$$

$$sO_{37} = -B_{66}I_{31} - B_{12}I_{12} + B_{22}I_{21} + c_1E_{66}I_{31} + c_1E_{12}I_{12} - c_1E_{22}I_{21},$$

$$O_{38} = B_{12}I_{11} - B_{22}I_{12} - c_1E_{12}I_{11} + c_1E_{22}I_{12},$$

$$\Delta = A_{11}A_{22} - A_{12}^2, j_{11} = \frac{A_{22}}{\Delta}, j_{12} = \frac{A_{12}}{\Delta}, j_{13} = \frac{B_{12}A_{12} - B_{11}A_{22}}{\Delta},$$

$$j_{14} = \frac{B_{22}A_{12} - B_{12}A_{22}}{\Delta},$$

$$j_{15} = \frac{E_{12}A_{12} - E_{11}A_{22}}{\Delta}, j_{16} = \frac{E_{22}A_{12} - E_{12}A_{22}}{\Delta}, j_{17} = \frac{A_{22}}{\Delta}, j_{18} = -\frac{A_{12}}{\Delta},$$

$$j_{21} = \frac{A_{11}}{\Delta}, j_{23} = \frac{B_{11}A_{12} - B_{12}A_{11}}{\Delta}, j_{24} = \frac{B_{12}A_{12} - B_{22}A_{11}}{\Delta},$$

$$j_{25} = \frac{E_{11}A_{12} - E_{12}A_{11}}{\Delta}, j_{26} = \frac{E_{12}A_{12} - E_{22}A_{11}}{\Delta}, j_{27} = -\frac{A_{12}}{\Delta},$$

$$j_{28} = \frac{A_{11}}{\Delta}, j_{31} = \frac{1}{A_{66}}, j_{32} = -\frac{B_{66}}{A_{66}}, j_{33} = -\frac{E_{66}}{A_{66}}.$$

Appendix 7

$$l_{11} = -k_1 - k_2\left(\lambda_m^2 + \delta_n^2\right) + X_{13}\lambda_m^4 + X_{14}\lambda_m^2\delta_n^2 + X_{15}\delta_n^4 + X_{110}Q_1\lambda_m^4 + X_{111}Q_1\lambda_m^2\delta_n^2 + X_{112}Q_1\delta_n^4,$$

$$l_{12} = -X_{11}\lambda_m + X_{16}\lambda_m^3 + X_{17}\lambda_m\delta_n^2 + X_{110}Q_2\lambda_m^4 + X_{111}Q_2\lambda_m^2\delta_n^2 + X_{112}Q_2\delta_n^4,$$

$$l_{13} = -X_{12}\delta_n + X_{18}\delta_n^3 + X_{19}\lambda_m^2\delta_n + X_{110}Q_3\lambda_m^4 + X_{111}Q_3\lambda_m^2\delta_n^2 + X_{112}Q_3\delta_n^4,$$

$$l_{14} = \frac{32Q_2\lambda_m\delta_n}{3ab}, l_{15} = \frac{32Q_3\lambda_m\delta_n}{3ab},$$

$$n_1 = -X_{11}\lambda_m^2 - X_{12}\delta_n^2, n_2 = \frac{32Q_1\lambda_m\delta_n}{3ab},$$

$$n_3 = -\frac{8X_{110}\lambda_m\delta_n}{3abI_{21}} - \frac{8X_{112}\lambda_m\delta_n}{3abI_{11}}, n_4 = -\frac{\lambda_m^4}{16I_{11}} - \frac{\delta_n^4}{16I_{21}}, n_5 = \frac{16}{mn\pi^2},$$

$$l_{21} = -\lambda_m^3(X_{22} + Q_1X_{27}) - \lambda_m\delta_n^2(X_{23} + Q_1X_{28}),$$

$$l_{22} = X_{21} - X_{24}\lambda_m^2 - X_{25}\delta_n^2 - X_{27}Q_2\lambda_m^3 - X_{28}Q_2\lambda_m\delta_n^2,$$

$$l_{23} = -X_{26}\lambda_m\delta_n - X_{27}Q_3\lambda_m^3 - X_{28}Q_3\lambda_m\delta_n^2,$$
$$n_6 = X_{21}\lambda_m, n_7 = \frac{8X_{27}\delta_n}{3abI_{21}},$$

$$l_{31} = -\delta_n^3(X_{33} + Q_1X_{38}) - \lambda_m^2\delta_n(X_{32} + Q_1X_{37}),$$

$$l_{32} = -X_{34}\lambda_m\delta_n - X_{38}Q_2\delta_n^3 - D_{37}Q_2\lambda_m^2\delta_n,$$
$$l_{33} = X_{31} - X_{35}\lambda_m^2 - X_{36}\delta_n^2 - X_{38}Q_3\delta_n^3 - X_{37}Q_3\lambda_m^2\delta_n,$$
$$n_8 = X_{31}\delta_n, n_9 = \frac{8X_{38}\lambda_m}{3abI_{11}},$$
$$J_0 = J_1 - \lambda_m\overline{J_7} - \delta_n\overline{J_7^*}.$$

Appendix 8

$$g_1 = \frac{(j_{21}a_1 + j_{12}a_4)}{ab\left(j_{12}^2 - j_{11}j_{21}\right)}\frac{4}{\lambda_m\delta_n}, g_4 = -\frac{1}{8}\frac{\left(j_{21}\lambda_m^2 + j_{12}\delta_n^2\right)}{\left(j_{12}^2 - j_{11}j_{21}\right)},$$
$$g_2 = \frac{(j_{21}a_2 + j_{12}a_5)}{ab\left(j_{12}^2 - j_{11}j_{21}\right)}\frac{4}{\lambda_m\delta_n}, g_3 = \frac{(j_{21}a_3 + j_{12}a_6)}{ab\left(j_{12}^2 - j_{11}j_{21}\right)}\frac{4}{\lambda_m\delta_n},$$
$$g_5 = \frac{(j_{17}j_{21} + j_{27}j_{12})}{\left(j_{12}^2 - j_{11}j_{21}\right)}, g_6 = \frac{(j_{18}j_{21} + j_{28}j_{12})}{\left(j_{12}^2 - j_{11}j_{21}\right)},$$
$$f_1 = \frac{(a_1j_{12} + j_{11}a_4)}{ab\left(j_{12}^2 - j_{11}j_{21}\right)}\frac{4}{\lambda_m\delta_n}, f_4 = -\frac{1}{8}\frac{\left(\lambda_m^2j_{12} + j_{11}\delta_n^2\right)}{\left(j_{12}^2 - j_{11}j_{21}\right)},$$
$$f_2 = \frac{(a_2j_{12} + j_{11}a_5)}{ab\left(j_{12}^2 - j_{11}j_{21}\right)}\frac{4}{\lambda_m\delta_n}, f_3 = \frac{(a_3j_{12} + j_{11}a_6)}{ab\left(j_{12}^2 - j_{11}j_{21}\right)}\frac{4}{\lambda_m\delta_n},$$
$$f_5 = \frac{(j_{17}j_{12} + j_{11}j_{27})}{\left(j_{12}^2 - j_{11}j_{21}\right)}, f_6 = \frac{(j_{18}j_{12} + j_{11}j_{28})}{\left(j_{12}^2 - j_{11}j_{21}\right)},$$
$$a_1 = -j_{11}\delta_n^2G_1 + j_{12}G_1\lambda_m^2 + c_1j_{15}\lambda_m^2 + c_1j_{16}\delta_n^2,$$
$$a_2 = (-j_{13} + c_1j_{15})\lambda_m + j_{12}G_2\lambda_m^2 - j_{11}G_2\delta_n^2,$$
$$a_3 = (-j_{14} + c_1j_{16})\delta_n + G_3j_{12}\lambda_m^2 - j_{11}G_3\delta_n^2,$$

$$a_4 = -G_1j_{21}\lambda_m^2 + \delta_n^2G_1j_{12} + c_1j_{26}\delta_n^2 + c_1j_{25}\lambda_m^2,$$
$$a_5 = (-j_{23} + c_1j_{25})\lambda_m + j_{12}G_2\delta_n^2 - j_{21}G_2\lambda_m^2,$$
$$a_6 = (-j_{24} + c_1j_{26})\delta_n + j_{12}G_3\delta_n^2 - j_{21}G_3\lambda_m^2.$$

References

1. Zhang GY, Qu YL, Gao XL, Jin F (2020) A transversely isotropic magneto-electroelastic Timoshenko beam model incorporating microstructure and foundation effects. Mech Mater 149:103412. https://doi.org/10.1016/j.mechmat.2020.103412
2. Vinyas M, Harursampath D (2020) Nonlinear vibrations of magneto-electroelastic doubly curved shells reinforced with carbon nanotubes. Compos Struct 253:112749. https://doi.org/10.1016/j.compstruct.2020.112749
3. Dat ND, Quan TQ, Mahesh V, Duc ND (2020) Analytical solutions for nonlinear magneto-electroelastic vibration of smart sandwich plate with carbon nanotube reinforced nanocomposite core in hygrothermal environment. Int J Mech Sci 186:105906. https://doi.org/10.1016/j.ijmecsci.2020.105906
4. Li M, Zhou L, Liu C (2020) The multiphysical cell-based smoothed finite element method for analysing transient behavior of functionally grade magneto-electroelastic thin-walled structures under thermal environment. Thin-Walled Struct 155:106876. https://doi.org/10.1016/j.tws.2020.106876
5. Li M, Liu M, Zhou L (2020) The static behaviors study of magneto-electroelastic materials under hygrothermal environment with multiphysical cell-based smoothed finite element method. Compos Sci Technol 193:108130. https://doi.org/10.1016/j.compscitech.2020.108130
6. ur Rehman N, Uzair M, Allauddin U (2020) An optical-energy model for optimising the geometrical layout of solar photovoltaic arrays in a constrained field. Renew Energy 149:55–65. https://doi.org/10.1016/j.renene.2019.12.040
7. Ling W, Zhou W, Liu C, Zhou F, Yuan D, Huang J (2020) Structure and geometric dimension optimisation of interlaced microchannel for heat transfer performance enhancement. Appl Therm Eng 170:115011. https://doi.org/10.1016/j.applthermaleng.2020.115011
8. Chen X, Ding H, Shao W (2020) Adaptive data-driven collaborative optimisation of both geometric and loaded contact mechanical performances of nonorthogonal duplex helical face-milling spiral bevel and hypoid gears. Mech Mach Theory 154:104028. https://doi.org/10.1016/j.mechmachtheory.2020.104028
9. Liu H, Hu M, Jiao J, Li Z (2020) Geometric optimisation of aerogel composites for high temperature thermal insulation applications. J Non Cryst Solids 547:120306. https://doi.org/10.1016/j.jnoncrysol.2020.120306
10. Bahiraei M, Rahimi Z, Nazari F (2021) A combined multicriterion optimisation to determine optimum geometrical parameters for flow of an ecofriendly Graphene-based nanofluid inside tube enhanced with twisted conical strip inserts. Powder Technol 377:336–349. https://doi.org/10.1016/j.powtec.2020.08.044
11. Quang VD, Quan TQ, Tran P (2022) Static buckling analysis and geometrical optimization of magneto-electro-elastic sandwich plate with auxetic honeycomb core. Thin Wall Struct 173:108935. https://doi.org/10.1016/j.tws.2022.108935
12. Dat ND, Quan TQ, Tran P, Lam PT, Duc ND (2020) A first-principle study of nonlinear large amplitude vibration and global optimization of 3D penta-graphene plates based on the Bees algorithm. Acta Mech 231:3799–3823. https://doi.org/10.1007/s00707-020-02706-7
13. Anh VM, Quan TQ, Tran P (2021) Nonlinear vibration and geometric optimization of nanocomposite multilayer organic solar cell under wind loading. Thin Wall Struct 158:107199. https://doi.org/10.1016/j.tws.2020.107199
14. Dat ND, Anh VM, Quan TQ, Duc PT, Duc ND (2020) Nonlinear stability and optimization of thin nanocomposite multilayer organic solar cell using Bees algorithm. Thin Wall. Struct. 149:106520. https://doi.org/10.1016/j.tws.2019.106520
15. Quyen NV, Thanh NV, Quan TQ, Duc ND (2021) Nonlinear forced vibration of sandwich cylindrical panel with negative Poisson's ratio auxetic honeycombs core and CNTRC face sheets. Thin Wall Struct 162:107571. https://doi.org/10.1016/j.tws.2021.107571
16. Gibson LJ, Ashby MF (1997) Cellular solids: structure and properties. Cambridge University Press, Cambridge, UK, Second Edition

17. Haghgoo M, Ansari R, Hassanzadeh-Aghdam MK, Darvizeh A (2019) Fully coupled thermo-magneto-electro-elastic properties of unidirectional smart composites with a piezoelectric interphase. Proc Inst Mech Eng, Part C: J Mech Eng Sci, pp 2813–2829. https://doi.org/10.1177/0954406218797976
18. Reddy JN (2004) Mechanics of laminated composite plates and shells: theory and analysis. CRC Press, Boca Raton
19. Pham DT, Castellani M (2009) The Bees algorithm: modelling foraging behavior to solve continuous optimization problems. Proc Inst Mech Eng Part C J Mech Eng Sci 223:2919–2938. https://doi.org/10.1243/09544062JMES1494
20. Xue Y, Jin G, Ma X, Chen H, Ye T, Chen M, Zhang Y (2019) Free vibration analysis of porous plates with porosity distributions in the thickness and in-plane directions using isogeometric approach. Int J Mech Sci 152:346–362. https://doi.org/10.1016/j.ijmecsci.2019.01.004
21. Ansari R, Hassani R, Gholami R, Rouhi H (2020) Nonlinear bending analysis of arbitrary-shaped porous nanocomposite plates using a novel numerical approach. Int J Nonlin Mech 126:103556. https://doi.org/10.1016/j.ijnonlinmec.2020.103556
22. Shen HS, Wang H (2017) Nonlinear vibration of compressed and thermally postbuckled nanotube-reinforced composite plates resting on elastic foundations. Aero Sci Tech 64:63–74. https://doi.org/10.1016/j.ast.2017.01.017

Integrating the Bees Algorithm with WSAR for Search Direction Determination and Application to Constrained Design Optimisation Problems

Adil Baykasoglu and Mumin Emre Senol

Abstract The bees algorithm (BA) is a novel swarm-based intelligent metaheuristic search algorithm that was proposed by Pham et al. in 2005. It has been applied to several complex optimisation problems successfully. One of the main mechanisms in BA is to direct search agents (bees) to the search sites that were discovered by the best-performing bees (elite bees). In the original BA, this mechanism is realised by generating random bees in the close neighbourhood of elite bees. In this chapter, a different approach borrowed from the weighted superposition attraction–repulsion algorithm (WSAR) is incorporated into BA for search direction determination. In this approach, attractive and repulsive superpositions are determined by considering "elite bees" and "nonselected site bees" (the worst performing bees), similar to WSAR. The performance of this new approach is tested on four different constrained engineering design optimisation problems. The obtained results are compared with the basic BA and some other metaheuristic algorithms from the literature. The performance of BA can be further improved by utilising such an approach.

Keyword Weighted superposition attraction-repulsion algorithm · Search direction · Superposition · Design optimization · WSAR · Random walk · Hybrid · Levy flight · Inversetangent-constrainthandling · No-free-lunch-theorem

---

A. Baykasoglu (✉) · M. E. Senol
Department of Industrial Engineering, Faculty of Engineering, Dokuz Eylül University, Izmir, Türkiye
e-mail: adil.baykasoglu@deu.edu.tr

M. E. Senol
e-mail: emre.senol@deu.edu.tr

D. T. Pham and N. Hartono (eds.), *Intelligent Engineering Optimisation with the Bees Algorithm*, Springer Series in Advanced Manufacturing,
https://doi.org/10.1007/978-3-031-64936-3_6

1 Introduction

Optimisation is at the heart of almost all engineering disciplines. Nonoptimal designs are costly to maintain and can result in market loss. However, most practical optimisation problems contain complex objective functions and constraints, usually with discrete decision variables, making them very difficult to solve with classical derivative-based optimisation approaches. Mainly due to this fact, researchers have developed a new generation of optimisation algorithms that are named metaheuristic search algorithms. Most of these metaheuristic search algorithms are inspired by nature, such as the Bees Algorithm (BA) of Pham et al. [1] and the weighted superposition attraction–repulsion algorithm (WSAR) of Baykasoğlu [2]. There are many other nature-inspired metaheuristic algorithms for a detailed review of some of these algorithms; one can refer to [3]. In this research, the focus is on the BA and WSAR algorithms and their integration.

BA was first proposed by Pham and his colleagues in 2005 [1]. It has been successfully applied to many diverse optimisation problems, including clustering [4], crack detection of beam-type structures [5], workload balancing in examination job assignment [6], automatic design of control systems for robot manipulators [7], interference suppression of linear antenna arrays [8], and design of two-dimensional recursive filters [9]. BA mimics the behaviour of natural bees for solving the aforementioned complex optimisation problems. BA can be classified as a nature-inspired swarm intelligence-based metaheuristic algorithm. On the other hand, WSAR was proposed by Baykasoglu [2] as a successor to his previous algorithm known as weighted superposition attraction (WSA) [10, 11]. WSAR has also been applied to many diverse optimisation problems with success, including cutting parameter optimisation problems [2], multiple objective optimisation problems [12], permutation flow shop scheduling problems [13], constrained optimisation problems [14], truss structure optimisation with multiple frequency constraints [15] and others. WSAR employs the superposition principle and attracted movements of particles that are inspired by classical physics for solving the aforementioned complex optimisation problems. WSAR is classified as a physics-inspired swarm-based metaheuristic algorithm.

The No-Free-Lunch-Theorem of Macready and Wolpert [16] indicates that no optimisation algorithm performs better than other optimisation algorithms for all optimisation problems, and if we try to solve all optimisation problems, different metaheuristic algorithms can perform statistically the same. However, there is still room for possible improvements by borrowing and sharing ideas between different metaheuristics for enhancing their capabilities in problem solving. Hybridisation is one such possible approach. There are many successful applications of hybridising several metaheuristic algorithms in the literature. For example, Zachariadis et al. [17] hybridised tabu search and guided local search algorithms for solving vehicle routing problems with simultaneous delivery and pick-up; Morasaei et al. [18] presented a multihybrid metaheuristic framework to predict structural characteristics of steel–concrete composite floor systems; and Kalayci et al. [19] hybridised continuous ant colony optimisation, artificial bee colony and genetic algorithms for solving cardinality constrained portfolio optimisation problems. Blum et al. [20] presented a good review and discussion on hybrid metaheuristics. Ting et al. [21] also presented a critical review of hybrid metaheuristics by discussing their past, present and possible future research directions. Pellerin et al. [22] presented a detailed survey on a hybrid metaheuristic algorithm for solving resource-constrained project scheduling problems by presenting several ways of possible hybridisations. Analyses of these previous studies show that performance enhancements are possible with suitable integration of several metaheuristics in solving complex problems. Based on this observation, we aim to integrate the BA and WSAR algorithms to search for possible performance improvements in the searching ability of the basic BA. In our approach, we specifically focused on the search direction determination mechanism of the basic BA. In one of the stages of the basic BA, search agents (bees) are directed to the search sites that were discovered by the best-performing bees (elite bees). There is no explicit mechanism in the basic BA for search direction determination, and exploring the good search sites is performed by generating random bees in the close neighbourhood of the previously determined good search sites. In this work, the search direction determination mechanism of WSAR is incorporated into the basic BA for possible performance improvement. The performance of this hybridisation is tested on four constrained engineering design optimisation problems collected from the literature. The computational results are also compared with the basic BA and some other metaheuristic algorithms. It is shown that the performance of the basic BA can be further improved by utilising such a hybrid approach.

The remainder of this chapter is organised as follows. In Sect. 2, the basic BA and WSAR algorithms are overviewed. Section 3 presents the hybridisation approach. In Sect. 4, test functions and experimental results are reported. Finally, concluding remarks are presented in Sect. 5.

2 A Brief Overview of the BA and WSAR Algorithms

2.1 An Overview of BA

The BA is a swarm-intelligent-based metaheuristic search algorithm that was proposed by Pham et al. [1]. The BA mimics the foraging behaviour of bees in nature and performs a neighbourhood search combined with a random search to find good quality solutions to complex optimisation problems. BA needs several algorithm-specific parameters to be set, namely,

- # scout bees (n),
- # sites selected for neighbourhood search (out of n visited sites) (m),
- # elite bees among m selected sites (e),
- # bees recruited for sites visited by e elite bees (nes),
- # bees recruited for the other (m-e) selected sites (nss),
- size of sites.

```
1. Initialise the population (generate random bees [scout bees]
and calculate their fitness)
2. While (stopping criterion is not met)//Forming new population
      2.1. Determine elite site bees [e], selected site bees [m-e]
           and nonselected site bees [n-m]
      2.2. Recruit bees by performing waggle dance for selected
           sites (the number of bees for a site is proportional
           to the type of that site) and evaluate their fitness.
      2.3. Select the fittest bee from each site
      2.4. Assign remaining bees to search randomly and evaluate
           their fitness.
3. End While
```

Fig. 1 Main steps of BA

The main steps of BA are shown in Fig. 1.

BA starts with randomly generated *n* scout bees. The fitness values of scout bees are calculated in step 1. The bees with the highest fitness values are determined as "elite bees", and sites visited by them are selected for neighbourhood search in step 2.1. BA recruits more bees with waggle dance to follow elite bees (*e*) than the other selected bees (*m-e*) for searching around the neighbourhood of the elite bees. Only one bee with the highest fitness value is selected for each site for generating the next bee population in step 2.3. Additionally, other bees (*n-m*) in the population perform a random search for exploring new potential sites in step 2.4. These steps repeat until predetermined stopping conditions are satisfied.

2.2 An Overview of WSAR

The WSAR algorithm is the successor of the WSA algorithm. Both of them were proposed by Baykasoglu [2, 10, 11] as new-generation swarm intelligence-based metaheuristic algorithms for solving complex optimisation problems. WSAR utilises the "superposition principle" and "attracted movements of particles" from classical physics. The WSAR algorithm generates two superpositions from the current population of Solution Vectors (SV). Attractive superposition (AS) is the first superposition, and it may attract SVs to it. On the other hand, repulsive superposition (RS) is the second superposition, and it may repel SVs from its position. Superpositions can inherently determine search directions towards good search spaces, which is one of the most anticipated search principles in developing metaheuristic algorithms. The WSAR algorithm has three main steps, namely, initialisation, superposition determination, and moving solution vectors. Figure 2 depicts the main steps of the WSAR algorithm.

The initialisation steps of the WSAR algorithm consist of parameter setting, creation/evaluation of initial random SVs and determination of the best SV (Steps 1–3). In the superposition determination stage, a specially developed "biased

```
1. Parameters initialisation for WSAR
2. Generate initial SVs randomly and determine, the best SV and the
   best fitness
3. Compute fitness of initial SVs and update the best SV
4. i = 1 //iteration counter
While i <= Max_iteration
   4.1 Rank SVs in relation to their fitness from "best to worst" &
       "worst to best"
   4.2 Assign weights to each SV in relation to their ranks
       (best and worst weights for each SV)
   4.3 Determine two superpositions based on the best and worst
       weights
   4.4 Calculate the fitness of the superpositions
   4.5 Determine AS and RS. If RS comes with better fitness replace
       AS with RS
   4.6 Decide search direction for each SV by considering AS and
       its fitness value
   4.7 Move each SV towards its determined direction
   4.8 Calculate fitness value of each SV & update the best SV
i = i + 1
5. End While
```

Fig. 2 Main steps of the WSAR

randomised sampling approach" that was proposed by Baykasoglu [2] is employed to determine AS and RS (Steps 4.1–4.5). More explanations about the superposition determination process in relation to the hybridisation approach that is proposed in this work are presented in the following subsection. In deciding the search directions and moving the SVs, the SVs' next positions are delineated according to their fitness values. If the AS has a better fitness value than the SV, the SV will move towards the AS and escape from the RS by employing Eq. 1. On the other hand, it can also move towards the AS and escape from the RS with a probability, which is calculated by using the following equation: $e^{(f(i)-f(AS))}$, where $f(i)$ represents the fitness of the SV i and $f(AS)$ represents the fitness of the AS. If the generated random number is less than the calculated probability, the corresponding SV will move towards the AS and escape from the RS. In addition, the solution vector will move randomly by utilising Eq. 2.

$$x_{ij}(t+1) = x_{ij}(t) + \left(rand(0,1)*\left(x_{AS}(t) - \left|x_{ij}(t)\right|\right) - rand(0,1)*\left(x_{RS}(t) - \left|x_{ij}(t)\right|\right)\right) \quad (1)$$

$$x_{ij}(t+1) = x_{ij}(t) + ss(t)*unifrnd(-1,1)*\left|x_{(r1)j}(t) - x_{(r2)j}(t)\right| \quad (2)$$

$ss(t)$ is defined in Eq. 3.

$$ss(t+1) = \begin{cases} ss(t) - e^{-\frac{t}{t+1}} * \varphi_{step} * ss(t) if rand(0,1) \leq \mu \\ ss(t) + e^{-\frac{t}{t+1}} * \varphi_{step} * ss(t) if rand(0,1) > \mu \end{cases} \quad (3)$$

where $x_{ij}(t)$ is the value of the position of SV i on dimension j at iteration t; $x_{AS}(t)$ and $x_{RS}(t)$ are the AS and RS at iteration t. $|.|$ shows the absolute value function; $x_{(r1)j}(t)$ and $x_{(r2)j}$ represent randomly selected SVs from the population at iteration t. $r1$ and $r2$ are random indexes and $r1 \neq r2 \neq i$; $ss(t)$ is the random walk step-setting function that begins with an initial step size $ss(0)$ for each variable and updates step sizes as the search evolves by means of a proportional rule [2]. The proportional rule requires a random number and two user-defined parameters, φ_{step} and μ.

3 Hybridising BA and WSAR Algorithms

As mentioned above, there is no explicit mechanism in the basic BA for search direction determination. In the basic BA, the selected site bees (*m-e*) make a waggle dance to explore its neighborhood. Making waggle dance actually means generating random bees in the very close neighbourhood of the selected site bees (*m-e*) in BA. In the existing implementations of the basic BA, the selected site bees (*m-e*) do not follow elite bees (*e*) with an explicitly defined mechanism. In the present work, the search direction determination mechanism of WSAR, which includes determining attractive and repulsive (AS and RS) superpositions, is incorporated into the basic BA for search direction determination. In other words, by employing the superposition determination mechanism along with move equations of WSAR (with slight modifications in Eq. 2), we enable selected site bees (*m-e*) to follow elite bees (*e*) and escape from the nonselected site bees (*n-m*).

The superposition determination mechanism is summarised as follows by incorporating BA's terminology. A biased randomised sampling algorithm is used to determine superpositions [2]. AS and RS are recomputed in each iteration as follows: first, an unoccupied vector is formed. Afterwards, a random number [0, 1] is assigned to each cell of this vector. Subsequently, these random numbers are compared with the formerly calculated rank-based weights (weight columns in Table 1, where τ is set to 0.8) of bees. Weights are computed as follows: $weight(i) = i^{-\tau}$, where i is the rank of the bee. Elite site bees are considered in determining AS, and nonselected site bees are considered in determining RS. If a bee's weight is larger than the formerly set random number, then the corresponding element of the related bee becomes a candidate for the empty position. The roulette wheel technique is used to select one of the candidates as an element of the AS/RS after determining the candidates. This procedure is performed iteratively until complete AS/RS vectors are determined. An illustration of AS/RS formation is shown in Table 1. Assume that the population comprises six elite bees. Bees are ranked from "best to worst" in determining AS. Similarly, nonselected site bees are ranked from "worst to best" in determining RS. The values selected from the corresponding dimensions undergo roulette wheel selection. These values are emphasised in grey in Table 1. For more details, one can refer to Baykasoglu [2]. A schematic representation for AS and RS determination is depicted in Fig. 3.

Table 1 Biased randomised sampling procedure for determining AS and RS

Ranking	Elite bees					*weight* $(i) = i^{-\tau}, \tau = 0.8$
1	3*	2.21	0*	4.32*	3.11	1.000
2	4	3.22*	1	4.22	4.65	0.574
3	5	4.54	1	3.65	7.65	0.415
4	6	5.65	1	9.67	9.65*	0.329
5	3	7.89	0	1.11	2.42	0.275
6	5	9.11	0	2.45	6.11	0.238
Random numbers	0.841	0.351	0.904	0.564	0.222	
AS	**3**	**3.22**	**0**	**4.32**	**9.65**	
Ranking	Non-Selected Site Bees					*weight* $(i) = i^{-\tau}, \tau = 0.8$
1	8	3.11*	0*	8.45*	2.11	1.00
2	4	7.89	0	1.11	4.82	0.574
3	1*	5.65	1	1.67	3.65	0.415
4	7	1.74	0	3.65	6.65	0.329
5	4	3.22	1	4.27	1.65	0.275
6	5	2.21	0	2.72	1.11*	0.238
Random numbers	0.203	0.434	0.876	0.934	0.124	
RS	**1**	**3.11**	**0**	**8.45**	**1.11**	

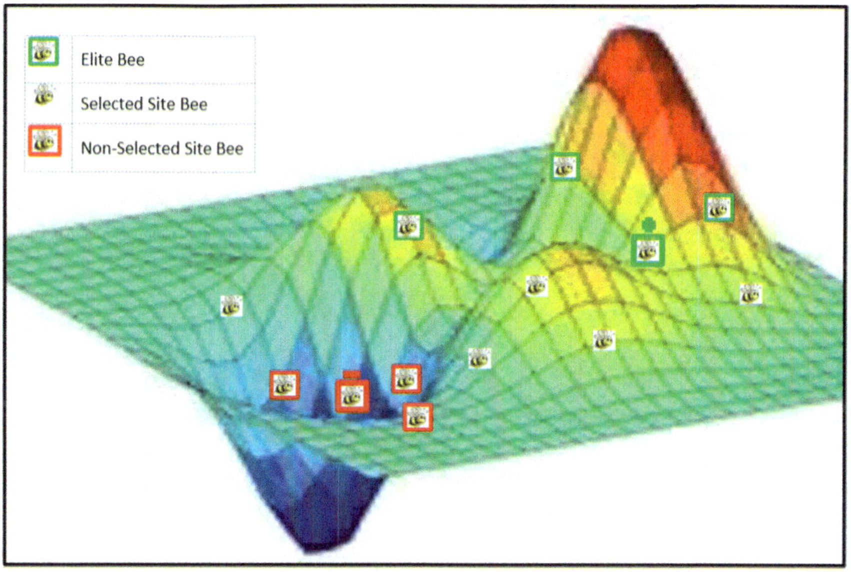

Fig. 3 Schematic representation of AS and RS determination in hBA

After computing AS and RS, bees' next moves are determined as follows in the hybrid BA (hBA): Elite bees make waggle dance as in the basic BA with some modifications, i.e., in the hBA, all of the solution elements of the elite bees are simultaneously modified by utilising the Levy Flight equation (see Eq. 4). On the other hand, if the AS has a better fitness value than the "selected site bee", the selected site bee will move towards the AS and escape from the RS by employing Eq. 1. Otherwise, it will move randomly by utilising Eq. 7. Nonselected site bees also utilise Eq. 7 to move randomly. After moving all of the bees, fitness values are computed, and improved moves are accepted. Figure 4 presents bees' next moves in hBA.

$$x_{ij}(t+1) = x_{ij}(t) + unifrnd(-r_j,\ r_j)*\left|Le'vy(t)\right| \tag{4}$$

$$Distribution_{\mathrm{Le'vy}} = \frac{U}{|V|^{1/\beta}} \tag{5}$$

$$\sigma = \left(\frac{\Gamma(1+\beta) * \sin\left(\frac{\Pi*\beta}{2}\right)}{\Gamma\left(\frac{1+\beta}{2}\right) * \beta * 2^{\frac{\beta-1}{2}}}\right) \tag{6}$$

$$x_{ij}(t+1) = x_{ij}(t) + unifrnd(-r_j,\ r_j)*|ss(t)| \tag{7}$$

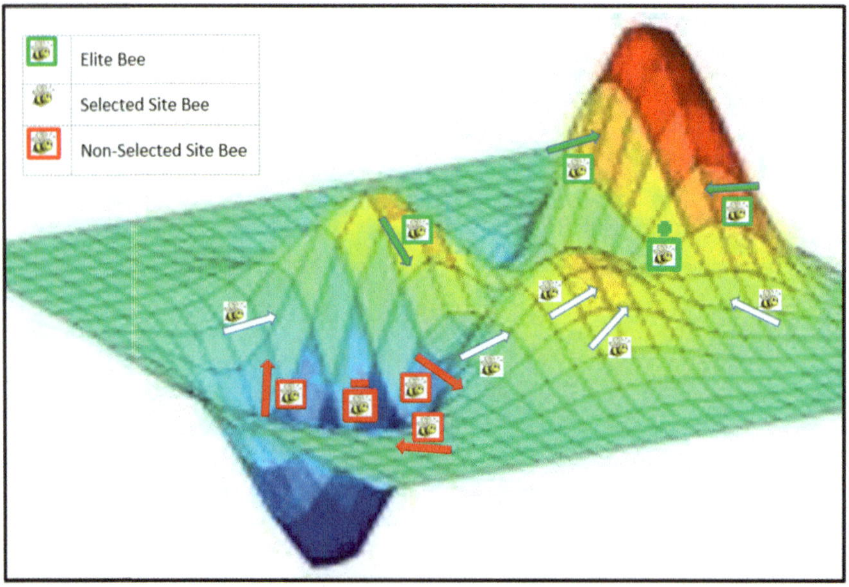

Fig. 4 Schematic representation of AS and RS determination in hBA with search directions

where $x_{ij}(t)$ is the value of the position of bee i on dimension j at iteration$t$; |.| shows the absolute value function;$-r_j$,r_j denotes the lower and upper bounds of dimension j; β denotes the power-law index; V denotes a random number sampled from the Gaussian distribution $N(0, 1)$; U is a random number sampled from the Gaussian distribution $N(0, \sigma 2)$; and σ is the standard deviation.

A schematic representation of the proposed hBA is given in Fig. 5.

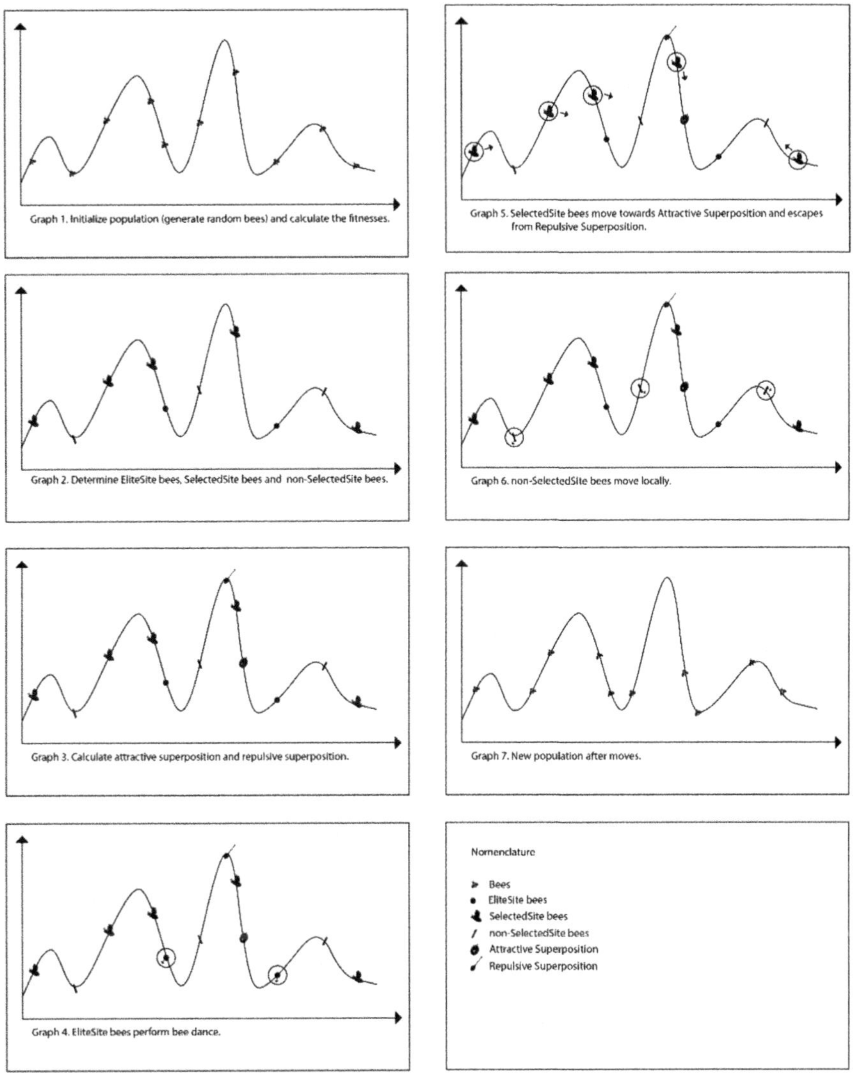

Fig. 5 Schematic representation of hBA

4 Computational Study

The performance of the proposed hBA algorithm is tested against BA and some of the other metaheuristic algorithms from the literature on four constrained engineering design problems. Thirty independent runs are performed for BA and hBA, and the best, worst and average results are reported. The hBA algorithm is programmed in MATLAB 2022a on a PC with 2.2 GHz CPU and 8 GB RAM.

Four engineering design problems, namely, the car crash side impact problem, pressure vessel design problem, speed reducer problem, and welded beam design problem, are solved. The car crash side impact problem considers the case of a car side impact. It aims to increase the side impact crash performance while reducing the overall car weight. The problem has 11 decision variables, nine of which are continuous variables x1, x2, x3, x4, x5, x6, x9, x10, x11) and two of which are discrete variables (x7, x8). The mathematical model of the problem and other details can be obtained from the following reference [23].

The pressure vessel design problem aims to design a pressure vessel with hemi-spherical heads on both sides at minimum cost. The total cost is the sum of the welding, material, and forming costs. Optimisation variables included the thickness of the cylindrical skin (Ts), the thickness of the spherical head (Th), the inner radius (R), and the length of the cylindrical segment of the vessel (L). Ts can only take discrete values with integer multiples of 0.625. The other variables can take continuous variables. The following study includes the details of this problem [24].

The speed reducer problem aims to minimise the weight of the speed reducer, which is subject to constraints such as bending and surface stresses of the gear teeth, transverse deflection of the two shafts caused by the transmitted force, and shaft stresses. The decision variables from x1 to x7 represent b (face width), m (tooth module), n (number of teeth on the pinion), l1 (length of shaft 1 between bearings), l2 (length of shaft 2 between bearings), d1 (shaft diameter 1), and d2 (shaft diameter 2), respectively (diameter of shaft 2). In this problem, x3 should be an integer, while the other variables are continuous. Details of the speed reducer problem can be found in reference [24].

Welded beam design is a common engineering optimisation problem. In this problem, four decision variables, h, l, t, and b, need to be optimised. The problem aims to minimise the cost of fabrication. Detailed information about this problem can be found in reference [24].

All of the abovementioned engineering design optimisation problems are constrained problems; therefore, a constraint handling mechanism is needed in the BA and hBA algorithms. In this study, the inverse-tangent-constraint-handling (ITCH) approach is employed to handle constraints. Kim et al. [25] proposed the ITCH technique, and they proved its efficacy on a range of constrained design optimisation problems. ITCH is given in Eq. (8).

$$(\vec{x}) = \begin{cases} \hat{g}(\vec{x}) = g_{\max}(\vec{x}) & if\, g_{\max}(\vec{x}) > 0 \\ \hat{f}(\vec{x}) = a\tan\left[f(\vec{x})\right] - \pi/2 & otherwise \end{cases} \tag{8}$$

where $g_{\max}(\vec{x}) = \max_{g_i(\vec{x})}\left[g_1(\vec{x}), g_2(\vec{x}), g_3(\vec{x}), \ldots, g_m(\vec{x})\right]$ and $a\tan[.]$ represents the inverse tangent. Note that $\hat{g}(\vec{x}) < 0$ for any $\vec{x}$, and thus, $\hat{f}(\vec{x}) < \hat{g}(\vec{x})$ is guaranteed.

4.1 Computational Results

In the case of the car crash side impact problem, the performance of the hBA is compared with four different metaheuristic algorithms, namely, Particle Swarm Optimisation (PSO), Differential Evolution (DE), Genetic Algorithm (GA), Firefly Algorithm (FA) and BA. The proposed hBA shows competitive performance with compared metaheuristic algorithms and shows better performance than the BA. All results are illustrated in Table 2.

The hBA's performance on the pressure vessel problem is compared with a hybrid of Particle Swarm Optimisation and Genetic Algorithm (PSO-GA), Harmony Search (HS), Derivative-free Filter Simulated Annealing (SA-DS), FA, PSO and BA. According to the computational results, which are presented in Table 3, PSO-GA and HS are not able to find feasible solutions, and hBA provides the best results in terms of the average of 30 independent runs.

In the speed reducer problem, the performance of the hBA algorithm is tested against the Sin Cosine Algorithm (SCA), FA, Arithmetic Optimiser Algorithm (AOA), Grey Wolf Optimiser (GWO), Moth Flame Optimiser (MFO) and BA. The

Table 2 Computational results of the car crash side impact problem

Method	PSO [26]	DE [26]	GA [26]	FA [26]	BA	hBA
x1	0.50000	0.50000	0.50005	0.50000	0.5758	0.5000
x2	1.11670	1.11670	1.28017	1.36000	1.1343	1.3492
x3	0.50000	0.5000	0.50001	0.50000	0.5164	0.5
x4	1.30208	1.30208	1.0302	1.20200	1.3777	0.9732
x5	0.50000	0.50000	0.50001	0.50000	0.875	0.8750
x6	1.50	1.50	0.50000	1.12	1.1247	0.40
x7	0.50000	0.50000	0.50000	0.50000	0.4	0.4
x8	0.34500	0.34500	0.34994	0.34500	0.345	0.345
x9	0.19200	0.19200	0.19200	0.19200	0.192	0.345
x10	– 19.54935	– 19.54935	103.119	887.307	– 10.362	22.096
x11	– 0.00431	– 0.00431	0.00167	– 18.99808	8.803	19.794
Feasibility	Feasible	Feasible	Infeasible	Infeasible	Feasible	Feasible
Best	22.84474	22.84298	22.85653	22.84298	24.1469	23.47212
Worst	23.21354	24.12606	26.240578	24.06623	26.7299	24.16836
Average	22.89429	23.22828	23.51585	22.89376	25.3076	23.56475

Table 3 Computational results of the pressure vessel design problem

Method	PSO-GA [27]	HS [28]	SA-DS [29]	FA [26]	PSO [30]	BA	hBA
x1(l)	221.365	221.365	207.225	221.365	221.365	12.5859	12.4492
x2(r)	38.860102	38.86010	39.80962	38.86010	38.860099	6.2131	6.1518
x3(t_s)	0.75	0.75	0.7683	0.75	0.75	40.809	40.4359
x4(t_h)	0.3750	0.3750	0.3797	0.375	0.375	193.292	199.481
Feasibility	Infeasible	Infeasible	Feasible	Feasible	Feasible	Feasible	Feasible
Best	5850.38	5849.76	5868.76	5850.383	5850.38	5891.579	5879.695
Worst	N/A	N/A	6804.328	6258.968	5850.591	5998.384	5931.1653
Average	N/A	N/A	6164.585	5937.33	N/A	5933.976	5905.572

hBA surpasses all of the other algorithms in the spring reducer problem in terms of the best, worst and average results. The results are given in Table 4.

The performance of the hBA in solving the welded beam design problem is compared with the novel differential evolution algorithm (NDE), multitrial vector-based differential evolution algorithm (MTDE), coevolutionary particle swarm optimisation (CPSO), GWO, AOA and BA. hBA shows competitive performance with other algorithms and better performance than BA. The results are presented in Table 5.

To further analyse the performances of hBA and BA comparatively, two nonparametric tests (Friedman and Wilcoxon signed-rank test) are carried out. According to the test results, which are presented in Table 6, hBA outperforms BA. In addition, the difference between hBA and BA was statistically significant at $p < 0.05$.

Table 4 Computational results of the speed reducer problem

Method	SCA [31]	FA [32]	AAO [33]	GWO [34]	MFO [35]	BA	hBA
× 1	3.521	3.507	3.50384	3.501	3.5	3.499	3.4999
× 2	0.7	0.7001	0.7	0.7	0.7	0.699	0.6999
× 3	17	17	17	17	17	16.999	17
× 4	8.3	7.7196	7.3	7.3	7.3	7.2999	7.3
× 5	7.923	8.080855	7.72933	7.811	7.8	7.7999	7.8
× 6	3.355	3.351512	3.3564	3.350	3.350	3.35	3.3502
× 7	5.300	5.2867	5.2867	5.287	5.286	5.286	5.2866
Feasibility	Feasible	Feasible	Feasible	Feasible	Feasible	Feasible	Feasible
Best	3026.837	3010.137	2997.058	2997.81965	2996.348	2996.348	2996.302
Worst	N/A	N/A	N/A	N/A	N/A	3044.103	2998.652
Average	N/A	N/A	N/A	N/A	N/A	3004.520	2997.713

Table 5 Computational results of the welded beam design problem

Method	NDE [36]	MTDE [37]	CPSO [38]	GWO [34]	AOA [39]	BA	hBA
h	0.20572	0.20573	0.20236	0.20567	0.19447	0.20169	0.2055
l	3.4707	3.4704	3.5444	3.47837	2.57092	3.52033	3.47262
t	9.0366	9.0366	9.04821	9.03681	10	9.13648	9.04436
b	0.2057	0.2057	0.20572	0.20578	0.20182	0.2052	0.20569
Feasibility	Feasible	Feasible	Feasible	Feasible	Infeasible	Feasible	Feasible
Best	1.72485	1.72485	1.72802	1.72624	1.7164	1.73877	1.7259
Worst	N/A	N/A	N/A	N/A	N/A	2.1041	1.7619
Average	N/A	N/A	N/A	N/A	N/A	1.9111	1.7347

Table 6 Friedman and Wilcoxon signed-rank test results

Average rankings achieved by the Friedman test		Wilcoxon signed-rank test between hBA and BA	
Algorithms	Sum of Ranks	hBA vs	*p* value
BA	2 (2)	BA	0.0425
hBA	1 (1)		

5 Conclusions

In this chapter, the bees algorithm and the WSAR algorithms are integrated through the superposition determination mechanism of the WSAR algorithm. Other small modifications are also made in the move equations of the basic bees algorithm. The newly proposed hybrid bees algorithm is applied to four very well-known constrained design optimisation problems from the literature. It is observed that the proposed hybrid bees algorithm can provide competitive results in comparison to existing metaheuristic algorithms. Moreover, the proposed hybrid bees algorithm is found to be more effective in all test problems than the basic bees algorithm. Statistical test results also show that the hybrid bees algorithm is able to provide better results than the basic bees algorithm.

References

1. Pham DT, Ghanbarzadeh A, Koc E, Otri S, Rahim S, Zaidi M (2005) The Bees Algorithm. Manufacturing Engineering Centre, Cardiff University, UK, Technical Note
2. Baykasoglu A (2021) Optimizing cutting conditions for minimizing cutting time in multi-pass milling via weighted superposition attraction-repulsion (WSAR) algorithm. Int J Prod Res 59(15):4633–4648

3. Ezugwu AE, Shukla AK, Nath R, Akinyelu AA, Agushaka JO, Chiroma H, Muhuri PK (2021) Metaheuristics: a comprehensive overview and classification along with bibliometric analysis. Artif Intell Rev 54(6):4237–4316
4. Pham DT, Darwish AH (2010) Using the bees algorithm with Kalman filtering to train an artificial neural network for pattern classification. J Syst Control Eng 224(7):885–892
5. Moradi S, Razi P, Fatahi L (2011) On the application of bees algorithm to the problem of crack detection of beam-type structures. Comput Struct 89:2169–2175
6. Baykasoglu A, Özbakir L, Tapkan P (2009) The bees algorithm for workload balancing in examination job assignment. Eur J Ind Eng 3(4):424–435
7. Fahmy AA, Kalyoncu M, Castellani M (2012) Automatic design of control systems for robot manipulators using the bees algorithm, proceedings of the institution of mechanical engineers. Part I: J Syst Control Eng 226(4):497–508
8. Guney K, Onay M (2010) Bees algorithm for interference suppression of linear antenna arrays. Expert Syst Appl 37:3129–3135
9. Pham DT, Koç E (2011) Design of a two-dimensional recursive filter using the bees algorithm. Int J Autom Comput 7(3):399–402
10. Baykasoglu A, Akpinar Ş (2017) Weighted superposition attraction (WSA): a swarm intelligence algorithm for optimization problems–Part 1: unconstrained optimization. Appl Soft Comput 56:520–540
11. Baykasoglu A, Akpinar Ş (2015) Weighted superposition attraction (WSA): a swarm intelligence algorithm for optimization problems–part 2: constrained optimization. Appl Soft Comput 37:396–415
12. Baykasoglu A (2022) Multiple objective optimization with weighted superposition attraction-repulsion algorithm (moWSAR). In: Book: modeling and advanced techniques in modern economics (1st ed), World Scientific Publishing. https://doi.org/10.1142/q0346
13. Baykasoglu A, Senol ME (2022) Parallel WSAR for solving permutation flow shop scheduling problem. Comput Sci Math Forum 2(1):10
14. Baykasoglu A, Senol ME (2022) WSAR with levy flight for constrained optimization. In: 7th International conference on harmony search, soft computing and applications (ICHSA 2022), Virtual Conference, Seoul, South Korea
15. Baykasoglu A, Baykasoglu C (2021) Weighted superposition attraction-repulsion (WSAR) algorithm for truss optimization with multiple frequency constraints. Structures 30:253–264
16. Wolpert DH, Macready WG (1997) No free lunch theorems for optimization. IEEE Trans Evol Comput 1(1):67–82
17. Zachariadis EE, Tarantilis CD, Kiranoudis CT (2009) A hybrid metaheuristic algorithm for the vehicle routing problem with simultaneous delivery and pick-up service. Expert Syst Appl 36(2):1070–1081
18. Morasaei A, Ghabussi A, Aghlmand S, Yazdani M, Baharom S, Assilzadeh H (2021) Simulation of steel–concrete composite floor system behavior at elevated temperatures via multi-hybrid metaheuristic framework. Eng Comput 1–16 (article in press)
19. Kalayci CB, Polat O, Akbay MA (2020) An efficient hybrid metaheuristic algorithm for cardinality constrained portfolio optimization. Swarm Evol Comput 54:100662
20. Blum C, Roli A, Sampels M (Eds) (2008) Hybrid metaheuristics: an emerging approach to optimization, vol 114. Springer
21. Ting TO, Yang XS, Cheng S, Huang K (2015) Hybrid metaheuristic algorithms: past, present, and future. Recent advances in swarm intelligence and evolutionary computation, vol 585. Springer, Cham, pp 71–83. https://doi.org/10.1007/978-3-319-13826-8_4
22. Pellerin R, Perrier N, Berthaut F (2020) A survey of hybrid metaheuristics for the resource-constrained project scheduling problem. Eur J Oper Res 280(2):395–416
23. Gu L, Yang RJ, Tho CH, Makowskit M, Faruquet O, Y Li YL (2001) Optimisation and robustness for crashworthiness of side impact. Int J Veh Des 26(4):348–360
24. Sandgren E (1990) Nonlinear integer and discrete programming in mechanical design optimization. J Mech Des 112(2):223–229

25. Kim TH, Maruta I, Sugie T (2010) A simple and efficient constrained particle swarm optimization and its application to engineering design problems. Proc Inst Mech Eng C J Mech Eng Sci 224(2):389–400
26. Gandomi AH, Yang X-S, Alavi AH (2011) Mixed variable structural optimization using firefly algorithm. Comput Struct 89:2325–2336
27. Dimopoulos GG (2007) Mixed-variable engineering optimization based on evolutionary and social metaphors. Comput Methods Appl Mech Eng 196:803–817
28. Mahdavi M, Fesanghary M, Damangir E (2007) An improved harmony search algorithm for solving optimization problems. Appl Math Comput 188:1567–1579
29. Hedar AR, Fukushima M (2005) Derivative-free filter simulated annealing method for constrained continuous global optimization. J Global Optim 35:521–649
30. Pulido GT, Coello CAC (2004) A constraint-handling mechanism for particle swarm optimization In: Proceedings of the 2004 congress on evolutionary computation, Portland, OR, USA, pp 1396–1403
31. Mirjalili S (2016) Sca: a sine cosine algorithm for solving optimization problems. Knowl Based Syst 96:120–133
32. Baykasoglu A, Ozsoydan FB (2015) Adaptive firefly algorithm with chaos for mechanical design optimization problems. Appl Soft Comput 36:152–164
33. Czerniak JM, Zarzycki H, Ewald D (2017) Aao as a new strategy in modeling and simulation of constructional problems optimization. Simul Model Pract Theory 76:22–33
34. Mirjalili S, Mirjalili SM, Lewis A (2014) Grey wolf optimizer. Adv Eng Softw 69:46–61
35. Mirjalili S (2015) Moth-flame optimization algorithm: a novel nature-inspired heuristic paradigm. Knowl-Based Syst 89:228–249
36. Mohamed AW (2018) A novel differential evolution algorithm for solving constrained engineering optimization problems. J Intell Manuf 29(3):659–692
37. Nadimi-Shahraki MH, Taghian S, Mirjalili S, Faris H (2020) Mtde: An effective multi-trial vector-based differential evolution algorithm and its applications for engineering design problems. Appl Soft Comput 97:106761
38. He Q, Wang L (2007) An effective co-evolutionary particle swarm optimization for constrained engineering design problems. Eng Appl Artif Intell 20(1):89–99
39. Abualigah L, Diabat A, Mirjalili S, Abd EM, Gandomi AH (2021) The arithmetic optimization algorithm, computational methods applied. Mech Eng 376:113609

Bees Algorithm-Based Optimisation of Welding Sequence to Minimise Distortion of Thin-Walled Square Al–Mg-Si Alloy Tubes

Chunbiao Wu and Chao Wang

Abstract The welding sequence has a significant influence on the magnitude and distribution of the final welding distortion, which often contributes negatively to the dimensional accuracy, structural instability, fabrication cost, and in-service performance of the welded assembly. Hence, it is of great significance to automatically plan optimal welding robot paths for minimum residual deformations in real-world engineering. In this study, an integrated optimisation approach combining an artificial neural network (ANN) model and intelligent optimisation algorithm was proposed to optimise welding sequences for minimising residual distortions in the assembly of thin-walled aluminium tubes. During the optimisation process, an ANN model was successfully established and implanted into this optimisation method to rapidly predict welding distortion within good computational precision. Subsequently, the Bees Algorithm (BA) was developed to systematically search for the optimal welding sequence. This algorithm takes advantage of performing a multi-neighbourhood search combined with a randomised global search to improve computational efficiency and solve premature convergence problems. Finally, a series of experiments were conducted to confirm the accuracy of the proposed BA. Verification experiments demonstrate that the acquired optimum welding sequence was in good agreement with the experimental results.

Keywords Welding distortion · Welding sequence optimisation · Finite element method · Back propagation neural network · Bees Algorithm

---

C. Wu
School of Mechanical Engineering, Changshu Institute of Technology, Jiangsu 215500, China

C. Wang (✉)
School of Mechanical Engineering, Suzhou University of Science and Technology, Jiangsu 215004, China
e-mail: cslgcw@mail.usts.edu.cn

D. T. Pham and N. Hartono (eds.), *Intelligent Engineering Optimisation with the Bees Algorithm*, Springer Series in Advanced Manufacturing,
https://doi.org/10.1007/978-3-031-64936-3_7

1 Introduction

Fusion welding is one of the most frequently used metal joining processes owing to its high productivity and good stability. It has been widely applied in many industries and fields, such as aerospace, automotive, construction, and shipbuilding. Despite its superior advantage, residual stresses concentrated near the weld are inevitably generated from localised and nonuniform heating and subsequent rapid cooling during the welding process, which often leads to undesirable distortions in whole welded structures. This welding-induced residual stress and distortion often contribute negatively to the welding quality and service lift of welded components. Therefore, it is of practical significance to accurately predict and control the residual stress and distortion beforehand.

1.1 Influence of the Welding Sequence

Many studies have indicated that factors affecting welding residual stress and distortion can be divided into three categories, namely, material properties, design-related parameters, and process-related variables [1, 2]. Among them, the welding sequence is considered one of the most crucial process-related factors. Over the last two decades, many researchers have devoted much effort to investigating the influence of the welding sequence on the residual stress and distortion of welded structures through experimental observation and numerical analysis based on the finite element method (FEM). For example, Fu et al. [3] developed a sequentially coupled thermo-mechanical finite element (FE) model to study the influence of welding sequence on residual stresses and distortions in T-joint welds with experimental validation. The comparative results indicate that the welding deposition sequence significantly influences the magnitude and deformation mode of the deflections. Azad et al. [4] used experimentally validated FEM numerical simulations to analyse the effect of eight welding sequences on weld distortion in a ship deck structure. They discovered that different welding sequences significantly affect both the magnitude and shape of weld distortions.

Based on the aforementioned studies, the FEM is a powerful tool for modelling the welding process with a high degree of accuracy. Furthermore, the welding deformation is significantly influenced by the welding deposition sequences. Hence, welding sequence optimisation (WSO) for minimising welding-induced distortion is a critical issue.

1.2 Welding Sequence Optimisation

When searching for an optimal welding sequence using full factorial designs, the total number of welding configurations (S) can be calculated using Eq. (1).

$$S = w! \times \tag{1}$$

Where w and d denote the number of weld seams and welding directions, respectively.

It is noted that the available combinations grow exponentially with the number of welds. In this study, the research object is an assembly of cowl cross bar systems with various types of functional brackets, as shown in Fig. 1. During the assembly process, to minimise welding-induced distortion and save weld metal, multiple short welds ($w \geq 20$) are adopted to avoid over-welding. Accordingly, the total welding configuration exceeds millions. The computational cost of using experiments or FE analysis with full factorial designs to select an optimal sequence becomes extremely high and unacceptable. Thus, the development of a rapid and reliable intelligence framework to automatically search for the optimal welding sequence is significant, and it is the main purpose of this study.

The process of WSO is a typical discrete optimisation problem that cannot be effectively solved by traditional techniques (e.g., analytic methods and gradient-based methods). Recently, with the development of evolutionary computation theory, the genetic algorithm (GA) has become the most popular optimisation approach for finding the best welding sequence. It is a global stochastic search optimal method inspired by natural evolutionary theory. The GA searches for the best solution through three basic operators, i.e., selection, crossover, and mutation.

In recent years, many applications of GA related to the aspect of WSO have been reported. For example, Islam et al. [5] developed an effective numerical approach

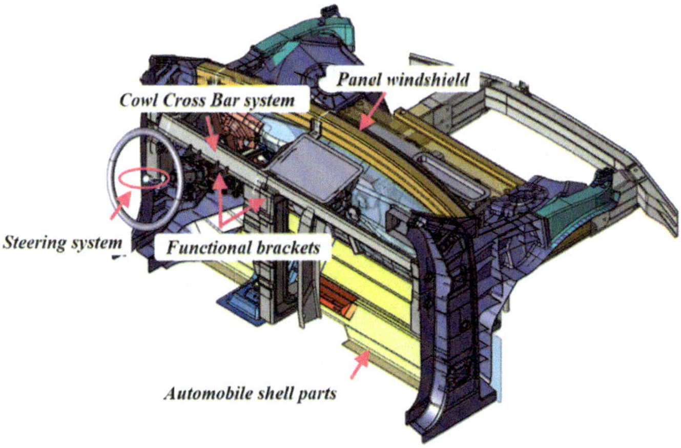

Fig. 1 Cowl cross bar system of automobile

based on coupled GA and FE analysis for welding process parameter optimisation to minimise weld-induced distortion. In this research, the welding speed, voltage, current, and welding direction were considered to be the process design variables. Jesus et al. [6] proposed a WSO framework based on the principles of an elitism GA, in which thermo-mechanical FE analysis was utilised to accurately predict welding distortion. They discovered that using GA significantly reduces the search space of the exhaustive search and enhances the computational efficiency. Choi et al. [7] presented a welding sequence and direction optimisation procedure based on the coupled GA-FEM to mitigate approximately 40% of the final welding deformation.

The abovementioned studies indicated that the complicated WSO problem can be successfully solved by the coupled GA-FE computing framework. However, the GA-based optimisation method has certain limitations. First, its local search ability and convergence speed are relatively poor [8, 9]. The GA cannot guarantee that the solution obtained is the actual global optimum due to the nature of the randomised search. Second, the selection of multiple controllable parameters in the GA significantly affects the performance of the algorithm, and there are no effective rules to analytically determine these values. Usually, the appropriate variables are acquired through the researcher's experience. Moreover, for the WSO issue, there are still several challenges, such as the chromosome coding mode and the operation of crossover and mutation.

In this chapter, to circumvent the limitation of the GA-based optimisation method, this study developed a novel numerical computational approach based on the bees algorithm (BA) for welding sequence optimisation to minimise the welding distortion of a welded structure. Moreover, the optimal results obtained by this proposed BA were validated by experimental measurements and FE numerical analysis. The comparative study demonstrated that the BA-based optimisation framework has the ability to optimise the welding sequence with good computational accuracy and efficiency in practical engineering.

2 Research Theory and Method

The aim of this study is to propose a novel optimisation methodology to automatically search for an optimal welding sequence to minimise welding distortion. Therefore, the total welding distortion in a welded structure is defined as an objective function, as expressed in Eqs. (2–3).

$$F = \min(d_{total}) \tag{2}$$

$$d_{total} = \max\left(\sqrt{d_{x_j}^2 + d_{y_j}^2 + d_{z_j}^2}\right) \tag{3}$$

Where d_{x_j}, d_{y_j}, and d_{z_j} are the displacement components in the x, y and z-axis directions at the j-th position within the entire welded structure. Accordingly, the solving process consists of two main stages: (i). calculation of welding distortion, and (ii). BA-based WSO algorithm.

2.1 Calculation of Welding Distortion

Despite the fact that the experimental measurements and FE numerical simulations can effectively investigate the influence of the welding sequence on the residual distortion, the computational cost and time consumption are extremely high, particularly for WSO problems, where these available welding combinations grow exponentially with the number of welds. Hence, there is a strong need to provide an available intelligent system for accurately predicting welding distortions under different combinations of input parameters.

To date, an artificial neural network (ANN) has been successfully applied in a variety of engineering applications to solve classification, recognition, and prediction problems owing to its powerful adaptive learning capability and nonlinear mapping ability. Moreover, related applications of ANNs to various aspects of welding engineering have also been reported in many studies. For instance, a variety of neural network models have been proposed to precisely predict and analyse the real-time geometric characteristics of weld seams during the welding process [10, 11]. Anand et al. [12] implemented ANNs based on five different training algorithms to predict the mechanical properties of Incoloy 800H friction-welded joints. The comparative results indicated that the ANN model with a GA may be able to predict the welding process parameters better. Dhas and Kumanan [13] developed intelligent predictive techniques (i.e., ANN) and fuzzy logic models for accurately computing welding residual stresses in a butt-welded joint. Tian et al. [14] and Rubio-Ramirez et al. [15] established ANN models to estimate the transverse shrinkage and angular distortions generated during the welding process with good computational accuracy.

In summary, the variety of ANNs has shown that they are capable of analysing the different characteristics of the welding process. Among them, a back propagation neural network (BPNN) is one of the most frequently used neural networks. It is a typical supervised multilayer feedforward neural network, which has a parallel-distribution architecture with a great number of interconnected neurons. Based on the back propagation algorithm, the connected weights and biases values are adjusted to minimise the prediction error of a BPNN. Therefore, in this study, a reliable BPNN model was established to rapidly and accurately estimate welding distortion for an arbitrary welding sequence.

2.2 BA-Based WSO Algorithm

BA is a typical population-based stochastic search algorithm, which is inspired by the intelligent foraging behaviors of honeybee swarms [16]. Compared to GA and particle swarm optimisation (PSO), the BA is more suitable for accurately searching narrow valleys and holes and for highly multimodal functions [17, 18]. Furthermore, the BA takes advantage of a multineighbourhood search combined with a randomised global search to improve computational efficiency and solve combinatorial or continuous optimisation problems.

In this BA, each nectar source in the enormous search space is regarded as a potential solution. In the initial stage, a number of initial scout bees are randomly sent out to collect nectar in multiple directions simultaneously. Sequentially, the quality of each nectar source is evaluated by the amount of nectar (i.e., fitness value). Afterwards, based on the fitness evaluation, two types of selected sites, i.e., best site and elite site, are selected to perform a multi-neighbourhood search. During the neighbourhood search, more recruited bees are assigned to these selected sites to further explore other potential solutions. A neighbourhood radius (λ) is set to limit the local search space. Accordingly, a next generation ($x_{id}\prime$) is addressed by Eq. (4).

$$x_{id}\prime = x_{id} + \phi(-\lambda, \lambda) \tag{4}$$

Where x_{id} is the location of the old nectar source. ϕ is the random number between $(-\lambda, \lambda)$.

Additionally, regarding unselected sites from scout bees, these nectar sources were abandoned due to their relatively low level of fitness. Then, the same number of scout bees are again randomly placed on the search space to execute a global search, which aims to find potential new regions of higher-ranking fitness in the next iteration. After that, the aforementioned steps are repeated by means of iterative computations until the stopping criteria are satisfied. Finally, a desired optimal solution was obtained using this proposed BA. Here, the flowchart of the BA-driven optimisation framework with a computer program for fitness evaluation is shown in Fig. 2.

3 Experimental Procedure

To verify the computational accuracy of the developed FE modes and the BA-based optimisation approach, a series of welding experiments were carried out to measure welding deformations. In this study, cold metal transfer (CMT) pulse welding technology was used to perform bead-on multi-seam welding of a thin-walled square Al–Mg–Si alloy (AA6063) tube with ER4043 wire (Ø1.2 mm).

The geometry and dimension of the welded square tube structure and distribution of multi-seam welds are shown in Fig. 3. The welding experiment was performed using a YASKAWA HP206 robotic welding system. The main welding parameters

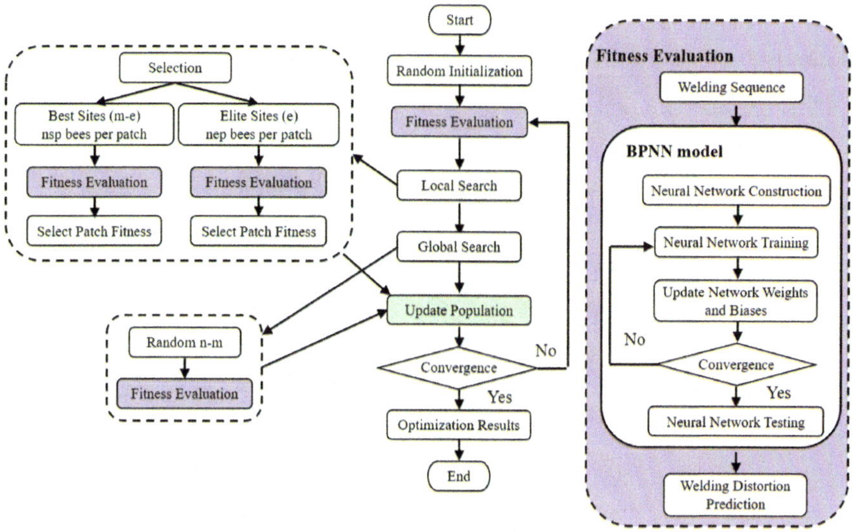

Fig. 2 The flowchart of the BA-driven optimisation framework

used in this study are presented in Table 1. During welding, laser displacement sensors were adopted to measure the welding-induced displacements in both the X-axis and Y-axis directions, as shown in Fig. 4.

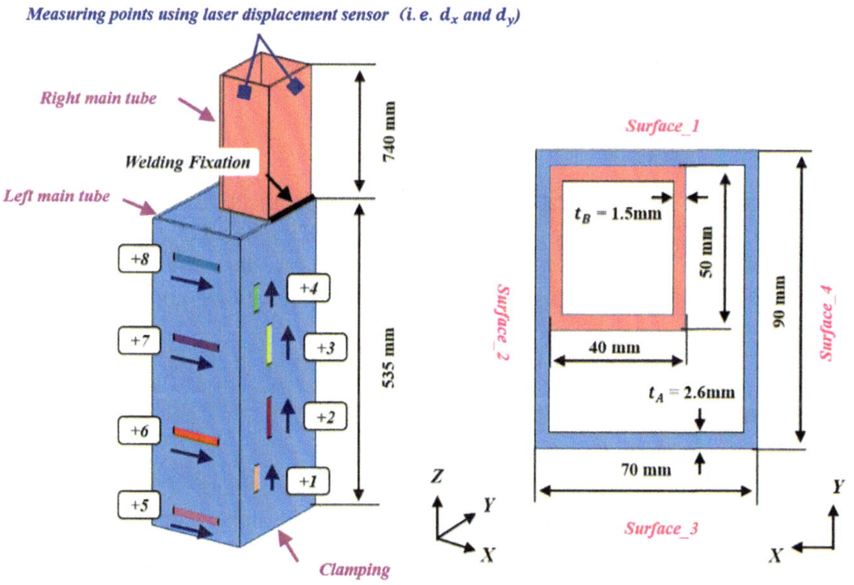

Fig. 3 Geometry and dimensions of the square tube structure with eight weld seams

Table 1 Welding conditions

Parameter	Voltage	Current	Welding speed	Wire feed rate	Shielded gas
Value	17.4 V	119A	1 m/min	5.2 m/min	Ar (15 L/min)

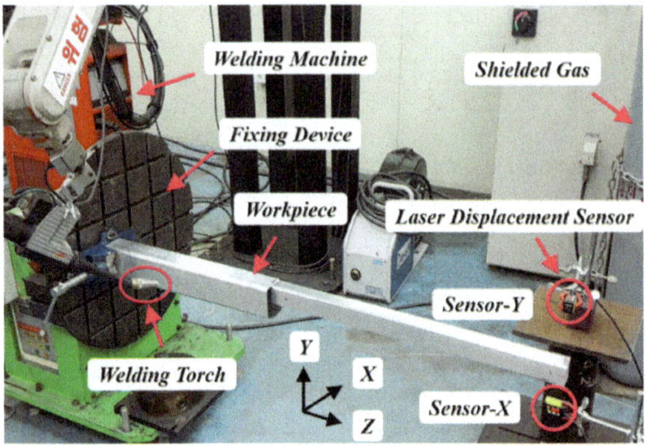

Fig. 4 Cold metal transfer welding robot system

After welding was performed, metal microstructure analysis was carried out in the middle cross-section of the welds to gather knowledge about the weld morphology and measure the weld shape. Accordingly, the numerical heat source model used in the FE simulation can be adjusted by the trial-and-error method to match well with the experimentally measured weld shape and size during heat source fitting analysis in the future.

4 Numerical Simulation

In this section, a 3D sequentially coupled thermo-mechanical FE analysis was developed using MSC-Marc software to simulate the CMT pulse welding process with multiple welds. Subsequently, the residual deformation caused by various welding sequences was evaluated by using this numerical computational approach.

4.1 Thermal Modelling

The shape and dimensions of the FE model are the same as those of the experimental specimen. Considering the balance of computation time and calculation accuracy, a fine grid was designed near the welds, whereas a relatively coarse mesh was utilised in the area away from the welds. In addition, considering the dynamic addition of filler material, the birth and death technique was used. During the FE analysis, the filler elements are not active initially. When the welding heat source reached these elements, it was activated, which was shown on the post file [19, 20].

During thermal modelling, a high non-uniform temperature field was produced by a moving heat source model. This model plays an important role in the accuracy of the entire welding simulation. To date, a variety of mathematical models including a surface heat source and volumetric heat source with Gaussian distribution have been widely used to simulate the welding process. In this study, to accurately model the heat input from the welding arc and molten metal droplets, the double-ellipsoidal volumetric heat source model suggested by Goldak et al. [21] was employed.

Throughout the welding process, the other boundary conditions should also be taken into account in the FE model, such as temperature-dependent material properties, initial temperature, and heat losses due to thermal convection and radiation. In addition, the birth and death element technique was used to model the dynamic addition of filler metal. For multi-seam welding, when the deposition of each weld is performed, an average 5-s cooling process is introduced into the FE model to simulate the moving process of a welding torch. Moreover, the latent heat effect and fluid flow in the weld pool are assumed to be negligible in this study.

4.2 Mechanical Modelling

Based on the assumption that mechanical responses are not affected by the previous thermal behaviors, a quasistatic mechanical analysis is conducted using the temperature cycles of all nodes obtained from the thermal analysis as the input load to calculate the stress–strain field and welding deformation. In the current FE model, the mechanical boundary condition was constrained at one end of the tube to prevent rigid body motion.

In the mechanical simulation, a nonlinear elastic-perfectly plastic material constitutive model with Von-Mises plasticity criteria was adopted to investigate the welding distortion formation behavior. The strains due to phase transformation and creep behavior are assumed to be negligible. Thus, the total strain is defined as the summation of the other three strain components, as expressed in Eq. (5).

$$\varepsilon_{total} = \varepsilon_e + \varepsilon_p + \varepsilon_{th} \tag{5}$$

Where ε_{total} is the total strain and ε_e, ε_p, and ε_{th} are the elastic strain, plastic strain, and thermal strain, respectively.

4.3 FE Model Validation

During the welding simulation process, the weld pool shape significantly affects the temperature distribution around the weld. Lindgren [22] concluded that if the weld

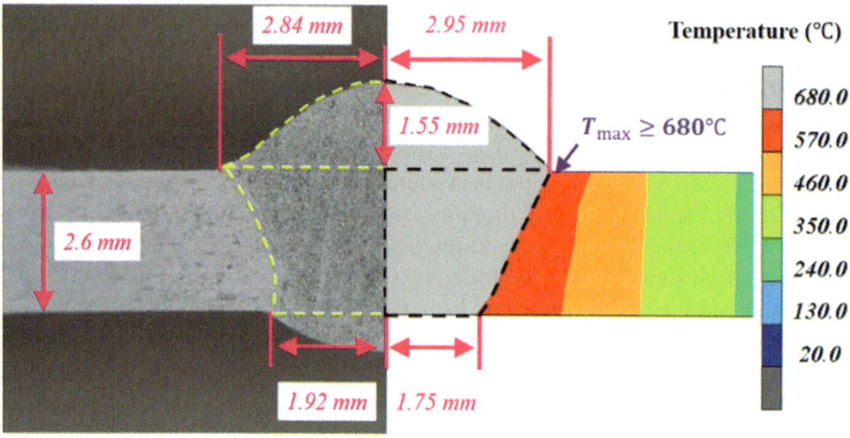

Fig. 5 Comparison of weld pool shape between experimental data and FE simulation

pool boundary is correct, then the temperature field outside this region will also be correct. Accordingly, Fig. 5 compares the overall boundary dimensions of the weld pool in the cross-section obtained by the experimental measurements and FE simulations. From this figure, it can be observed that the weld pool shape predicted by numerical simulation agrees well with experimentally measured values.

After validating the heat source model, the welding distortion predicted by the proposed FE model should be compared with the experimentally measured data. In this study, two simulation cases (Case A and Case B) with different welding sequences are performed to extract the welding deformation, as presented in Table 2. The order welding sequence with a positive direction is adopted in Case A. However, the symmetry and intermittent welding sequence with two directions are considered in Case B. Table 2 shows the final welding deformations of Case A and Case B in the X-axis and Y-axis directions computed by FE analysis and experiments. Obviously, a close agreement between the results from FE analysis and experimental measurements is obtained.

Small differences between the FE simulation and experiment results may be due to welding torch moving time inconsistency, experimental random error, and the lack of enough accurate material property data for the welding. Hence, the prediction error

Table 2 Comparison between the experimental and FE simulation results

Case	Welding Sequence								Experimental data		FE simulation	
									d_x (mm)	d_y (mm)	d_x (mm)	d_y (mm)
Case A	+ 1	+ 2	+ 3	+ 4	+ 5	+ 6	+ 7	+ 8	1.242	− 3.892	1.440	− 3.784
Case B	+ 1	+ 5	− 3	− 7	+ 2	+ 6	− 4	− 8	1.078	− 3.027	1.207	− 3.212

of the FE simulation is reasonable and acceptable. The aforementioned comparison result implies that the developed FE model has demonstrated its ability to accurately investigate the influence of the welding sequence on the welding deformation.

5 The Proposed ANN Model

In this section, based on the database obtained from the FE analysis, a supervised multilayer feedforward BPNN model was established to rapidly and accurately evaluate the welding distortion. The input variables used for developing the ANN model depend on the features of the welding sequences and directions with eight weld seams, whereas the bending deformation in the X-axis and Y-axis directions are considered as the output variables. The performance of the developed ANN model was evaluated by comparing its predicted values with the FE simulation results.

5.1 Database Generation

To generate the database, a total of 400 FE numerical simulations with randomly selected welding sequences are conducted to extract the welding-induced welding deformation in the X-axis and Y-axis directions. Except for the welding sequence, all other parameters, i.e., mesh size, constraint and boundary conditions, and analysis steps, remain the same during FE analysis.

5.2 BPNN Model Design

During the training of a BPNN model, the architecture and function/algorithm used in the network significantly affect the network performance. However, there is still no exact rule for analytically determining the appropriate network structure, selecting a reasonable activation function and learning algorithm. Generally, the parameters needed for optimal network performance are obtained by the trial-and-error method.

Overfitting is considered a crucial problem that may occur during neural network training. It often contributes negatively to the generalisation capability of the trained ANN model. In the present study, the Bayesian regularisation backpropagation training algorithm, denoted by 'trainbr' in MATLAB software, was adopted to solve the overfitting problem and enhance the generalisation ability of the designed network. Based on these numerical simulations described above, a total of 400 FE results were classified into 300 datasets (75%) for training and the remaining 100 datasets (25%) for testing. The training set was used to update the weight and bias values, whereas the testing set was utilised to evaluate the generalisation capability of the trained network.

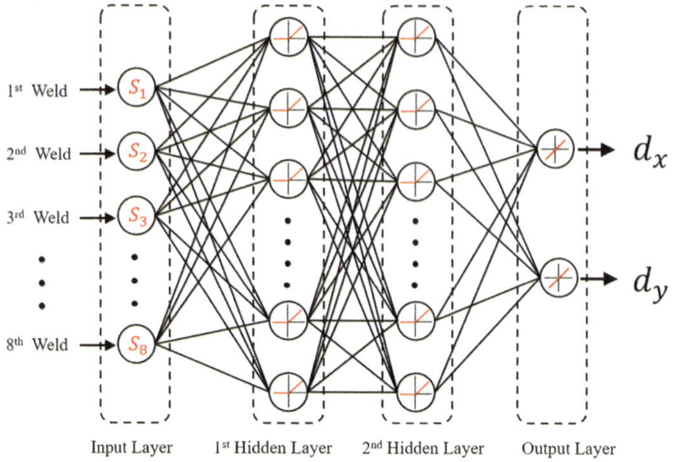

Fig. 6 Structure of the proposed BPNN model

During the designed BPNN training, the initial network adaptive parameters (weights and biases) of each layer are selected as arbitrary values between −1 and 1. Then, the rectified linear unit (ReLU) function and linear function were applied as the activation functions in the hidden layer and output layer, respectively. The comparative results demonstrated that the BPNN model with two hidden layers can obtain the desired generalisation capability with sufficient precision. Finally, in this study, an 8-32-32-2 architecture of the BPNN model was suggested to predict welding distortion induced by various welding sequences. Figure 6 illustrates the structure of the proposed BPNN model.

6 Optimisation Results and Discussion

6.1 Related Input Parameters

To perform a BA-based WSO algorithm, the most important issue is generating the new welding sequence that occurs during the neighbourhood search. In this study, a schematic diagram of the mechanism for generating a new welding sequence is illustrated in Fig. 7. A random number defines the positions of the operating point (P_j) and the adjacent elements $(P_{j-1}$ and $P_{j+1})$. Next, the positions of these selected elements are randomly swapped to control the welding order. Simultaneously, to change only welding directions, an operation is implemented to randomly determine the position and number of elements.

The proposed BA-based optimisation system comprises several important input parameters, as presented in Table 3. In the initial phase, a total of 20 welding

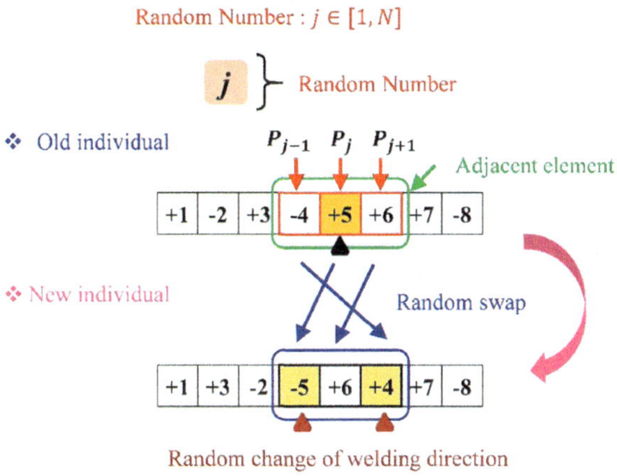

Fig. 7 Schematic diagram of the mechanism for generating a new individual

Table 3 Input parameters used in the optimisation system

Parameters	Value
Maximum number of iterations	50
Number of initial scout bees	20
Number of selected best sites	10
Number of recruited bees for best sites	30
Number of selected elite sites	5
Number of recruited bees for elite sites	10

sequences were generated randomly. Meanwhile, the repetition of the same welding sequence should be avoided. A large number of local search bees are assigned to search for the potential optimal solutions around nectar sources with higher-ranking fitness. For each iteration, the maximum number of possible calculations in the optimisation system is approximately 94. As the iterations are completed, the total welding configuration that needs to compute the welding deformation exceeds 4700. However, because the ANN model was used to compute fitness values in this study, the response speed of the proposed optimisation system is very fast, and the computational cost is relatively low.

6.2 Optimisation Results

In this study, two optimisation procedures were carried out to search for the best welding sequence (minimum distortion) and worst welding sequence (maximum distortion). Figure 8 shows the evolution of fitness values in the proposed BA-based optimisation system. From this figure, it was observed that the optimisation process converged within 50 iterations. Subsequently, the corresponding optimal welding

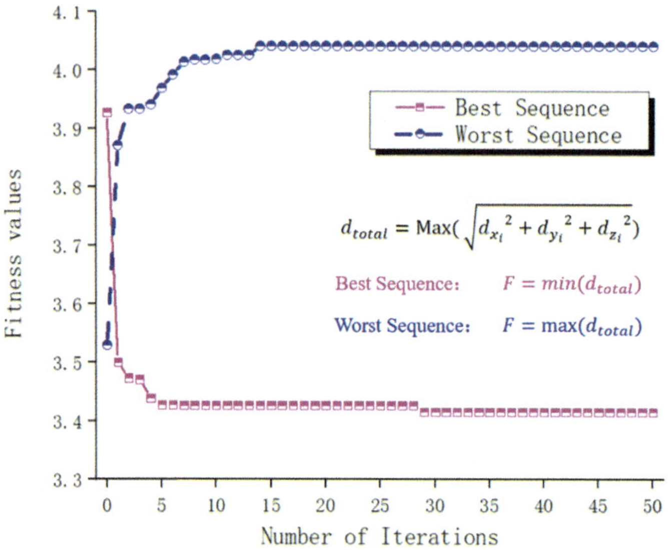

Fig. 8 Evolution of fitness values in the proposed optimisation system

sequence is presented in Table 4. Moreover, the welding distortion obtained from experimental measurements, FE numerical simulations, and the proposed BPNN model were compared. It is clear that the welding sequence can significantly affect the welding distortion. In this study, the contribution of the best and worst welding sequences to welding distortion differs by approximately 30%.

Moreover, from the results in Table 3, a close agreement was observed between the welding distortions obtained from experiment measurements, FE numerical simulations, and the proposed BPNN model. The comparative results indicate that the BA-based optimisation framework can be successfully utilised to search for the optimal welding sequence with good computational accuracy.

Due to the characteristic of multiple neighbourhood search combined with global search, a BA was used to greatly reduce the search space during the optimisation process. Compared with conventional optimisation techniques, the significant advantages of the proposed BA are high efficiency and simple operation. In addition, the proposed approach can avoid falling into the local optimum.

Table 4 Optimal welding sequence and corresponding welding deformation

Case	Welding sequence								Experimental data		FE simulation		BPNN model	
									d_x (mm)	d_y (mm)	d_x (mm)	d_y (mm)	d_x (mm)	d_y (mm)
Best	+ 1	− 4	− 3	− 7	− 5	− 6	+ 2	− 8	0.685	− 3.172	0.781	− 3.298	0.888	− 3.264
Worst	+ 7	+ 8	+ 6	+ 5	+ 2	− 1	+ 3	+ 4	1.556	− 3.807	1.778	− 3.783	1.824	− 3.701

7 Conclusion

An optimisation framework based on BA was created in this study to search for the best and worst welding sequences in thin-walled, square Al–Mg–Si alloy tubes. In this method, an implanted trained BPNN model was used to compute the fitness function to assess the adaptability of each welding sequence. The findings of the study allow for the following conclusions to be made.

- The BA-based optimisation search technique was successfully proposed to find the optimal welding sequences. The best outcomes demonstrate how the welding sequence can have a large impact on welding distortion.
- Due to the trait of multiple neighbourhood search mixed with global search, the suggested BA has considerable benefits in terms of high efficiency and ease of use.

References

1. Deng D (2013) Influence of deposition sequence on welding residual stress and deformation in an austenitic stainless steel J-groove welded joint. Mater Des 49:1022–1033
2. Kumanan S, Vaghela SJ (2017) Investigations on effect of direction of welding on distortion in combined butt and filled joint using finite element analysis. Indian J Eng Mater Sci 24:201–206
3. Fu G, Lourenco MI, Duan M, Estefen SF (2016) Influence of the welding sequence on residual stress and distortion of fillet welded structures. Mar Struct 46:30–55
4. Azad N, Iranmanesh M, Darvazi AR (2019) A study on the effect of welding sequence on welding distortion in ship deck structure. Ships Offshore Struct 15:1–13
5. Islam M, Buijk A, Rais-Rohani M, Motoyama K (2014) Simulation-based numerical optimisation of arc welding process for reduced distortion in welded structures. Finite Elem Anal Des 84:54–64
6. Romero-Hdz J, Aranda S, Toledo-Ramirez G, Segura J, Saha B (2016) An Elitism based genetic algorithm for welding sequence optimisation to reduce deformation. Res Comput Sci 121:17–36
7. Choi M, Wu C, Kim J-W (2020) Numerical optimisation of the welding sequence for mitigating welding deformation in aluminium pipe structures by using a genetic algorithm. Int J Precis Eng Manuf 21:2323–2333
8. Li G, Yang X, Wu D (2017) A novel collaborative optimization algorithm in solving complex optimization problems. Soft Comput 21:4387–4398
9. Katoch S, Chauhan SS, Kumar V (2021) A review on genetic algorithm: past, present, and future. Multimedia Tools Appl 80:8091–8126
10. Lei Z, Shen J, Wang Q, Chen Y (2019) Real-time weld geometry prediction based on multi-information using neural network optimized by PCA and GA during thin-plate laser welding. J Manuf Process 43:207–217
11. Chang Y, Yue J, Guo R, Liu W, Li L (2020) Penetration quality prediction of asymmetrical fillet root welding based on optimized BP neural network. J Manuf Process 50:247–254
12. Anand K, Barik BK, Tamilmannan K, Sathiya P (2015) Artificial neural network modeling studies to predict the friction welding process parameters of Incoloy 800H joints. Eng Sci Technol Int J 18:394–407
13. Dhas JER, Kumanan S (2011) Weld residual stress prediction using artificial neural network and fuzzy logic modeling. Indian J Eng Mater Sci 18:351–360

14. Tian L, Luo Y, Wang Y, Wu X (2014) Prediction of transverse and angular distortions of gas tungsten arc bead-on-plate welding using artificial neural network. Mater Design 54:458–472
15. Rubio-Ramirez C, Giarollo DF, Mazzaferro JE, Mazzaferro CP (2021) Prediction of angular distortion due GMAW process of thin-sheets Hardox 450® steel by numerical model and artificial neural network. J Manuf Process 68:1202–1213
16. Pham DT, Ghanbarzadeh A, Koç E, Otri S, Rahim S, Zaidi M (2006) The Bees algorithm—a novel tool for complex optimisation problems. Intell Prod Mach Syst 454–459.
17. Pham DT, Castellani M (2015) A comparative study of the Bees algorithm as a tool for function optimization, Cogent Engineering, vol 2 (online). https://doi.org/10.1080/23311916.2015.1091540
18. Li J, Gu Y, Wang C, Liu M, Zhou Q, Lu G, Pham DT, Xu H (2022) Pedestrian-aware supervisory control system interactive optimization of connected hybrid electric vehicles via fuzzy adaptive cost map and bees algorithm. IEEE Trans Transp Electrification 8(2):2959–2970
19. MSC Software, Marc and Mentat (2016) Installation and operation guide. MSC Software Corporation
20. Wu C, Kim J-W (2020) Numerical prediction of deformation in thin-plate welded joints using equivalent thermal strain method. Thin-Walled Struct 157:107033
21. Goldak J, Chakravarti A, Bibby M (1984) A new finite element model for welding heat sources. Metall Trans B 15:299–305
22. Lindgren L-E (2007) Computational welding mechanics: thermomechanical and microstructural simulations. Woodhead Publishing Limited, Cambrige England

Hybrid Genetic Bees Algorithm (GBA) for Continuous and Combinatorial Optimisation Problems

Tran Duc Vi, Vu Hoai Anh Thu, Truong Tran Mai Anh, and Nguyen Dao My Vy

Abstract The Bees Algorithm (BA) has proven its effectiveness in performing optimisation in many problems in recent years with many applications, but the global search of the BA is not its strength. This chapter summarised reports on a hybrid model called the Genetic Bees Algorithm (GBA) to enhance the BA for both continuous and combinatorial optimisation problems. For continuous problems, GBA in training a multilayer perceptron (MLP) is reported for the first time to solve real-world problems. Experimental results show that GBA provides significantly better performance than Particle Swarm Optimisation (PSO) in training MLP with higher accuracy. For combinatorial problems, in the two-stage flow shop scheduling problem (TFSSP), three methods, such as Particle Swam Optimiser (PSO), Genetic Algorithm (GA) and GBA, are compared. GBA is competitive and better than PSO and GA in a majority of instances in terms of results, proving that GBA is a realistic and efficient solution to the TFSSP. Finally, GBA proves its strength in solving the Just-in-Time Job Shop Scheduling Problem (JITJSSP) by solving a set of 36 benchmark instances ranging from 20 to 200 operations, the outcomes obtained from that are then compared to an exact method, two recent studies, and best-known solutions. The results show that the GBA has its strengths and weaknesses in solving the JITJSSP, and it performs well in some instances but does not perform well in others.

Keywords Metaheuristics · Hybrid genetic Bees Algorithm (GBA) · Continuous optimisation problems · Combinatorial optimisation problems

T. D. Vi (✉) · V. H. A. Thu · T. T. M. Anh · N. D. M. Vy
School of Industrial Engineering and Management, International University—Vietnam National University HCM City, Block 6, Linh Trung Ward, Thu Duc District, HCM City, Vietnam
e-mail: tdvi@hcmiu.edu.vn

D. T. Pham and N. Hartono (eds.), *Intelligent Engineering Optimisation with the Bees Algorithm*, Springer Series in Advanced Manufacturing,
https://doi.org/10.1007/978-3-031-64936-3_8

Table 1 The parameters of the Bees Algorithm [1]

ns	number of scout bees
nb	number of best sites selected out of ns visited site
ne	number of elite sites out of nb selected sites
nre	number of bees recruited for the best ne sites
nrb	number of bees recruited for the other (nb − ne) selected sites
stlim	number of cycles after which a site is abandoned

1 Introduction

1.1 Bees Algorithm (BA)

Before presenting the hybrid Genetic Bees Algorithm (GBA), this section first briefly describes the Bees Algorithm (BA) proposed by Pham et al. [1]. More details about the BA and its many applications can be found in [1, 2]. In general, the BA is a search algorithm that uses honeybees' food gathering technique to find the best solution to a given optimisation issue, mimicking the bees' behaviours, including process scouting, waggle dancing, and foraging in the search for the best flower patch. The parameters and general flow chart of BA are shown in Table 1 and Fig. 1, respectively.

The following process is employed for the search in the BA:

1. Calculate the fitness function of the ns visited site
2. Select the best nb sites for local search
3. Recruit nre forager bees for the ne sites and nrb bees for the best nb-ne sites, evaluate their fitness
4. Select the fittest bee from the local site as the new centre (the dominant one)
5. If the site fails to improve in a single local search, the neighbourhood size decreases (neighbourhood shrinking) (this step is applied in other problems with continuous decision variables).
6. If the site fails to improve for *stlim* cycles, the site is abandoned (site abandonment).
7. Allocating other bees to search for the whole space and evaluate fitness
8. If the stopping criterion is not met, return to step 2.

1.2 Genetic Bees Algorithm (GBA)

For the BA, although it has been implemented successfully on a variety of optimisation problems, the algorithm has weak global search capability [3]. This flaw arises from the search process's use of a simple global random search procedure. Recently, several works have proposed improving the performance of BA using the

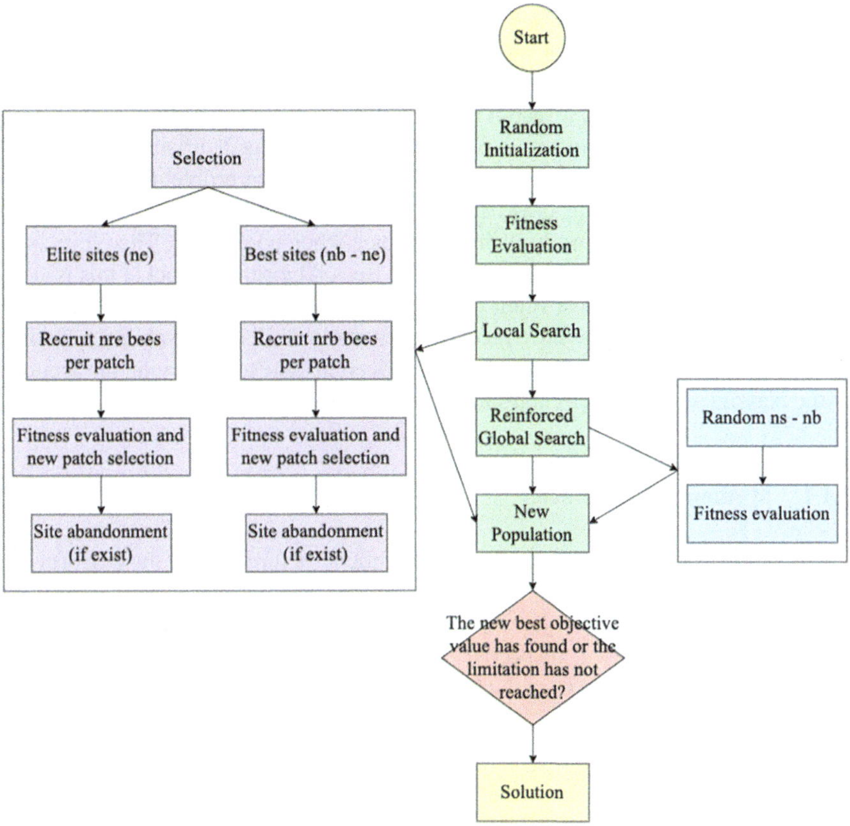

Fig. 1 Flowchart of the basic Bees Algorithm [1]

hybrid Genetic Bees Algorithm (GBA) algorithms. More details about the algorithm can be found in [3–5]. Applications of GBA for both continuous and combinatorial problems will be presented in the following sections.

2 Applications

2.1 *GBA for Continuous Problem: Training Multilayer Perception (MLP)*

Neural networks (NNs) have been widely known for their efficiency in solving classification problems. The concept of mimicking the neurons of the human brain was initially proposed in McCulloch and Pitts [6]. NNs often refer to learning, which is the ability to extract meaningful information from complex and inaccurate data. In this way, this can cluster data with similarities within a number of items with labels. Because of this usefulness, it can be applied in several fields, including pattern recognition, voice recognition, etc.

The Bees Algorithm is a search algorithm that combines an exploitative neighbourhood search with a random explorative search. Its metaheuristic has proven its effectiveness in exploring global solutions to optimisation problems [1]. However, the current method also has drawbacks. The disadvantage of random search in the Bees Algorithm is that it produces high variance when computing. Therefore, other meta-heuristics should be combined with the Bees Algorithm to tackle this issue.

These two methods have often progressed in different directions. Attempts to merge the two technologies have recently been made. Utilising the ability to find the global optimum of the Bees Algorithm, we will contribute to improving the accuracy of the classification problem in training the MLP.

2.1.1 Mathematical Models

The coding strategy is important to organise the structure so that the performance of the metaheuristic can be improved. In this work, weights and biases are encoded as {iw, hw, hb, ob}, where iw, hw, hb, ob are input weights, hidden weights, hidden biases, and output biases, respectively. Figure 2 illustrates how we encode our solution. We follow the process to obtain the output of MLPs:

$$s_j = \sum_{i=1}^{n} \left(W_{ij} x X_{ij}\right) - \theta_j \tag{1}$$

The output of the hidden node is computed as follows:

$$S_j = sigmoid\left(s_j\right) = \frac{1}{1 + exp(-s_j)} \tag{2}$$

The following are the final outputs based on the hidden nodes' computed outputs:

$$o_{\mathrm{k}} = \sum_{\mathrm{j}=1}^{\mathrm{h}} \left(w_{\mathrm{ij}} \times S_{\mathrm{j}}\right) - \theta'_{\mathrm{k}} \tag{3}$$

$$O_k = sigmoid(o_k) = \frac{1}{1 + exp(-o_k)} \tag{4}$$

After being trained, the ANN models will be evaluated using statistical metrics: mean squared error (MSE). MSE represents the difference between the actual and predicted data. The lesser the mean squared error is, the closer the fit is to the dataset. RMSE is the error rate by the square root of MSE.

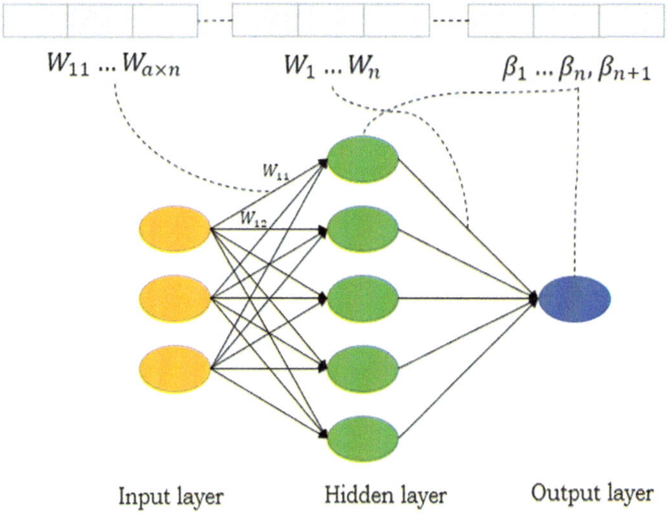

Fig. 2 Weights and biases mapping for MLP

$$\text{MSE} = \frac{1}{n}\sum_{i=1}^{n}(y_i - \hat{y}_i)^2 \quad (5)$$

- y_i is the desired output, and $\hat{y}_i$ is the neural network output.

The GBA used for training the MLP is presented in the (Fig. 3).

2.1.2 Experimental Settings

In this work, we benchmark the efficient performance of HGBA by testing on 4 classification datasets. The whole training process of each testing case will be performed 30 times to generalise the results. The parameters of the HGBA algorithm and classification datasets are as follows (Table 2):

The classification datasets are summarised as follows (Table 3):

We also consider the Particle Swarm Optimisation for training MLP in the work of Kenedy and Eberhart [7]. In this work, we will adopt the following suggestions for classification problems: 2 × na + 1, where na is the number of attributes (Table 4).

2.1.3 Result and Analysis

(a) Mean Squared Error

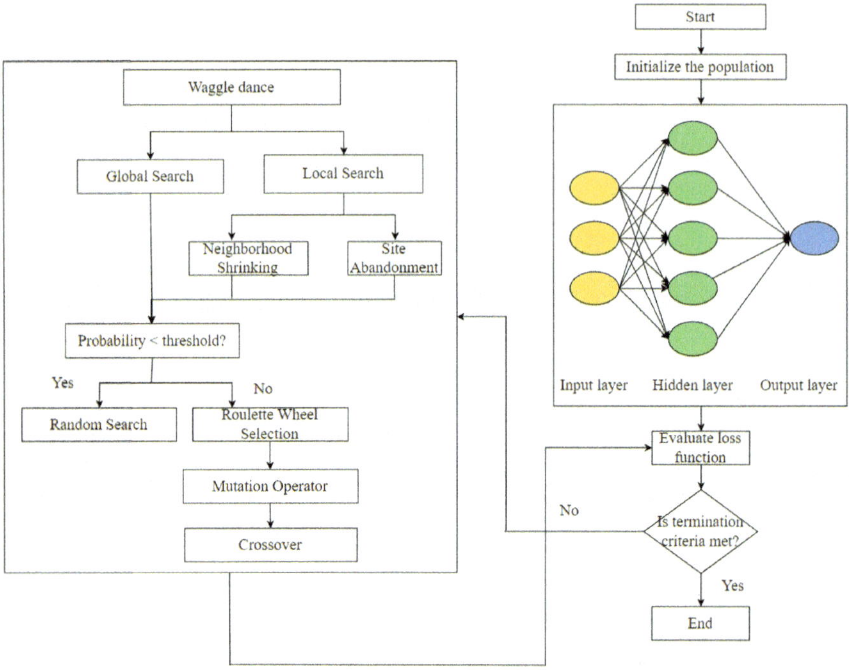

Fig. 3 Training the MLP network by GBA

Table 2 Parameter setting for HGBA for MLP

Parameter (GBA)	Value
Number of scouts bees (ns)	100
Number of best sites (nb)	5
Number of elite sites (ne)	3
Number of foragers in elite sites (nre)	20
Shrink factor (sf)	0.1
Mutation rate	0.1
Crossover rate	0.8
Max iterations	100
Number of iterations unchanged	20

Table 3 Classification datasets

Dataset	Attribute number	Training samples	Test samples	Classes number
XOR	3	8	8	2
Breast cancer	9	599	100	2
Heart	22	80	80	2
Tic-Tac-Toe	9	637	300	2

Table 4 MLP structure for classification datasets

Dataset	Number attributes	MLP structure
XOR	3	3-7-1
Breast cancer	9	9-19-1
Heart	22	22-45-1
Tic-Tac-Toe	9	9-19-1

The following five figures illustrate the convergence curve of four different types of datasets. They were performed by two different metaheuristic algorithms: HGBA-MLP and PSO-MLP. The dataset we used includes the "Breast cancer" dataset, "Heart" dataset, "Tic Tac Toe Endgame" dataset and "XOR" dataset. Overall, in all four datasets, HGBA-MLP always performs better than PSO-MLP in minimising the objective value (Fig. 4).

(b) Shrink Factor (Sf)

The neighbourhood size of patches was decreased by using the following equation:

$$nghi = nghi - 1 \times sf \tag{6}$$

The shrinking factor is assumed to be from 0 to 1. If sf = 0, then this figure will eliminate the effect of the neighbourhood size because the size will approach 0. If sf = 1, then there is no effect on the size. In the case that sf > 1, the size will become larger, which contradicts our purpose.

Table 5 illustrates how different the objective value is when we apply different values of the shrink factor. The bold values indicate the smallest values in this

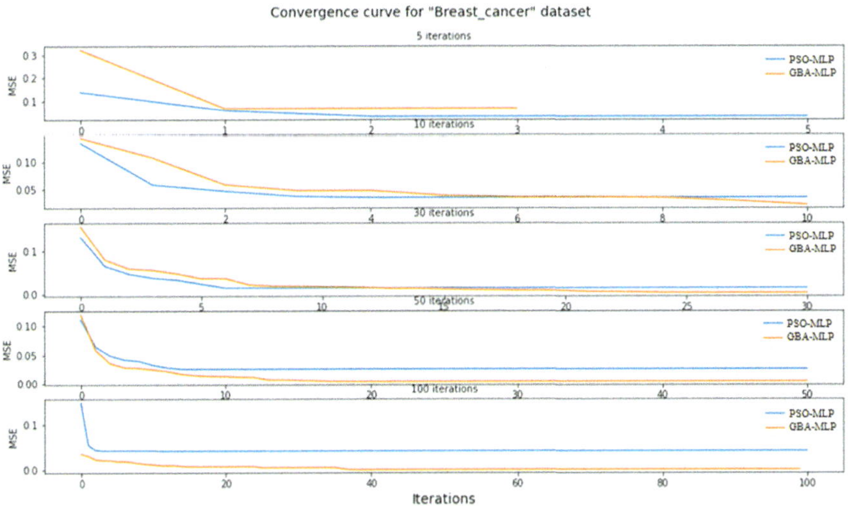

Fig. 4 Convergence curve for the "Breast cancer" dataset

dataset. These bold values are minimal objective values achieved at the corresponding shrinkage factors. It can be seen that parameter tuning is relatively important because with different shrink factor values, the objective value for each dataset differs. For example, a shrink factor of 0.2 will bring the best value for the "Breast cancer" dataset, while for XOR, the shrink factor should set at 0.001.

(c) MSE and Accuracy

Table 6 compares the performance of the two algorithms when applied to four different classification problems. It can be seen from the table that the higher accuracy rate belongs to our proposed algorithm. Moreover, the average of HGBA-MLP is also smaller than that of PSO-MLP for all cases except for the "Heart" dataset. The HGBA-MLP gives a higher deviation compared to the figure for PSO-MLP, but the difference is not large enough to consider.

Table 5 Sensitivity analysis of the shrinkage factor

Shrinkage factor	Breast cancer	Heart	Tic Tac Toe endgame	XOR
0.001	0.039887534	0.248317181	0.255780597	**0.023857062**
0.002	0.042906719	0.245772466	0.258271766	0.030685722
0.005	0.042080124	0.245691644	0.258035425	0.037550701
0.01	0.047253134	**0.24216748**	0.255754658	0.03707171
0.05	0.046908873	0.244228577	**0.245897409**	0.036551992
0.1	0.042367223	0.252891685	0.252114593	0.025976337
0.2	**0.038874115**	0.247698323	0.253178375	0.033243973
0.5	0.047651649	0.250066689	0.250787427	0.03265171

Table 6 Comparisons between HGBA-MLP and PSO-MLP

Dataset	HGBA-MLP			PSO-MLP		
	Mean	Std	Accuracy	Mean	Std	Accuracy
Breast cancer	0.043491171	0.040366654	92.30%	0.044877396	0.03001	92.10%
Heart	0.247104256	0.049575268	66.70%	0.246917228	0.03046	58.10%
Tic Tac Toe endgame	0.253727531	0.057619606	66.90%	0.259985173	0.03321	60.00%
XOR	0.032198651	0.0589139	93.50%	0.053474034	0.061656	32.00%

2.2 *GBA for Combinatorial Problems*

2.2.1 GBA for the Two-Stage Flow Shop Scheduling Problem (TFSSP)

The scheduling problem is one of the most studied problems in manufacturing systems, and it is important for efficient production planning and management. Over the last few years, a variety of scheduling algorithms have been developed to address different production systems. Because of its application in real-life problems, the TFSSP, like many other problems, has received much attention in the literature [8–12]. The bees make no assumptions about the continuity of the search space, requiring only problem domain knowledge in addition to the needed solution performance. In this section, the Genetic Bees Algorithm (GBA) [3] is employed to solve the TFSSP.

Mathematical Model

The mathematical model considered by Tian et al. [13] is presented below. The objective is to minimise the makespan ($C\max$) and mean completion time (MCT).

The notations of the two-stage assembly scheduling problem are defined as follows:

- n is the number of jobs;
- m is the number of machines at the first stage;
- i is the index of jobs, i {1, 2,…, n};
- j is the index of machines, j {1, 2,…, m};
- k is the coefficient of setup time, k {0.3, 0.6, 0.9, 1.2}; α is the weight given to makespan, α {0.2, 0.5, 0.8};

Parameters

- $s[i, j]$ is the setup time of job i on machine j at stage 1
- $t[i, j]$ is the processing time of job i on machine j at stage 1;
- $s[k]$ is the setup time of job i on the assembly machine at stage 2;
- $p[k]$ is the processing time of job i on the assembly machine at stage 2;
- $C[n]$ is the completion time of job n;

The objective of minimising makespan is:

$$C\max = C[n] \tag{7}$$

Then, the completion time of each job can be calculated using the following formula:

$$MCT = n1\sum ni = 1C[n] \tag{8}$$

$$C[k] = \max\left\{\left\{\sum ki = 1\left(s[i,j] + t[i,j]\right)\right\}, C[k-1] + s[k]\right\} + p[k] \tag{9}$$

Table 7 Data used to generate instances [10]

Data	
Number of jobs (n)	20, 40, 60, 80
Number of machines at the first stage (m)	2, 4, 6, 8
Coefficient of setup time (k)	0.3, 0.6, 0.9, 1.2
Weights (α)	0.2, 0.5, 0.8
Processing time at the first stage ($t[i, j]$)	Unidrnd(1, 100)
Processing time at the assembly machine ($p[k]$)	Unidrnd(1, 100)
Setup time at the first stage ($s[i, j]$)	Unidrnd(1, 100)
Setup time at the assembly machine ($s[k]$)	Unidrnd(1, 100 k)

Table 8 Data of the Bee Algorithm

Parameter	Value
Population (ns)	400
Number of best sites (nb)	10
Number of elite sites (ne)	6
Number of recruited bees for remaining best sites (nrb)	10
Number of recruited bees for elite sites (nre)	10
Initial size of neighbourhood (ngh)	10

In this study, these two objectives are combined to generate a new objective *FO*max. Thus, the final measure *FO*max is considered as the objective function to be minimised and computed by:

$$FO\max \; = \alpha \times C\max + \beta \times MCT \tag{10}$$

Where α and β are the weights (α + β = 1) assigned to the makespan and mean completion time, respectively, and express the relative importance of the makespan and mean completion.

Experimental Settings

(a) Data processing

Problem data were generated in accordance with the literature [10] and are listed as follows (Tables 7 and 8):

For convenience, we use n_m_α_k_u to denote one test instance, where u denotes the number of instances and u = 3 denotes the third instance. Because each combination has ten instances with different n, m, α and k, u is between 1 and 10. For example, 20_4_0.2_0.9_2 describes the second instance (u = 2) of the combination with 20 jobs, 4 machines, α = 0.2 and k = 0.9.

Table 9 Computation results for n = 20, 40 and k = 0.3

n	m	a	PSO	GA	GBA
20	2	0.2	0.2775	0.3040	0.0621
20	2	0.5	0.4543	0.2931	0.2026
20	2	0.8	0.3629	0.3677	0.1140
20	4	0.2	0.2510	0.2622	0.0722
20	4	0.5	0.2858	0.2716	0.1465
20	4	0.8	0.2956	0.4250	0.1195
20	6	0.2	0.2753	0.2708	0.1162
20	6	0.5	0.2768	0.2766	0.1160
20	6	0.8	0.3074	0.3655	0.1500
20	8	0.2	0.3116	0.2590	0.0947
20	8	0.5	0.2949	0.2690	0.0504
20	8	0.8	0.3376	0.3289	0.0906
40	2	0.2	0.1609	0.1654	0.0400
40	2	0.5	0.1539	0.1765	0.0289
40	2	0.8	0.2170	0.2404	0.0546
40	4	0.2	0.1313	0.1538	0.0525
40	4	0.5	0.1644	0.1486	0.0500
40	4	0.8	0.1977	0.2151	0.0924
40	6	0.2	0.1526	0.1417	0.0846
40	6	0.5	0.1188	0.1836	0.0450
40	6	0.8	0.2046	0.1991	0.0976
40	8	0.2	0.1335	0.1328	0.0804
40	8	0.5	0.1851	0.1764	0.0691
40	8	0.8	0.2025	0.1941	0.1246

We tested the performance of the GBA using measures of the average percentage error (APE). The percentage error (PE) is defined as follows:

$$PE = \frac{(\text{Soli} - \text{BestSol}) \times 100}{\text{BestSol}} \tag{11}$$

Where Sol$i$ is the best solution found for the ith run and BestSol is not the known best solution but the best solution obtained from the total run. The best possible performance is clearly obtained when Sol$i$ = BestSol, that is, PE = 0. Examples of computational results are shown in Table 9.

Figure 5 shows that when the job number is 60, GA is close to PSO in the APE and is slightly better than the latter. In all cases, GBA is the best, GA is the second and PSO is the worst.

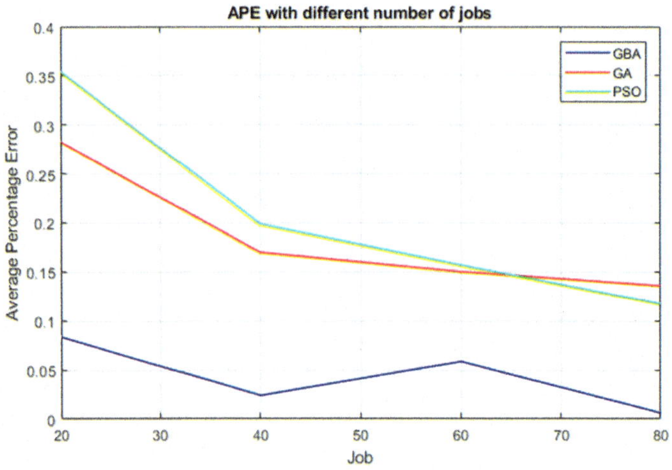

Fig. 5 APE with different job numbers

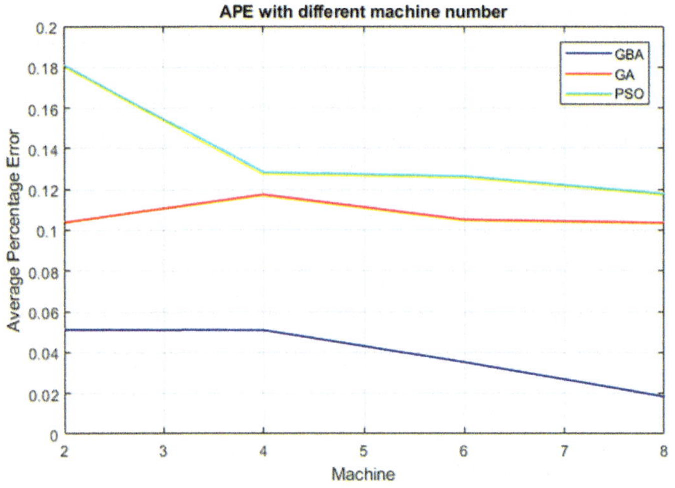

Fig. 6 APE with different machine numbers

As seen from Fig. 6, no matter how the machine number changes, the results of the GBA are the best, while those of PSO are the worst.

Figure 7 describes the comparative results with different weights α. When the weight α is 0.5 and 0.8, the GBA achieved the smallest APE and is the best of the three algorithms.

Figure 8 illustrates the runtimes (averaged) of the algorithms with different job numbers. When the job numbers are 40 and 60, the GBA runtime is 105 s. The

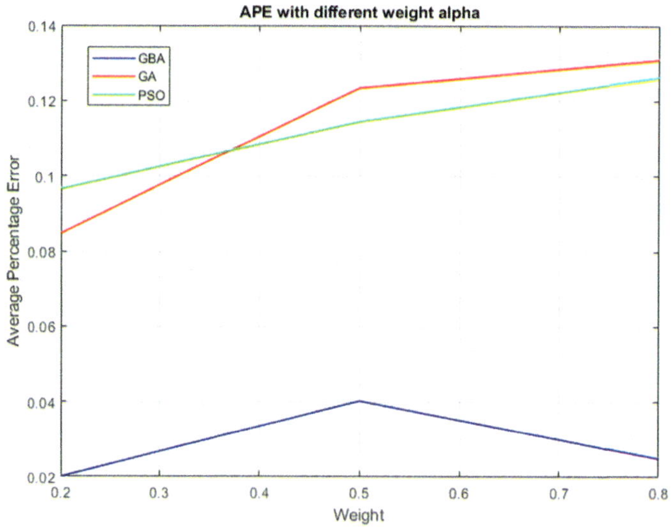

Fig. 7 APE with different weights α

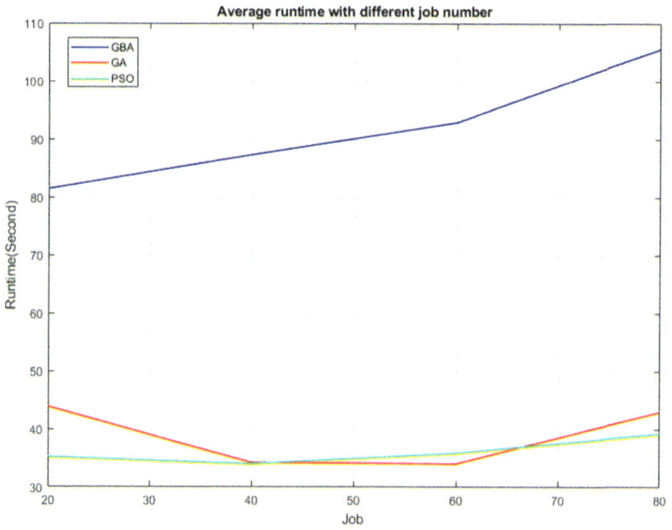

Fig. 8 Average runtime with different job numbers

runtimes of the GBA increase proportionally to the problem's scale (number of jobs).

Based on the experimental results, we can conclude that the GBA proposed in this chapter can achieve better performance for almost all problems, and the advantage becomes more apparent as the number of jobs increases. GBA, on the other hand, has a much longer runtime than GA and PSO.

2.2.2 GBA for Just-In-Time Job Shop Scheduling Problem (JITJSSP)

A brief introduction of current approaches for solving the JITJSSP is presented as follows. Baptiste et al. [14] investigated the subject of job-shop scheduling with early and tardy penalties. The problem is described in terms of two Lagrangian relaxations. The first is based on precedence constraint relaxation, while the second is based on machine constraint relaxation. The relaxations allow them to develop lower bounds and heuristic solutions in both circumstances. Finally, to improve the best answer identified, they offered a simple local search strategy. The benchmark data provided by them are now widely used for the purpose of comparing the effectiveness of the methods in solving JIT-JSSP. Santos et al. [15] used a hybrid method to solve the problem. They presented a combination of an evolutionary algorithm and mathematical programming with a new local search proposal, which was said to be suitable for the JIT JSSP. Wang and Li [16] used a mix of variable neighbourhood search (VNS) and mathematical programming to reduce the sum of all operations' earliness and tardiness penalty costs. The hybrid algorithm combines a VNS procedure to efficiently explore the large feasible solution spaces by alternating the swap and insertion neighbourhood structures and a mathematical programming model to optimise the operation completion times for a given solution in each iteration procedure.

To solve JIT-JSS, Ahmadian et al. [17] also used the VNS technique. The method employs numerous neighbourhood structures, including new relaxation neighbourhoods. Several processes are partially destroyed (relaxed) and then reconstructed (sequenced) by the relaxation neighbourhoods. Unlike traditional neighbourhoods, where changes are performed randomly, the new neighbourhoods' moves are undertaken in relation to other operations, so their effects on the overall sequence are properly studied. In general, most of the studies were based on a single metaheuristic or a hybrid of a metaheuristic and a mathematical program. It is proven that they have achieved remarkable results by their incredible work. However, obviously, the GBA has not been implemented to solve the problem, and the results may not be optimal since the global search is not strong enough in the above method.

Mathematical Model

This study takes the model from Ahmadian et al. [17] and describes it as follows (Table 10):

Set:

- Job j has nj operations, where Oj = {1,...,nj} nj $\leq$ m is the set of operations of job j.

Table 10 Objective function and constraint explanation

Objective/Constraints	Explanation
Objective function	Minimising the total penalty of earliness and tardiness
Equation (13)	The start time of the first operation of each job must not be earlier than time zero
Equation (14)	Defining the job precedence relationships among a job's operations
Equation (15)	Machine precedence relations constraint ensures that each machine can only process one operation at a time
Equation (16) and (17)	Earliness and Tardiness definition, respectively
Equation (18) and (19)	Earliness and Tardiness are nonnegative

- Machine: processes operation Ojk (the kth operation of job j) as M(Ojk) and M(Ojk) in M.
- Mi in M, O(Mi): set of operations that must be processed by machine Mi.
- N: Set of all the jobs, N = {1,…,n}.
- M: Set of all the machines, M = {1,…,m}.
- Oj: Set of operations of job j, Oj = {1,…,nj}, nj $\leq$ m.

Index:

- j: Job index, j $\in$ N.
- k: Operation index for job j, k $\in$ Oj.
- i: Machine index, i $\in$ M.

Parameters:

- pjk: Processing time of operation Ojk, pjk $\geq$ 0, j $\in$ N, k $\in$ Oj.
- dk: Due date of operation Ojk, djk > 0, j $\in$ N, k $\in$ Oj.
- αjk: Earliness penalty coefficient associated with operation Ojk, αjk $\geq$ 0, j $\in$ N, k $\in$ Oj.
- βjk: Tardiness penalty coefficient associated with operation Ojk, βjk $\geq$ 0, j $\in$ N, k $\in$ Oj.

Variables:

- Cjk: Completion time of operation Ojk, Cjk $\geq$ 0, j $\in$ N, k $\in$ Oj.
- ESjk: Earliest start time of operation Ojk, ESjk $\geq$ 0, j $\in$ N, k $\in$ Oj.
- ECjk: Earliest finish time of operation Ojk, ECjk $\geq$ 0, j $\in$ N, k $\in$ Oj.
- Tjk: Tardiness of operation Ojk, Tjk $\geq$ 0, j $\in$ N, k $\in$ Oj.

If Cjk $\leq$ djk, earliness penalty αjkEjk incurs, Ejk = max {djk − Cjk, 0}.
If Cjk $\geq$ djk, tardiness penalty βjkTjk incurs, Tjk = max {Cjk − djk, 0}.
Objective function:

$$Z = \min \sum_{j=1}^{n} \sum_{k=1}^{n_j} \alpha_{jk} E_{jk} + \beta_{jk} T_{jk} \tag{12}$$

Table 11 BA parameters of the model [18]

Parameter	Value
Number of scout bees (ns)	40
Number of elite bees (nre)	40
Number of best bees (nrb)	20
Number of elite sites (ne)	16
Number of best sites (nb)	20

Subject to:

$$C_{j1} - p_{j1} \geq 0, j \in N \tag{13}$$

$$C_{jk} \leq C_{j,k+1} - p_{j,k+1}, j \in N, k \in O_j - \{n_j\} \tag{14}$$

$$C_{jk} \geq C_{lh} - p_{jk} or C_{lh} \geq C_{jk} - p_{lh}, i \in M, O_{jk}, O_{lh} \in \text{O}(M_i) \tag{15}$$

$$E_{jk} \geq d_{jk} - C_{jk} j \in N, k \in O_j \tag{16}$$

$$T_{jk} \geq C_{jk} - d_{jk} j \in N, k \in O_j \tag{17}$$

$$E_{jk} \geq 0, j \in N, k \in O_j \tag{18}$$

$$T_{jk} \geq 0, j \in N, k \in O_j \tag{19}$$

Experimental Settings

(a) **Parameters**

Apart from setting the mutation rate equal to 0.1 and the crossover rate equal to 0.8, the set of parameters is given below, while the number of runs is modified (equal to 5), and the number of iterations is not constant since there are small-size and large-size instances (Table 11).

(b) **Data Processing**

The data used are the benchmark data from Baptiste et al. [14]. There are eight instances for each (n, m) ∈ ({10, 15, 20} × {2, 5, 10}), ranging from 20 to 100 operations. Each instance follows the format DD-W-ID-nxm, where n and m are denoted as the number of jobs and number of machines alternatively. The definitions of DD, W, and ID are given below.

DD: The distance between due dates of consecutive operations is equal to the processing time (DD = Tight), or the distance between due dates of consecutive operations is equal to the processing time plus a random number taken in [0, 10] (DD = Loose).

W: Both the earliness and tardiness costs (αjk, βjk) are chosen at random in [0.1, 1] (W = equal), or αjk is chosen at random in [0.1, 0.3], while βjk is chosen at random in [0.1, 1] (W = tard). The latter corresponds to a situation in which the cost of tardiness outweighs the cost of earliness, as is usually the case in real-world circumstances.

The total number of instances is 72 but as it is stated, because the limit of time, this study can only test 36 instances, which have ID equals to 1.

(c) **Encoding**

This study uses operation-based encoding, which is popularly known as was created for the JSSP [19]. A single sequence represents an execution order of the operations on the machines. For the problem of ten jobs and two machines (one of the instances), this sequence gives ten multiplied by two equal to exactly 20 elements (20 operations) in which each job appears exactly two times, with the first-time appearance of each job representing its first operation and so on. In the meantime, this study simply uses the left-shift decoding strategy.

(d) **Results and Analysis**

The result (z, which is the cost value) is now compared with other literature reviews, and the exact method given in the (Table 12) with numbers in bold indicates the best solutions found between models being compared. To see the difference in changes between the results from the proposed model and the other model being compared, the change in percentage is taken into consideration and is calculated by the following equation:

$$\%Change = \frac{zbest_{othermodels} - zbest_{GBA}}{zbest_{othermodels}} \tag{20}$$

According to Ahmadian et al. [17], the best-known solutions are compared with three metaheuristic models and an exact method. The number of best solutions are 9, 9, 30, and 10 from CPLEX, Wang and Li [16], Ahmadian et al. [17] and GBA from this study, respectively. While the exact method solved by CPLEX studio running on a powerful computer configuration has obtained 1 new best solution, GBA has 5. The changes are recorded above.

GBA Compares with Other Methods in Terms of Time (S)

In general, the model is best fit with certain instances, while it is clearly not a good choice for others, for instance (Fig. 9):

Table 12 Computational result comparison with best-known solutions

Instance	Best-known [6]	CPLEX	VNS [16]	VNS [17]	GBA (this study)	New best solution	% Change
Tight-equal-1-10 × 2	416.44	461.96	461.96	461.96	467.66	–	–
Tight-equal-1-10 × 5	689.11	689.11	689.11	689.11	850.69	–	–
Tight-equal-1-10 × 10	1276.23	1412.07	1277.44	1276.23	1409.98	–	–
Loose-equal-1-10 × 2	224.84	224.84	224.84	224.84	224.84	–	–
Loose-equal-1-10 × 5	1719.63	1738.05	1738.05	1719.63	1819.1	–	–
Loose-equal-1-10 × 10	364.39	360.74	365.81	364.39	1022.31	360.74	1.00
Tight-tard-1-10 × 2	179.46	181.1	179.46	179.46	179.46	–	–
Tight-tard-1-10 × 5	379.19	380.82	380	379.19	445.31	–	–
Tight-tard-1-10 × 10	668.14	668.14	746.03	668.14	770.42	–	–
Loose-tard-1-10 × 2	416.44	416.44	416.44	416.44	427.18	-	–
Loose-tard-1-10 × 5	175.08	175.08	175.08	175.08	246	-	–
Loose-tard-1-10 × 10	374.2	375.71	380.26	374.2	615.79	–	–
Tight-equal-1-15 × 2	3344.54	3366.66	3344.54	3344.54	3344.54	–	–
Tight-equal-1-15 × 5	1342.3	1412.71	1345.21	1342.3	1403.88	–	–
Tight-equal-1-15 × 10	6820.87	7388.3	8120.43	6820.87	6997.45	–	–
Loose-equal-1-15 × 2	1041.33	1057.75	1041.33	1041.33	1041.7	–	–
Loose-equal-1-15 × 5	3267.79	3244.19	3321.54	3272.46	3237.99	3237.99	0.91
Loose-equal-1-15 × 10	906.49	903.35	977.43	906.49	1218.74	–	–
Tight-tard-1-15 × 2	790.5	795.63	790.66	790.5	790.5	–	–
Tight-tard-1-15 × 5	1359.18	1384.32	1426.97	1359.18	1378.76	–	–
Tight-tard-1-15 × 10	767.26	782.48	832.08	767.26	1059.03	–	–
Loose-tard-1-15 × 2	654.84	654.84	654.84	654.84	654.84	–	–
Loose-tard-1–15 × 5	1281	1318.55	1315.47	1281	1317.2	–	–

(continued)

Table 12 (continued)

Instance	Best-known [6]	CPLEX	VNS [16]	VNS [17]	GBA (this study)	New best solution	% Change
Loose-tard-1-15 × 10	277	277	296.6	277	521.65	–	–
Tight-equal-1-20 × 2	1936.45	1940.06	1939.11	1936.45	1933.72	1933.72	0.14
Tight-equal-1-20 × 5	2853.31	3013.79	2935.95	2865.95	2975.34	–	–
Tight-equal-1-20 × 10	10,378.8	11,561.17	11,073.1	10,378.8	11,034.65	–	–
Loose-equal-1-20 × 2	2548.95	2565.51	2551.22	2548.95	2547.68	2547.68	0.05%
Loose-equal-1-20 × 5	7524.13	8271.34	7758.02	7524.13	7648.22	–	–
Loose-equal-1-20 × 10	4651.31	5920.05	5476.47	4651.31	5688.17	–	–
Tight-tard-1-20 × 2	1674.6	1685.19	1682.94	1674.6	1671.87	1671.87	0.16
Tight-tard-1-20 × 5	3620.61	3821.29	3669.09	3620.61	3633.81	–	–
Tight-tard-1-20 × 10	4445.59	5183.71	5536.64	4445.59	5016.86	–	–
Loose-tard-1–20 × 2	1204.92	1265.23	1205.59	1206.4	1203.88	1203.88	0.09%
Loose-tard-1-20 × 5	2931.11	3548.14	3517.2	2931.11	2973.53	–	–
Loose-tard-1-20 × 10	4817.32	6494.34	5478.21	4817.32	5480.77	–	–
Number of best solutions		9	9	30	10		

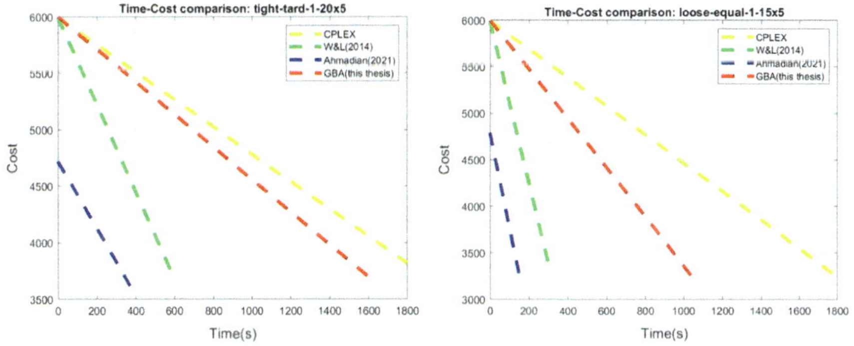

Fig. 9 Time–cost trade-off comparison: tight-tard-1-20 × 5 and loose-equal-1-15 × 5

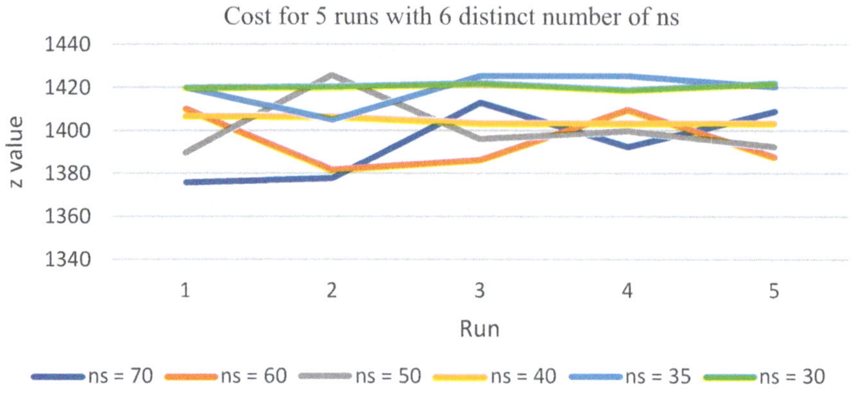

Fig. 10 Cost for 5 runs with 6 distinct ns: tight-equal-1-15 × 5

Table 13 Cost for 5 runs with 6 distinct ns: tight-equal-1-15 × 5

ns	Cost							% Change	
	1	2	3	4	5	Best	Average	Best	Average
70	1376.1	1377.98	1413.22	1392.69	1409.13	1376.1	1393.824	1.98%	0.81%
60	1410.63	1382.02	1386.72	1410.08	1388.12	1382.02	1395.514	1.56%	0.69%
50	1389.64	1425.63	1395.99	1399.73	1392.47	1389.64	1400.692	1.01%	0.32%
40	1407.21	1406.99	1403.88	1403.88	1403.88	1403.88	1405.168	0.00%	0.00%
35	1420.22	1405.24	1425.63	1425.63	1420.55	1405.24	1419.454	– 0.10%	– 1.02%
30	1420.22	1420.55	1422.24	1419.06	1422.24	1419.06	1420.862	– 1.08%	– 1.12%

As the fitness value of the initial population is not recorded in this study, as well as Wang and Li [16], the assumption of the initial cost has been made to be more illustrative. The chart shows that time is not this model's strong point. However, the instance "loose-equal-1-15 × 5" is one of the four instances that have new best solutions, while the GBA takes longer than Ahmadian's model and Wang and Li's model to solve. Nevertheless, when working with small problems, the model performs well, and this is the case of the instance "loose-equal-1-20 × 2". The time needed is only a few seconds, while it gives the best solutions among the four methods.

Sensitivity Analysis

To test the effectiveness of the model, this thesis changes the number of scout bees (ns) to determine whether it creates any differences. The instance "tight-equal-1-15 × 5" is evaluated with five other distinct numbers of scout bees, which are 70, 60, 50, 35 and 30, and the result is recorded by tables and charts (Fig. 10; Table 13):

The more scout bees give a better result (z best and z average as well) since the global search is stronger. However, it is a trade-off that the time needed also increases as it takes more time to perform the search in the reinforced global search.

3 Conclusions

GBA has proven its effectiveness in solving both continuous and combinatorial problems. The goal of optimisation algorithms is to find the optimal solution to a problem while meeting one or more objective functions, which are subject to some sets of constraints [20]. In this chapter, the hybrid method of GA and BA is used to obtain the optimal weights and biases in a continuous problem and to obtain the best job allocation in a combinatorial problem. Based on the training result and analysis, the findings from the experiment are shown below:

- The application of the GA in the BA can optimise the high variance of the random global search in the BA to improve the solution exploration.
- Of the four test datasets, the HGBA has proven to be a successful strategy for training the MLP in terms of biases and weights.
- The robust exploitation and exploration ability of our proposed method contributes to a lower error rate.

In the combinatorial problem, for the TSFSSP, the GBA achieves a lower percentage of error than the Genetic Algorithm and Particle Swarm Optimisation algorithm as well as the convergence rate in a certain short time. However, regarding future work, this is still a promising field to research more, such as addressing the issue of machine breakdown and a lack of buffer space between the two stages and application in the three-stage assembly scheduling problem. In JITJSSP, the computational result obtained shows that the model outperforms in the majority of scenarios compared to the exact method but is not truly a competitive model when compared with two recent studies and best-known solutions. Nevertheless, it still achieves 4 new best solutions among 36 instances tested. For future work, it is crucial to research the fields listed above to improve the model. Some of the suggestions can be listed as follows: decoding strategy (try backwards strategy), local search: try again the above operators but focus on earliness operations, global search (work on another local search operators or tun parameters).

Acknowledgements We greatly acknowledge Professor Duc Truong Pham from the University of Birmingham for the advice and comments on our work on the Bees Algorithm (BA) as well as his continuous support during our research for this work.

References

1. Pham DT, Ghanbarzadeh A, Koc E, Otri S, Rahim S, Zaidi M (2005) The Bees algorithm. Manufacturing Engineering Centre, Cardiff University, UK, Technical Note
2. Pham DT, Castellani M (2009) The bees algorithm: modelling foraging behaviour to solve continuous optimisation problems. Proc Inst Mech Eng C J Mech Eng Sci 223(12):2919–2938
3. Yuce B, Fruggiero F, Packianather MS, Pham DT, Mastrocinque E, Lambiase A, Fera M (2017) Hybrid genetic Bees algorithm applied to single machine scheduling with earliness and tardiness penalties. Comput Ind Eng 113:842–858
4. Anh TTM, Tran DV (2022) Hybrid genetic Bees algorithm in multilayer perceptron optimization. In: Proceedings of the 3rd International conference on data science and application (ICDSA 2022). India
5. Packianather MS, Yuce B, Mastrocinque E, Fruggiero F, Pham DT, Lambiase A (2014) Novel genetic Bees algorithm applied to single machine scheduling problem. In: 2014 World automation congress (WAC). IEEE, pp 906–911
6. McCulloch WS, Pitts W (1943) A logical calculus of the ideas immanent in nervous activity. Bull Math Biophys 5(4):115–133
7. Kennedy J, Eberhart R (1995) Particle swarm optimization. In: Proceedings of ICNN'95-international conference on neural networks, vol 4. IEEE, pp 1942–1948
8. Potts CN, Sevast'janov SV, Strusevich VA, Van Wassenhove LN, Zwaneveld CM (1995) The two-stage assembly scheduling problem: complexity and approximation. Oper Res 43(2):346–355
9. Lee CY, Cheng TCE, Lin BMT (1993) Minimizing the makespan in the 3-machine assembly-type flowshop scheduling problem. Manage Sci 39(5):616–625
10. Allahverdi A, Al-Anzi FS (2008) The two-stage assembly flowshop scheduling problem with bicriteria of makespan and mean completion time. Int J Adv Manuf Technol 37:166–177
11. Tozkapan A, Kirca O, Chung CS (2003) A branch and bound algorithm to minimize the total weighted flowtime for the two-stage assembly scheduling problem. Comput Oper Res 30(2):309–320
12. Madronero MD, Peidro D, Vasant P (2010) Vendor selection problem by using an interactive fuzzy multi-objective approach with modified S-curve membership functions. Comput Math Appl 60(4):1038–1048
13. Tian Y, Liu D, Yuan D, Wang K (2013) A discrete PSO for twostage assembly scheduling problem. Int J Adv Manuf Technol 66:482–499
14. Baptiste P, Flamini M, Sourd F (2008) Lagrangian bounds for just-in-time job-shop scheduling. Comput Oper Res 35(3):906–915
15. Santos AGD, Araujo RP, Arroyo JE (2010) A combination of evolutionary algorithm, mathematical programming, and a new local search procedure for the just-in-time job-shop scheduling problem. In: International conference on learning and intelligent optimization, Springer, Berlin, Heidelberg, pp 10–24
16. Wang S, Li Y (2014) Variable neighbourhood search and mathematical programming for just-in-time job-shop scheduling problem. Math Probl Eng
17. Ahmadian MM, Salehipour A, Cheng TCE (2021) A meta-heuristic to solve the just-in-time job-shop scheduling problem. Eur J Oper Res 288(1):14–29
18. Ismail AH, Hartono N, Zeybek S, Pham DT (2020) Using the Bees algorithm to solve combinatorial optimisation problems for TSPLIB. In: IOP conference series: materials science and engineering, vol 847, no 1. IOP Publishing, p 012027
19. Gen M, Tsujimura Y, Kubota E (1994) Solving job-shop scheduling problems by genetic algorithm. In: Proceedings of IEEE international conference on systems, man and cybernetics, vol 2, pp 1577–1582. IEEE
20. Yuce B, Packianather MS, Mastrocinque E, Pham DT, Lambiase A (2013) Honeybees inspired optimization method: the bees algorithm. Insects 4(4):646–662

Optimisation of Surface Roughness in 3D Printing Using the Bees Algorithm

Shafie Kamaruddin, Arman Hilmi Ridzuan, and Nor Aiman Sukindar

Abstract Additive manufacturing (AM) is renowned for its capability to produce parts that are low-cost and have less manufacturing time. One of the main challenges in this additive manufacturing technology is selecting proper input process parameters to achieve good quality of the 3D printed model. The focus of this study is to determine the optimum input parameter of the 3D printer using the Bees Algorithm (BA). This study uses the Bees Algorithm to predict the best combination parameters to optimise the surface roughness of parts printed by a fused deposition modelling (FDM) machine. The predicted results are compared with the experimental 3D model sample and previous findings of other optimisation methods. Comparative analysis between predicted and actual surface roughness measurements showed good agreement with differences of less than 2%, indicating a significant prediction method. The result also shows that the Bees Algorithm found a better combination of parameters compared to other algorithms. This research provides another alternative optimisation approach for industries that utilise 3D printing.

Keywords Additive manufacturing · Bees Algorithm · Intelligent optimisation

1 Introduction

Additive manufacturing, also known as 3D printing, is a process of making complex parts from CAD models. The 3D printed model is made by using additive processes by creating layer-by-layer materials until the desired shape model is created [1]. Additive manufacturing is the opposite of subtractive manufacturing, which removes materials by cutting or machining processes. The innovation of 3D printing has helped and

S. Kamaruddin (✉) · A. H. Ridzuan · N. A. Sukindar
Department of Manufacturing and Materials Engineering, Faculty of Engineering, International Islamic University Malaysia, 53100 Gombak, Malaysia
e-mail: shafie@iium.edu.my

N. A. Sukindar
e-mail: noraimansukindar@iium.edu.my

D. T. Pham and N. Hartono (eds.), *Intelligent Engineering Optimisation with the Bees Algorithm*, Springer Series in Advanced Manufacturing,
https://doi.org/10.1007/978-3-031-64936-3_9

become very useful for the industry, especially given the need to create a prototype for study purposes. This would help manufacturing industries or companies to cut costs when developing a new technology, as 3D printing leads to a reduction in waste materials. Apart from that, this process helps to reduce manufacturing time because manufacturers can easily make complex models, as it requires less effort from the manufacturers. It also provides benefits to both the industry and society by either producing good quality products or reducing the price of the products [2].

Fused deposition modelling (FDM) is a well-known 3D printing technology that is widely used in industry. A complex part is fabricated by depositing thermoplastic materials in the form of thin layers that move down a distance equal to the thickness of one layer after depositing a layer. The FDM machine is the most commonly used because it can withstand heat, chemicals, and mechanical stress, which makes it an ideal material for prototypes [3].

The Bees Algorithm is an optimisation algorithm mimicking the natural behaviour of honeybees in nature [4]. In recent years, researchers have investigated a variety of approaches to optimise the parameters of 3D printers, but to the best of the authors' knowledge, no researchers use the Bees Algorithm as an optimisation algorithm. Thus, the main purpose of this study is to apply the Bees Algorithm for optimisation of the surface roughness of a 3D model printed using an FDM machine. This study shows that the Bees Algorithm has the potential to be an alternative optimisation method. The best combination of parameters found by the Bees Algorithm is compared with the 3D printed model's surface roughness and the results from previous studies using other methods. There are four main parameters in the FDM machine: layer height, print speed, print temperature and outer shell speed. This study focuses on the study of parameters affecting the surface roughness in 3D printing and finding the optimal setup.

2 The Bees Algorithm

The Bees Algorithm is an optimisation algorithm that is influenced by the food foraging behaviour of honeybees to find the optimal solution. This algorithm has parameters to be set at the beginning [5]. The algorithm begins by sending several scout bees randomly across the search space. Then, the best sites are chosen based on the fitness value associated with them. The places they visited are designated "neighbourhood search sites". In each selected best site, several recruited bees are sent to these sites. The recruited bee with the highest fitness value from each patch is chosen to create a new population in the next iteration. The remaining bees in the colony are sent randomly across the search space to search for new possible solutions. The colony's new population consists of two parts at the end of each iteration, which are representatives from the chosen patches and scout bees allocated to perform random searches. These steps are repeated until the stopping condition is satisfied. The basic steps of the Bees Algorithm are as follows [6–8]:

1. Random Initialisation. Evaluate sites visited by scout bees.
2. While (stopping criterion not met)//
3. Rank the fitness values.
4. Send recruited bees to selected sites. The best recruited bee from each site is selected to form a new population.
5. Send the remaining unselected scout bees randomly across the search space.
6. New population.
7. End while.

3 Methodology

3.1 Parameter Initialisation

The parameters in this study are based on an article titled "Optimisation of Surface Roughness in an FDM 3D Printer Using Response Surface Methodology, Particle Swarm Optimisation, and Symbiotic Organism Search Algorithm" [9]. In addition to this article, a considerable amount of literature has also been published on optimising the parameters of a 3D Printer using optimisation algorithms [10–13]. The parameters are separated into two groups: fixed parameters and changeable parameters. The constant parameters are the FDM machine's specifications, which are listed in Table 1. The variable parameters are the printed model's layer thickness, print speed, print temperature and outer shell speed, as shown in Table 2. These settings are limited by the capabilities of the FDM machine.

Table 1 Specification of the Ender-3 max 3D printer

Technical specification	Ender-3 max
Layer thickness (mm)	0.1–0.4
Printing speed (mm/s)	30–60 (recommended), $\leq$ 180
Print precision (mm)	$\pm$ 0.1
Build size (mm3)	300 $\times$ 300 $\times$ 340
Material type	PLA, ABS, PETG, TPU, TPE, Wood-fil
Material diameter (mm)	1.75
Nozzle diameter (mm)	0.4
Maximum hot end temperature (°C)	260
Maximum bed temperature (°C)	100
Bed levelling	Manual
Extrusion system	Bowden-style extrusion

Table 2 Range of parameter settings for the 3D Printer

Parameters	Range
Layer height (mm)	0.18–0.30
Print speed (mm/s)	36–60
Print temperature (°C)	185–205
Outer shell speed (mm/s)	29–40

3.2 Implementation of the Bees Algorithm

The Bees Algorithm was run using R Software. It is a software for statistical computing developed for a statistical computing and graphics programming language as well as a free software environment. In the algorithm, the equation of the regression model for surface roughness (R_a) presented in Eq. (1) was used as the fitness function [9].

$$\begin{aligned} R_a = & -86.28432 + 43.75348\text{A} - 0.86231\text{B} + 2.75590\text{D} \\ & + 3.84465\text{AB} - 0.89363\text{AC} - 7.83291\text{AD} + 3.35859\text{E} \\ & - 003\text{BC} + 5.40956\text{E} - 003\text{BD} - 0.015738\text{CD} - 36.51104\text{A2} \\ & - 6.13478\text{E} - 004\text{C2} - 0.014970\text{ABC} - 0.022230\text{ABD} + 0.046619\text{ACD} \end{aligned} \quad (1)$$

In this Eq. (1), *A* represents the printed model's layer thickness, *B* is the print speed, *C* is the print temperature and *D* is the outer shell speed. The Bees Algorithm's sequence or flow started with a set of random parameters within the FDM machine's range. The surface roughness values are determined by computing the Ra value. Then, the parameters are ranked based on their surface roughness values. Several combinations of parameters with a low value of R_a are selected as the best sites. Recruited bees are sent to these selected best sites (local search). Next, a representative of recruited bees from each selected site is selected to form a new population. The remaining unselected parameters are set randomly again within the FDM machine's range (global search). These parameters from the local search and global search are merged to form a new population. This iteration process continues until the stopping condition is satisfied. As shown in Table 3, there are eight tunable parameters to run the Bees Algorithm. The parameters of the Bees Algorithm are the number of scout bees (*ns*), the number of elite sites (*ee*), the number of recruited bees for elite sites (*nep*), neighbourhood size (*ngh*), the number of sites selected for exploitation (*m*), the number of recruited bees for the remaining selected sites (*nsp*), iteration number (*it*), and stagnation limit (*stlim*). These parameters are important in finding the optimal combination of parameters [14, 15].

Table 3 Parameters of the Bees Algorithm

Parameters	Value
Scout Bees (ns)	100
Number of elite sites (ee)	5
Recruited Bees for elite site/s (nep)	3
Neighbourhood size (ngh)	0.1
Number of best sites (m)	15
Recruited Bees for remaining best sites (nsp)	2
Iteration number (it)	750
Stagnation limit (stlim)	5

3.3 Experiment Setup

A set of experiments was performed to verify the predicted R_a values found by the Bees Algorithm. As illustrated in Fig. 1, *SolidWorks* software was used to create the CAD model sample with dimensions of 3.2 mm (height) × 12.7 mm (width) × 125 mm (length). *Ultimaker Cura* software was used to slice the CAD model file into layers and produce printer-specific g-code from the CAD model sample. The software was used to adjust the layer height, print speed, print temperature, and outer shell speed. The Bees algorithm found three sets of combination parameters (optimal solutions), as illustrated in Table 4. Thus, this combination of parameters, as in Table 4, was used to print the CAD model sample.

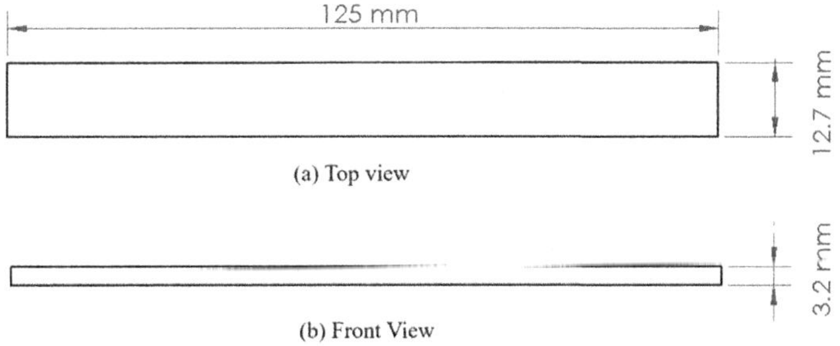

Fig. 1 CAD design specimen

Table 4 Parameters obtained by the Bees Algorithm

	Layer height (mm)	Print speed (mm/s)	Print temperature (°C)	Outer shell speed (mm/s)
Sample 1	0.18	36.02	204.93	39.92
Sample 2	0.18	38.12	203.49	37.98
Sample 3	0.18	39.88	204.52	39.93

Manual bed levelling is needed for the *Ender-3 Max 3D* printer, and bed levelling is necessary to keep the bed and nozzle in the same position throughout the 3D printing process. The following were the procedures for levelling the printing bed:

1. The 3D printer's bed and nozzle were heated, and then the print head was set to autohome.
2. The stepper motors were disabled, and the print head and the bed were moved manually to one corner of the print bed.
3. A piece of paper was slid between the print head and bed.
4. The bed corners were raised or lowered by turning the wheels to a point where the nozzle was lightly pressing on the paper.
5. The paper should be able to slightly move or wiggle with the nozzle on top.
6. Steps 3–5 were repeated for each corner and the middle of the bed.

The bed levelling procedure was performed before printing each sample. The printer bed should be fully levelled and calibrated so that when the sample is being printed, the filament would extrude evenly across the entire build surface. Figure 2 shows the calibration process of the printing bed. PLA, a biodegradable thermoplastic material obtained from natural sources, was used as the material to print the model. Table 5 shows the mechanical properties of the PLA.

The 3D model was printed to verify the predicted surface roughness using the optimal parameter settings (layer height, print speed, print temperature, and outer shell speed) found by the Bees algorithm. The surface roughness of the experimental 3D model was measured using an *Optical 3D measurement machine*. The samples were measured at three distinct places, as indicated in Fig. 3, on the left, middle, and right sides of the samples. This was done to ensure that the measurement of surface

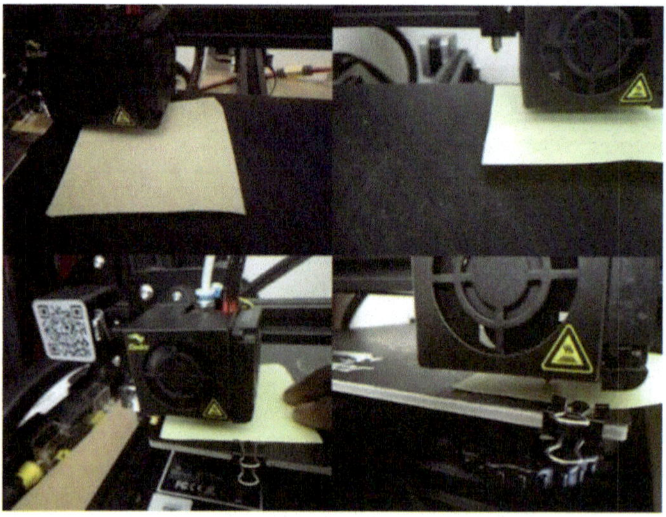

Fig. 2 Bed levelling using a piece of paper

Table 5 Typical range of mechanical properties of PLA

Mechanical properties	PLA
Tensile strength (MPa)	15.5–72.2
Tensile modulus (GPa)	2.020–3.550
Elongation break (%)	0.5–9.2
Flexural strength (MPa)	52.0–115.1
Flexural modulus (GPa)	2.392–4.930

roughness was accurate. The average measured R_a value of the samples is recorded in Table 6. The effect of input parameters such as layer thickness, print speed, print temperatures, and outer shell speed on the output response and surface roughness was analysed. The R_a value obtained by the Bees algorithm was compared with other predicted results in the literature and the experimental R_a value. The results comparison verified the ability of the Bees Algorithm to provide a better alternative for industries that utilise 3D printers to produce quality products.

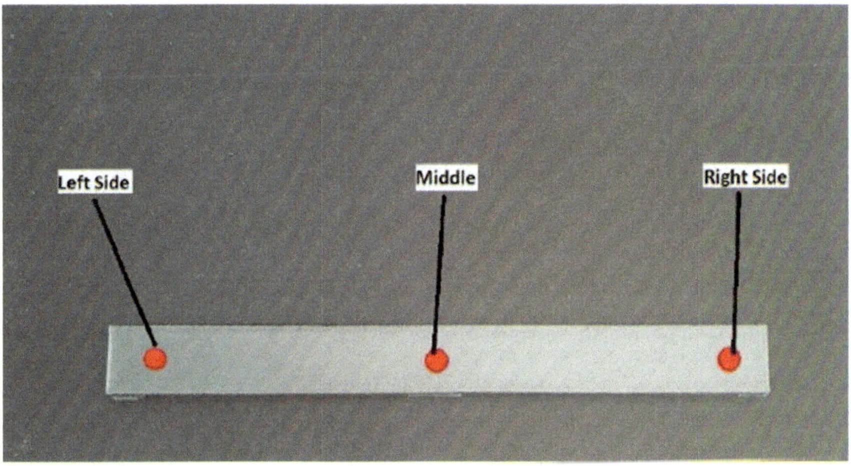

Fig. 3 Points on samples where R_a measured

Table 6 Measurement of surface roughness on printed samples

Sample	Left (μm)	Middle (μm)	Right (μm)	Average (μm)
Sample 1	2.330	2.339	2.376	2.348
Sample 2	2.399	2.371	2.538	2.436
Sample 3	2.687	2.687	2.708	2.694

4 Results and Discussion

4.1 Predicted Results Using the Bees Algorithm

The optimisation was performed by applying a range of variable parameter combinations indicated in Table 2. The predicted parameters obtained by the Bees Algorithm are recorded in Table 7, where the optimal value of surface roughness (R_a) is 2.3073 μm. The combination of parameters for this surface roughness is 0.180313 mm layer thickness, 36.212343 mm/s print speed, 204.982997 °C print temperature, and 39.957314 mm/s outer shell speed.

The convergence of the search progress in Fig. 4 shows that the optimal value of fitness or surface roughness in this optimisation is 2.3073 μm. The search progress started to converge between 6300 and 6800 function evaluations or iterations. After that, the Bees Algorithm was unable to find a better surface roughness value of less than 2.3073 μm. Based on this observation, the Bees Algorithm has successfully found the optimal solution.

Table 7 Optimal parameters found by the Bees Algorithm

Parameter	Value
Layer height, mm	0.18
Print speed, mm/s	36.21
Temperature, °C	204.98
Outer shell speed, mm/s	39.96
Predicted surface roughness, R_a	2.3073

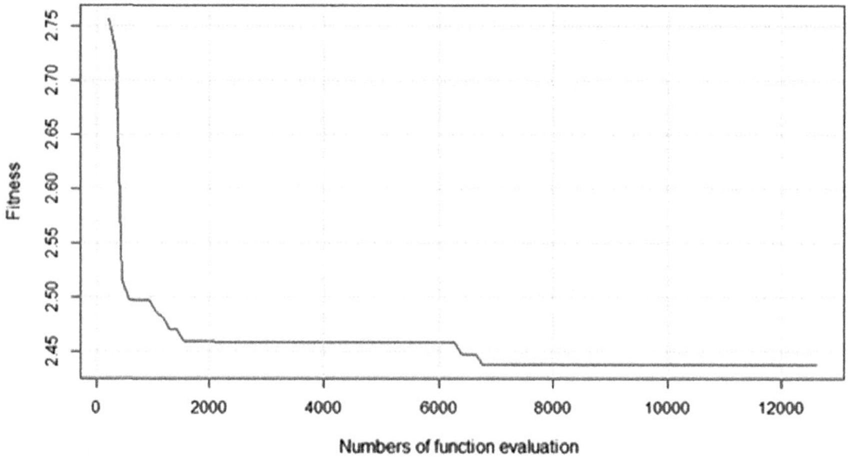

Fig. 4 Convergence graph of the Bees Algorithm for surface roughness

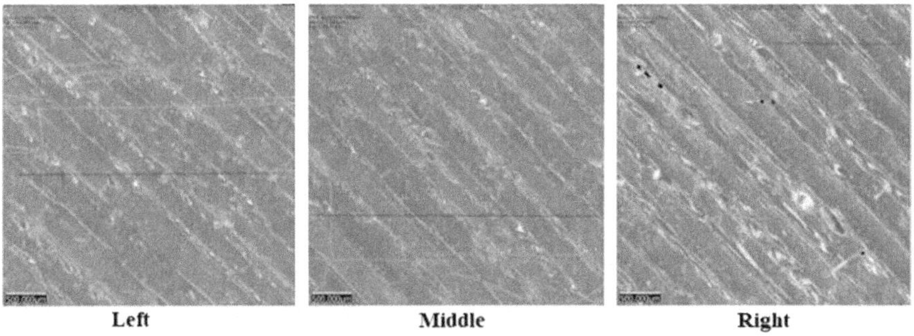

Fig. 5 Image of sample 1 Taken using optical 3D measurement machine

4.2 Experimental Validation Test

The experimental confirmation test was performed to validate the results of the Bees Algorithm approach in finding the best FDM process parameters. Three samples were printed using the best process parameters predicted by the Bees Algorithm, as shown in Table 4, and the surface roughness or the values of R_a were measured. Table 6 shows all the R_a values measured at three different points for each of the samples, and the average values of R_a were calculated.

The image surface of sample 1 projected using *Optical 3D measurement machine* is shown in Fig. 5. This figure shows that the left side of sample 1 has better surface roughness, followed by the middle and right sides of the sample. The right side of the sample has more small craters and holes than the left side. The variation in surface roughness on this sample might be due to a poorly calibrated 3D printer. As a result, a slight change in the process parameters may alter the quality of the sample's surface roughness during the printing process. The surface roughness of samples 2 and 3 is comparable to that of sample 1.

4.3 Result Comparison

The best solution found by the Bees Algorithm is compared with the results of previous studies using response surface methodology (RSM), the particle swarm optimisation (PSO) algorithm and symbiotic organism search (SOS) [9]. From Table 8, it is observed that the predicted surface roughness of the Bees Algorithm is higher than that of the other optimisation methods. However, the experimental surface roughness validation of the Bees Algorithm is 2.348 μm compared to the predicted value, which is 2.307 μm. The figure shows that the error between the experimental and predicted values is 1.73%.

The values of the process parameters of the Bees Algorithm and RSM are similar, but when compared to the PSO algorithm and SOS optimisation method, the most

Table 8 Results comparison of the Bees Algorithm with other algorithms

Parameter	BA	RSM	PSO	SOS
Layer height, mm	0.18	0.18	0.15	0.10
Print speed, mm/s	36.21	36.00	37.77	37.84
Temperature, °C	204.98	205.00	202.09	192.7
Outer shell speed, mm/s	39.96	40.00	39.69	38.59
Predicted surface roughness, R_a	2.307	2.295	**2.176**	2.18
Experimental surface roughness (μm)	2.348	2.444	2.236	**2.229**
Error (%)	**1.73**	6.10	2.68	2.21

significant differences are the print speed and print temperature. The PSO algorithm and SOS optimisation method predicted faster print speeds, which were 37.77 mm/s and 37.85 mm/s, respectively. The Bees Algorithm predicted only a print speed of 36.21 mm/s. The print temperatures found by the Bees Algorithm and PSO algorithm are 204.98 °C and 202.09 °C, respectively. Meanwhile, the print temperature of the SOS algorithm is 92.7 °C. Thus, these two process parameters are the main factors contributing to each algorithm having a different predicted surface roughness.

In this study, the optimisation of FDM machine parameters to optimise surface roughness was implemented using the Bees Algorithm approach. Overall, the Bees Algorithm shows good performance in terms of finding the best combination of parameters (optimise R_a). However, comparing the value of R_a found by the Bees Algorithm with previous studies using other methods shows that it is slightly higher. Nevertheless, the results are considered acceptable, as the response target range for surface roughness in a 3D printer is between 2.46 and 22.48 μm [16]. It is suggested to fine-tune the parameters of the Bees Algorithm in Table 6 to find better process parameters with lower surface roughness values.

The main finding of this study is that the comparison of the predicted surface roughness found by the Bees Algorithm with the experimental surface roughness is considered the most accurate compared to other methods in the literature. This study found the most accurate prediction of surface roughness, as shown in Table 8. This is significant for manufacturing companies that utilise 3D printers to improve the quality of printed products and save manufacturing costs by reducing the number of printed models.

5 Conclusion

In conclusion, the purpose of the current study was to determine the optimal parameters of the FDM using the Bees Algorithm. This study has identified that the Bees Algorithm obtained similar predicted surface roughness values with other algorithms. From the experimental confirmation test, it is observed that the Bees Algorithm produced the lowest error in predicting the surface roughness, which is 1.73%. These findings suggest that in general, it offers an alternative approach to finding the optimal parameters of the FDM. Further research should be undertaken to explore the capability of the Bees Algorithm to predict other outputs of 3D Printer, such as mechanical strength, build time, and dimensional accuracy.

Acknowledgements This study was supported by FRGS grant No. FRGS/1/2019/TK03/UIAM/02/3 from the Ministry of Higher Education Malaysia (MOHE). The authors are also grateful to the International Islamic University of Malaysia (IIUM), which made this study possible.

References

1. Boschetto A, Giordano V, Veniali F (2013) Surface roughness prediction in fused deposition modelling by neural networks. Int J Adv Manuf Technol 67:2727–2742
2. Wong KV, Hernandez A (2012) A review of additive manufacturing. Int Sch Res Not 2012:10. Article ID 208760
3. Mankar S, Kale C, Kanchan J (2019) Technology. 3D printing technology-a computer aided design-a review. Res J Sci Tech 11(3):217–224
4. Pham DT, Ghanbarzadeh A, Koc E, Otri S, Rahim S, Zaidi M (2005) The Bees algorithm. Manufacturing Engineering Centre, Cardiff University, UK, Technical Note
5. Pham DT, Ghanbarzadeh A, Koç E, Otri S, Rahim S, Zaidi M (2006) The Bees Algorithm—a novel tool for complex optimisation problems. Intell Prod Mach Syst 454–459
6. Pham DT, Castellani M, Le Thi H (2014) Nature-inspired intelligent optimisation using the Bees algorithm. Trans Comput Intell XIII 38–69
7. Pham DT, Castellani M (2015) A comparative study of the Bees algorithm as a tool for function optimisation. Cogent Eng 2:1091540
8. Pham DT, Castellani M (2019) The Bees algorithm: modelling foraging behaviour to solve continuous optimization problems. Proc Inst Mech Eng C J Mech Eng Sci 223(12):2919–2938
9. Saad MS, Nor AM, Baharudin ME et al (2019) Optimization of surface roughness in FDM 3D printer using response surface methodology, particle swarm optimization, and symbiotic organism search algorithms. Int J Adv Manuf Technol 105:5121–5137
10. Rao RV, Rai DP (2015) Optimization of fused deposition modeling process using teaching-learning-based optimization algorithm. Eng Sci Technol Int J 19(1):587–603
11. Deshwal S, Kumar A, Chhabra D (2020) Exercising hybrid statistical tools GA-RSM, GA-ANN and GA-ANFIS to optimize FDM process parameters for tensile strength improvement. CIRP J Manuf Sci Technol 31:189–199
12. Raju M, Gupta MK, Bhanot N et al (2019) A hybrid PSO–BFO evolutionary algorithm for optimization of fused deposition modelling process parameters. J Intell Manuf 30:2743–2758
13. Rayegani F, Onwubolu GC (2014) Fused deposition modelling (FDM) process parameter prediction and optimization using group method for data handling (GMDH) and differential evolution (DE). Int J Adv Manuf Technol 73:509–519
14. Kamaruddin S, Azmi N, Sukindar NA (2022) Assembly sequence optimization using the Bees algorithm. In: Enabling Industry 4.0 through advances in mechatronics. Lecture notes in electrical engineering, vol 900. Springer, Singapore
15. Kamaruddin S, Rosdi MN, Aiman Sukindar N (2021) Optimization of drilling path using the Bees algorithm. Manuf Technol 21(6): 788–792
16. Pérez M, Medina-Sánchez G, García-Collado A, Gupta M, Carou D (2018) Surface quality enhancement of fused deposition modelling (FDM) printed samples based on the selection of critical printing parameters. Materials 11(8):1382

The Bees Algorithm for Robotics-Enabled Collaborative Manufacturing

Wenjun Xu, Hang Yang, Zhenrui Ji, Zhihao Liu, and Jiayi Liu

Abstract Robotics-enabled collaborative manufacturing is vital to improving the efficiency and flexibility of industrial manufacturing processes and realising the digitalisation and intelligentisation of industry. In recent years, enabled by intelligent optimisation algorithms, robotics-enabled collaborative manufacturing approaches have been steadily developed as promising solutions to support industries. However, each robotics-enabled collaborative manufacturing approach has some optimisation problems, limiting its applicability in practice. To address this issue, in this chapter, the Bees Algorithm and robotics-enabled collaborative manufacturing are integrated as an effective approach. In this approach, the efficiency and flexibility of robotics-enabled collaborative manufacturing are significantly improved. Robotics-enabled collaborative manufacturing is composed of robotic collaborative manufacturing and human–robot collaborative manufacturing. In terms of robotic collaborative manufacturing, the improved multiobjective discrete Bees Algorithm and the improved discrete Bees Pareto algorithm are utilised to solve the robotic disassembly line balancing optimisation problem and multiobjective robotic collaborative manufacturing service aggregation optimal selection problem. In terms of human–robot collaborative manufacturing, the Pareto-based modified discrete Bees Algorithm is used to solve the task sequence planning optimisation problem. Afterwards, two case studies are presented to show the performance of the Bees Algorithm in

W. Xu · H. Yang · Z. Ji · Z. Liu · J. Liu (✉)
School of Information Engineering, Wuhan University of Technology, Wuhan 430070, China
e-mail: jyliu@whut.edu.cn

W. Xu
e-mail: xuwenjun@whut.edu.cn

H. Yang
e-mail: h_yang@whut.edu.cn

Z. Ji
e-mail: jizhenrui@whut.edu.cn

Z. Liu
e-mail: zhhliu@whut.edu.cn

D. T. Pham and N. Hartono (eds.), *Intelligent Engineering Optimisation with the Bees Algorithm*, Springer Series in Advanced Manufacturing,
https://doi.org/10.1007/978-3-031-64936-3_10

robotics-enabled collaborative manufacturing. The results demonstrate that the Bees Algorithm is superior to some other comparative approaches.

Keywords Bees algorithm · Robotics-enabled collaborative manufacturing · Robotics collaborative manufacturing · Human–robot collaborative manufacturing

1 Introduction

Collaborative manufacturing has received much attention in recent years due to its high flexibility and high efficiency [1]. As the key executors of collaborative manufacturing, industrial robots are advanced manufacturing equipment that have been widely used in the fields of assembly, welding and so on [2]. Recently, robotics-enabled collaborative manufacturing has been widely explored and mainly includes robotic collaborative manufacturing (RC-Mfg) [3] and human–robot collaborative manufacturing (HRC-Mfg) [4]. For RC-Mfg, several industrial robots collaboratively execute the given manufacturing tasks. HRC-Mfg covers the shortcomings of the full automation manufacturing paradigm by including human operators. It has more flexibility than RC-Mfg [5].

To promote the digitalisation and intelligentisation development of industrial manufacturing processes, in terms of perception, modelling, optimisation and decision making, much research has been finished in robotics-enabled collaborative manufacturing [6–9]. Pérez et al. built a digital twin for multirobot manufacturing cell commissioning to maximise the manufacturing quality and efficiency [6]. Shen et al. studied additive manufacturing using multirobot collaboration technology. In this research, the printing layer of additive manufacturing was divided into several safe areas. Multirobot collaborative printing experiments were completed to verify the proposed method [7]. Wang et al. overviewed the symbiotic human–robot collaborative assembly method and its enabling technologies [8]. After that, to undertake hazard manufacturing tasks, remote human–robot collaboration (HRC) was investigated by combining and verifying case studies that contain remote robot control systems and display systems [9].

The Bees Algorithm (BA) is inspired by the foraging behaviour of honey bees and can be utilised to solve many engineering problems [10]. It mainly contains elite bees and selected bees, which are used to search the local neighborhood, and scout bees, which are used to search the global space and can be utilised to solve continuous [11] and combinatorial optimisation problems [3]. Together with the other metaheuristic optimisation algorithms, its performance has also been verified in various application areas [12].

For robotics-enabled collaborative manufacturing, many research problems need to be solved, such as multirobot manufacturing optimisation, decision-making problems in HRC-Mfg, and robotics-related manufacturing service reconfigurations. These problems could also be solved by BA. In recent years, the BA has been utilised to solve robot disassembly line balancing problems [13], cross-layer optimisation problems [14], manufacturing service optimisation [12] and so on.

In this chapter, our research work on robotics-enabled collaborative manufacturing using BA is summarised. In Sect. 2, the related work of robotics-enabled collaborative manufacturing is reviewed. Afterwards, the framework of robotics-enabled collaborative manufacturing using BA is proposed in Sect. 3, and its enabling technologies are described in Sect. 4. Two case studies are carried out in Sect. 5, and a discussion is given in Sect. 6. Finally, Sect. 7 concludes this chapter.

2 Related Work

Much research has been conducted in the field of RC-Mfg. Towards the assembly problem for industrial applications, multirobots, which include autonomous mobile platforms, industrial robots and dexterous robots, collaboratively work to complete assembly operations [15]. To optimise the energy consumption of a multirobot system, a convex mixed integer nonlinear optimisation algorithm is utilised to solve the energy optimisation problem by choosing the optimal parameters [16]. To solve the coverage path planning problem of multirobot systems, Kapanoglu et al. utilised a pattern-based genetic algorithm to find the optimal sequence for each robot and then optimise the coverage completion time. P3-DX mobile robots were also used to verify the proposed method [17]. To evaluate the performance of multirobot systems, Boschetti et al. developed a simulation environment to simulate the assembly process and evaluate the degree of collaboration for multirobot collaborative assembly systems [18]. To integrate physical industrial robots with production sequence planning methods, Liu et al. developed a service platform for robotic disassembly planning. In this service platform, the disassembly command could be sent to the physical industrial robot, the status of the physical industrial robot could also be sent to the service platform, and then the remote monitoring could be realised [19].

There are many publications in the field of HRC-Mfg. Rahman proposed the cognitive cyber-physical system for HRC-Mfg, and a case study of assembling components under the HRC-Mfg process was carried out to verify the performance [20]. For the human–robot collaborative disassembly process, the disassembly task assignment method was proposed, and then the HRC disassembly sequence planning problem was solved. A case study based on a simplified computer was also carried out [21]. Towards the aerospace manufacturing industry, to maximise the productivity and the level of safety, Pérez et al. studied the symbiotic HRC approach and used the aircraft rib manufacturing process to verify the proposed method [22]. Towards a proactive HRC-Mfg process, Li et al. proposed a foreseeable cognitive manufacturing paradigm that includes intercollaboration cognition, spatiotemporal cooperation prediction and self-organising teamwork [23]. In addition, Liu et al. studied the collision-free method for the HRC-Mfg process considering context awareness.

The human pose recognition algorithm based on the transfer learning algorithm was adapted to verify the proposed method [24].

In addition, BA has been utilised to solve many engineering problems. Zeybek et al. used BA to train a recurrent neural network for sentiment classification. Its performance was also verified together with the long short-term memory method and recursive neural tensor network [25]. Laili et al. solved the robotic disassembly replanning problem using a two-pointer detection strategy and BA. This method could generate the optimal solution of given productions online [26]. Liu et al. studied robotic disassembly sequence planning using the enhanced discrete Bees Algorithm (DBA). In this research, case studies based on gear pumps were carried out [27]. Apart from the manufacturing fields, BA could also be applied to solve the optimisation problem of PID controllers under earthquake excitation [28] and the player's frame rate allocation problem in a cloud gaming environment [29].

From the aforementioned analysis, it is obvious that BA could be utilised to solve many engineering problems, and there are also many publications in the fields of robotics-enabled collaborative manufacturing. In this chapter, we mainly focus on the robotics-enabled collaborative manufacturing solved by BA.

3 Framework

As shown in the proposed framework in Fig. 1, robotics-enabled collaborative manufacturing includes two branches, namely, robotic collaborative manufacturing (RC-Mfg) and human–robot collaborative manufacturing (HRC-Mfg). RC-Mfg aims to integrate different types of robots with different manufacturing capabilities into an efficient manufacturing system to accomplish production tasks involving multi-type processes with various requirements. To achieve efficient collaboration among multiple robots, the BA-based cross-layer optimisation method is developed to address problems including energy efficiency optimisation, task scheduling, manufacturing service optimisation, etc. As a new production paradigm that allows robots to work alongside humans to complete the common task, HRC-Mfg requires collaborative robots to have adaptability to the dynamic environment with uncertainties, especially uncertain human factors. To address the task sequence planning problem in such HRC-Mfg, an HRC task model is built based on human fatigue and is used to replan the task sequence using BA. Towards the above optimisation problems in RC-Mfg and HRC-Mfg, some novel components of BA are also proposed, such as the more efficient searching strategy, Bees selection strategy, multiple objective optimisation, and combination with other algorithms.

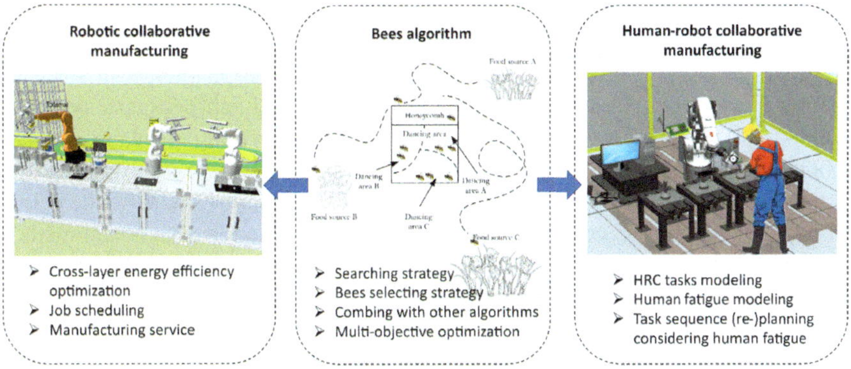

Fig. 1 Framework of robotics-enabled collaborative manufacturing using BA

4 Enabling Technologies

4.1 Robotic Collaborative Manufacturing

RC-Mfg enables more flexible and efficient manufacturing solutions and enhances the adaptability of manufacturing processes. However, such solutions require manufacturing processes with advanced industrial sensing and control technologies. To address this issue, BA is utilised to solve RC-Mfg optimisation problems. Common RC-Mfg optimisation problems can be divided into three layers: the manufacturing cell layer, production process layer, and manufacturing service layer [30, 31]. BA can not only efficiently solve complex optimisation problems in a single layer of RC-Mfg but also solve complex cross-layer optimisation problems [14] (Fig. 2).

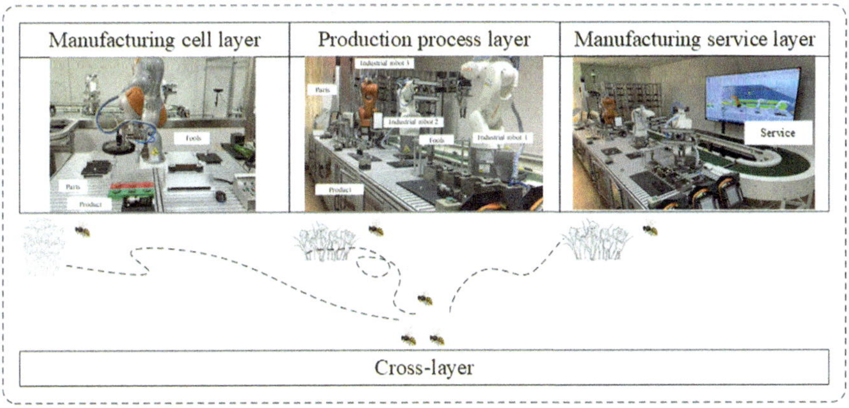

Fig. 2 Framework of RC-Mfg using BA

(1) Production process layer

In the production process layer, successful production planning can solve the robotic disassembly line balancing problem (RDLBP) and improve capacity utilisation [32, 33]. The major goal of RDLBP is to ensure the workload balance between robotic workstations [34]. However, it is complex and costly to find the optimal solution of RDLBP of an end-of-life product (EoLP) with n parts [35]. With the help of an improved space precedence matrix and robotic workstation assignment method, the improved multiobjective discrete Bees Algorithm (IMODBA) is presented to solve RDLBP [13]. It mainly contains three key components, namely, (1) RC-Mfg disassembly modelling for EoLP, (2) multiobjective formulation of RDLBP, and (3) IMODBA for robotic disassembly line balancing solution optimisation.

Many methods are used to build disassembly models for an EoLP [36–38]. To provide a feasible disassembly sequence and disassembly direction for industrial robots, in this part, the space interference matrix method (SIMM) [38] is utilised to generate the disassembly sequence and disassembly direction of an EoLP. Six interference matrices and the feasible disassembly sequence of each part can be obtained from SIMM.

RDLBP is an optimisation problem of how to evenly assign the workload to different robotic workstations. The multiobjective RDLBP consists of the following three objectives: (f_1) minimise the number of robotic workstations; (f_2) balance the workload of robotic workstations; and (f_3) disassemble high-value parts as early as possible [13]. Among them, (f_3) is the most important index in RDLBP [39].

Then, IMODBA is used to minimise three objectives f_1, f_2, and f_3. The specific workflow of IMODBA for optimising the RDLBP is shown in Fig. 3. By adding crowding-distance computation and improving the neighborhood search strategy, the effectiveness and computational efficiency of IMODBA are significantly improved. A gear pump and a camera are used to construct the case studies to verify the effectiveness of IMODBA. The IMODBA is compared with the multiobjective discrete Bees Algorithm (MODBA), multiobjective artificial Bees colony (MOABC) and multiobjective genetic algorithm (MOGA) in terms of the average running time and the optimisation performance. The results show that IMODBA can find a better solution in less time.

(2) Manufacturing service layer

Manufacturing service aggregation, which can significantly improve manufacturing efficiency, plays an important role in the manufacturing service layer [12]. The quality of service (QoS) is considered to evaluate the performance of manufacturing service aggregation. However, the existing works about the manufacturing service aggregation optimal selection problem assume that the services are independent from each other [12]. Most existing methods suggest that services are independent of each other, which leads to service performance degradation. To address the multiobjective RC-Mfg service aggregation optimal selection (MO-CMSAOS) problem, a manufacturing service aggregation optimisation algorithm based on BA is proposed [12]. It mainly contains three key components, including (1) three kinds of

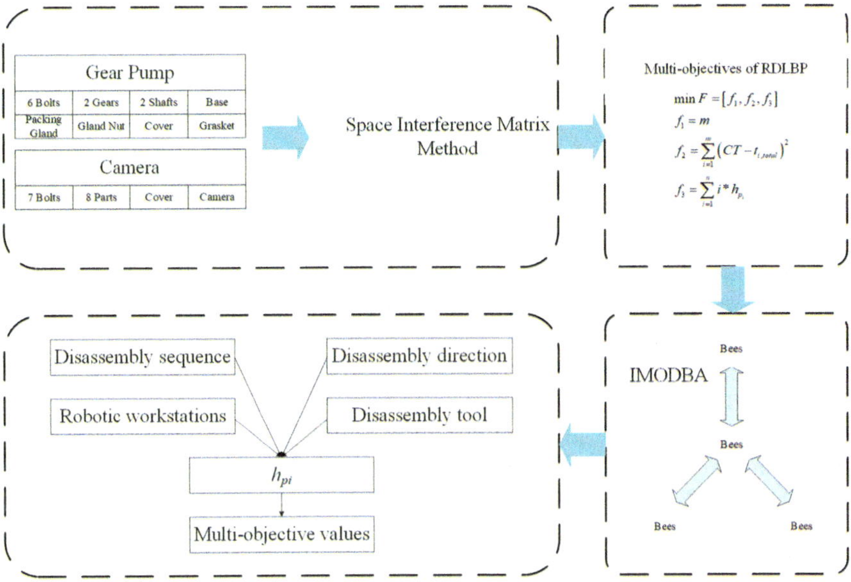

Fig. 3 Framework of RDLBP using IMODBA [13]

correlations between services, (2) a building correlation-aware QoS model, and (3) the improved discrete Bees Pareto algorithm (IDBA-Pareto) for the MO-CMSAOS problem (Fig. 4).

Three kinds of correlations between services are defined. Interface correlation (InC), business correlation (BuC), and statistical correlation (StC) define whether

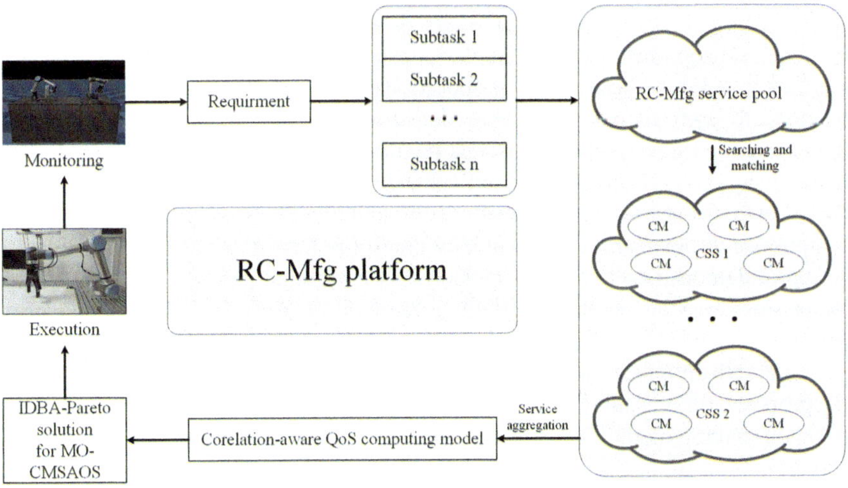

Fig. 4 Service aggregation optimisation using IDBA-Pareto [12]

data logical correlation, provider correlation, and spatiotemporal correlation exist between services, respectively.

In the correlation-aware QoS model of aggregation service, each component service has a set of service candidates with similar functions but differs in QoS. Four QoS criteria are considered: execution cost (C), response time (T), availability (*Ava*), and reliability (*Rel*) [12]. The four basic composite models of sequence, parallel, selection and circulation constitute the correlation-aware QoS model of aggregation service.

Aiming at the problem of how to select the optimal solution of manufacturing service aggregation, which has multiobjective and multiconstraint, IDBA-Pareto based on Pareto-optimal and BA is proposed. IDBA-Pareto improves the BA in terms of the neighborhood structure and search strategy. IDBA-Pareto designs three kinds of neighborhood structures, which not only expand the neighborhood search space but also better eliminate local optimisation. In addition, a search strategy named the dynamic adjustment mechanism is proposed, which can obtain a better solution distribution. IDBA-Pareto is compared with other optimisation algorithms [40–42] to emphasise its advantages in effectiveness, running time and ability to maintain population diversity.

(3) Cross-layer

Service-oriented robotic manufacturing systems (SORMS) include cloud manufacturing (CMfg) and industrial robotic systems [14]. The self-optimisation of industrial robots in the manufacturing cell layer and production process layer are two forms of optimisation in SORMS that pay little attention to mutual interference between each layer [33, 43]. Therefore, a cross-layer optimisation method at the manufacturing cell layer and production process layer based on DQN and adaptive BA (ABA) is proposed [14]. It mainly contains three key components, namely, (1) self-optimisation of industrial robot manufacturing cell service, (2) self-optimisation of production process service, and (3) cross-layer communications.

In the self-optimisation of a single industrial robot manufacturing cell service, the functional modules of the proposed optimisation method in the manufacturing cell layer include the event detection module, simulations and experiments module (SEM), subdecision-making module, and self-optimisation execution module (SOEM). Towards robot cell energy efficiency, a Markov decision process (MDP) model for a velocity controller is built, and a DQN controller is established to optimise the industrial robot velocity control policy. A simulated experiment is set up to demonstrate the utility of the DQN controller, and the result shows that the proposed velocity controller could decrease the energy consumption of the given path.

In the self-optimisation of production process service, the functional modules of the proposed optimisation method in the production process layer include the main-decision-making module, service composition optimisation simulation module and service composition optimisation execution module (SCOEM). A QoS-based industrial robot service assessment model for the multistructure industrial robot service

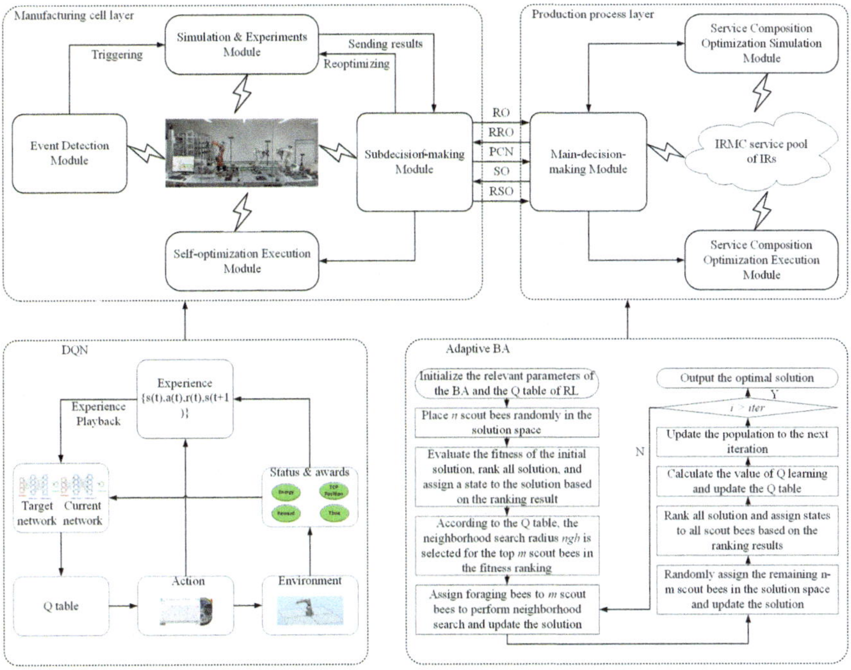

Fig. 5 Cross-layer optimisation using DQN and ABA [14]

composition is built to form the objective function of ABA. An adaptive neighborhood search strategy based on MDP is presented to improve the search efficiency, and the result shows that the ABA has a higher performance in the improvement of the energy efficiency (Fig. 5).

There are five communication instructions: request for optimisation (RO), reply to request for optimisation (RRO), performance change notification (PCN), suggestion for optimisation (SO), and reply to suggestion for optimisation (RSO). Based on five communication instructions, a cross-layer optimisation mechanism is proposed to achieve collaborative optimisation. There are three characteristics in the cross-layer optimisation mechanism: (1) self-optimisation at different layers affects each other; (2) self-optimisation will be executed only when it benefits both layers; and (3) the current execution composition scheme is replaced only when a better service composition scheme exists.

4.2 Human–robot Collaborative Manufacturing

HRC-Mfg is a novel paradigm that aims to combine humans and industrial robots into a united workforce, which makes full use of the strengths of humans and robots [44,

45]. It shows the capability to enhance productivity, energy and material efficiency in some complex production tasks, such as assembly and disassembly. [8]. In HRC-Mfg, humans and industrial robots are allowed to work side-by-side to accomplish common production goals with shared resources, where the activity of humans and robots usually overlaps in the temporal and spatial dimensions [46]. This feature of such a paradigm makes the production environment dynamic and uncertain, which brings a challenge to task sequence planning and decision-making.

In this part, leveraging the power of DBA, a branch of BA, as shown in Fig. 6, we present a task sequence planning framework for HRC-Mfg. It mainly contains three key components, namely, (1) HRC-Mfg task modelling, (2) solution generation and evaluation of the HRC-Mfg task sequence and (3) DBA for task sequence optimisation.

In HRC-Mfg task modelling, taking disassembly tasks as an example, the HRC disassembly tasks can be divided into three categories: (1) Type_Human: the human-specified tasks that need high flexibility, e.g., unscrewing the rusted screw. (2) Type_Robot: robot-specified tasks that need high repeatability, e.g., pick-and-place operations. (3) Type_Common: the tasks suitable for both humans and robots. To classify the disassembly tasks, a task difficulty function d_i^m is defined to evaluate the difficulty for the human or robot according to the attributions a_k of the i th disassembly tasks (e.g., perception ability, operation area, movement complexity, workload, hazardous level).

In addition to task difficulty, human fatigue is another key factor that affects the task sequence planner. Naturally, with the progress of the disassembly process, the

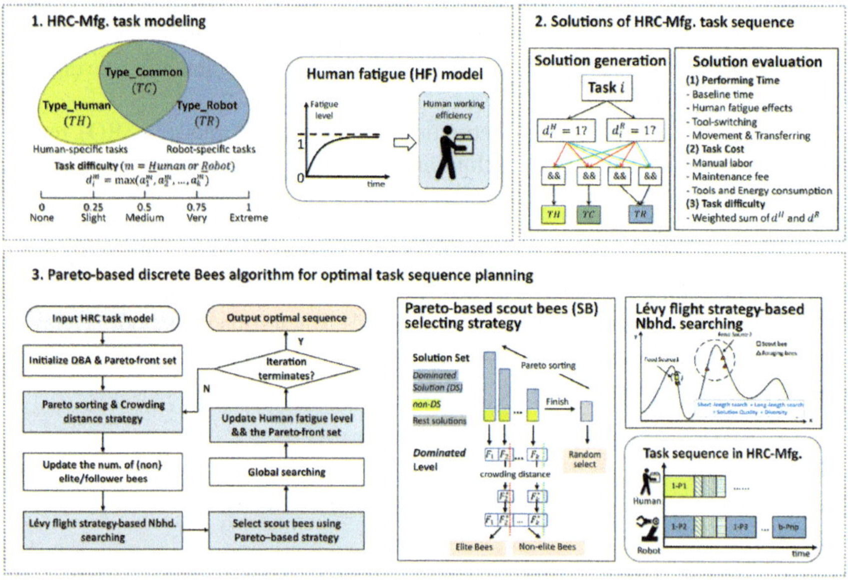

Fig. 6 Task sequence planning using DBA for HRC-Mfg [21]

accumulated human fatigue will lead to a reduction in the working efficiency, which should be considered in HRC disassembly task sequence planning. To this end, as shown in Eq. (1), a time-related human fatigue model is built based on the assumption of the exponentially increased fatigue level [47],

$$HF(t) = 1 - e^{\lambda t} \tag{1}$$

where λ is the accumulation rate of fatigue. The relationship between human fatigue and working efficiency (represented by task-performing time) can be described by Eq. (2), where T_0 is the baseline time (when $HF = 0$), and δ represents the effect of the human fatigue level on the working efficiency [48].

$$T = T_0(1 + \lambda\delta \ln[1 + HF(t)]) \tag{2}$$

With the built HRC-Mfg model, a feasible task sequence solution is generated for the actual production tasks, such as disassembly. The tasks are classified and assigned according to the task difficulty model w.r.t. human and robot. For the Type_ Common tasks, four following assign strategies are randomly chosen to assign them: (1) randomised strategy; (2) equilibrium assignment strategy; (3) preference strategy based on task difficulty; and (4) preference strategy based on disassembly time [21]. The generated solutions are evaluated in terms of task-performing time, task cost, and task difficulty. For task-performing time, the effects of human fatigue level and the robots' tool-switching are taken into consideration. For task cost, the manual labor, maintenance fee, tools, and energy consumption are the key factors, and the weighted sum of the task difficulty of humans and robots forms the third evaluation criterion.

To minimise the task-performing time, task cost, and task difficulty in the actual production tasks, a Pareto-based modified discrete Bees Algorithm (MDBA-Pareto) [21] is proposed to obtain the optimal solutions, whose workflow can be summarised as follows.

First, the task model input and the initialised parameters of MDBA-Pareto. Second, *nb*-sized scout bees and a Pareto front set *P* are initialised. The performance of these scout bees is ranked by the Pareto fast nondominated sorting method [49]. Among these *nb* scout bees, the best bees with a size of *mb* are then selected, where the best *eb* bees are seen as elite bees and the remaining $(mb - eb)$ bees are seen as nonelite bees. For each selected bees, there are *nep*-sized follower bees (or *nsp* for the nonelite bees) dispatching to search its neighborhood. The $(nb - mb)$-sized nonselected bees are dispatched to perform a global search. Then, the human fatigue level and the Pareto front set *P* are updated in each iteration. The above procedures will be repeated until the iteration name reaches the input parameter *MaxIter*. The Pareto front set *P* after the final iteration is the output optimal disassembly solution.

5 Case Study

5.1 Cross-Layer Optimisation Model for Industrial Robots in Assembly

To verify the proposed method, which can optimise the industrial robot in the cross-layer in Sect. 4.1, in this part, a motorcycle is assembled by industrial robots. As shown in Fig. 7, in the manufacturing cell layer, industrial robotic assembly for motorcycles includes 10 services [12]. Execution cost (*C*), response time (*T*), availability (*Ava*), and reliability (*Rel*) are considered in the assembly process. The initial parameters of each QoS are randomly generated, as shown in Table. 1. Equation (3) and (4) are computation methodologies for QoS and W_{QoS} of the assembly process. Equation (5) is the constraint of the assembly process. To avoid possible conflicts during the optimisation process, the maximum and minimum values of *C*, *T*, *Ava*, and *Rel* are limited. Then, the necessary parameters in this case study are initialised, as shown in Table. 2.

$$\begin{cases} T(cs) = \text{Min}(T(S1), T(S2), T(S3), T(S4), T(S5), T(S6)) + \sum_{i=7}^{10} T(Si) \\ C(cs) = \sum_{i=1}^{10} C(Si) \\ Ava(cs) = \prod_{i=1}^{10} Ava(Si) \\ Rel(cs) = \prod_{i=1}^{10} Rel(Si) \\ T_{max}(cs) = \text{Max}(T(S1), T(S2), T(S3), T(S4), T(S5), T(S6)) + \sum_{i=7}^{10} T(Si) \\ C_{total}(cs) = \sum_{i=1}^{10} C(Si) \end{cases} \tag{3}$$

$$\begin{cases} W_{QoS}(cs) = w_1 * T(cs) + w_2 * C(cs) + w_3 * Ava(cs) + w_4 * Rel(cs) \\ En(cs) = \sum_{i=1}^{10} En(S1) \\ F = min\left[1/W_{QoS}(cs), En(cs)\right] \end{cases} \tag{4}$$

$$s.t. \begin{cases} Rel(Si) \geq R_{min}, i \in \{1, \ldots, 10\} \\ Ava(Si) \geq A_{min}, i \in \{1, \ldots, 10\} \\ T_{max} \geq T(cs) \\ C_{max} \geq C(cs) \end{cases} \tag{5}$$

The proposed optimisation model is applied to S1 to show its effectiveness in the manufacturing cell layer. The distributions of only self-optimisation adopted ($QoS_1(S1)$) and the cross-layer optimisation model adopted ($QoS_2(S2)$) are shown in Fig. 8. Figure 8 shows that the proposed method optimises the objective in an integrated manner, rather than on only one metric. Figure 9 shows the QoS distribution of only service composition optimisation adopted ($QoS_1(cs)$) and the cross-layer optimisation model adopted ($QoS_2(cs)$). It can be concluded that the QoS of the production process layer is improved significantly. Then, the result of carrying out the cross-layer optimisation model on both layers (situation B) is compared with that of carrying out the optimisation on both layers without the cross-layer optimisation

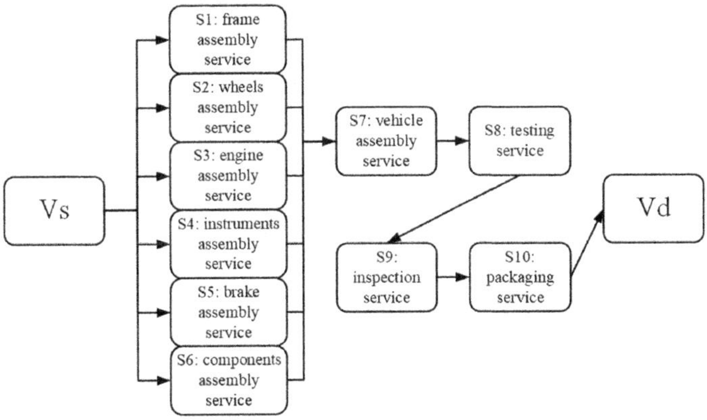

Fig. 7 The motorcycle assembly process model [12]

Table 1 Range of each QoS parameters [12]

QoS attribute	Range
Execution cost	[1,100]
Response time	[1, 10]
Availability	[0.5,1]
Reliability	[0.5,1]

Table 2 The initial values of necessary parameters [12]

Parameters	Value
Number of candidate service	20
T_{max}	15
C_{max}	400
w_1	0.5
w_2	0.5
R_{min}	0.6
A_{min}	0.6

mechanism (situation A) in Fig. 10. We can see that situation B is superior to situation A in terms of W_{QoS}, which can provide the usefulness and effectiveness of the cross-layer optimisation model.

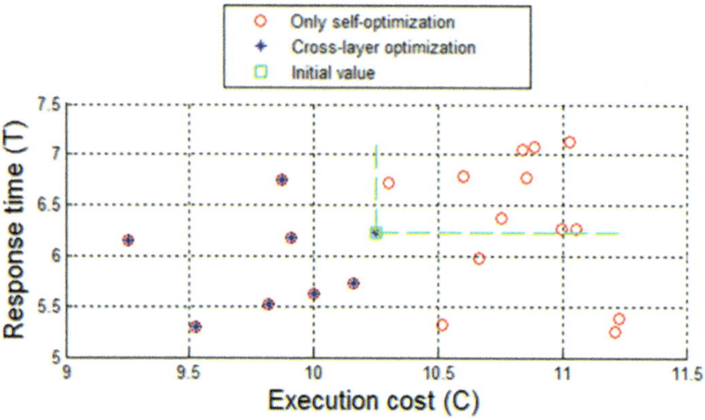

Fig. 8 QoS distribution [14]

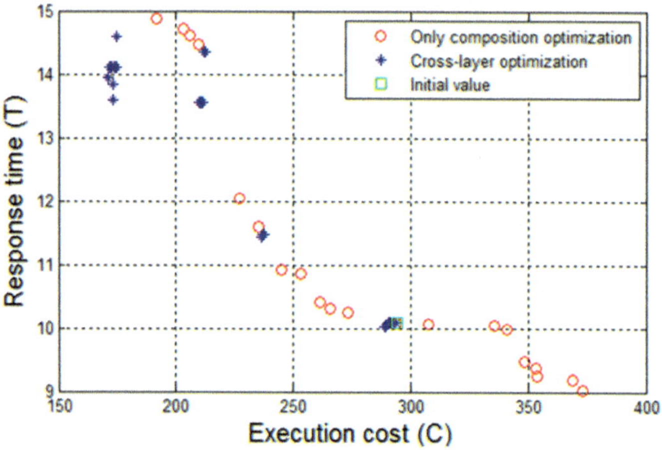

Fig. 9 QoS distribution of composition service [14]

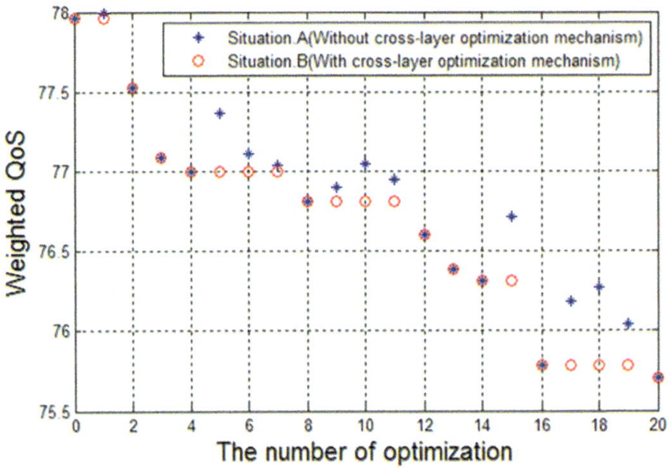

Fig. 10 Weighted QoS [14]

5.2 *Sequence Planning Considering Human Fatigue for HRC in Disassembly*

To verify the proposed method mentioned in Sect. 4.2, we developed a case study of batch disassembly tasks based on a bearing coupler consisting of 21 parts, as shown in Fig. 11a, and its explosive view is shown in Fig. 11b.

Fig. 11 Case study based on a bearing couple [50]

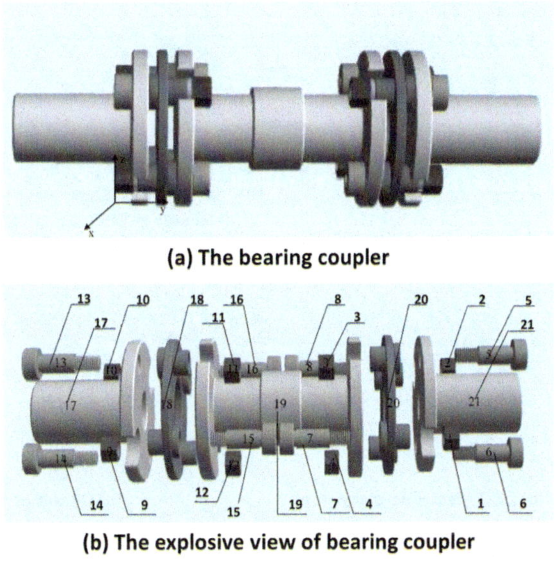

To build the human fatigue model, as shown in Fig. 12a, we collected the electroencephalogram (EEG) signal via the electrode placed at the human worker's forehead with a 1000-Hz sampling rate, which is shown in Fig. 12b. The total duration of the disassembly task is 2 h, where 10 s of data are collected every 5 min. Ten times the same experiments were conducted, and the average data among these experiments were recorded. To extract fatigue features from the denoising EEG signals, we selected the relative power ratio $R_{fatigue}$ of the three subbands of the EEG signal to represent fatigue:

$$R_{fatigue} = \frac{P_\alpha + P_\theta}{P_\beta} \tag{6}$$

where P_α, P_θ, and P_β are the average powers of the α, θ, and β waves in the EEG signal, respectively. The extracted feature is then used to fit the fatigue model based on Eq. (6), whose result is shown in Fig. 12c. The parameter $\lambda = 8.891 \times 10^{-3}$, followed by setting the fatigue effect parameter in Eq. (2) $\delta = 0.3/(\lambda \cdot \ln 2)$.

To optimise the task sequence in such disassembly tasks, we deployed the proposed MDBA with the number of selected scout bees $m = 20$, elite bees $n = 4$, follower bees led by nonelite scout bees $mb = 8$, and follower bees led by elite scout bees $nb = 8$. Combined with the fitted fatigue model, we test the performance of MDBA for batch disassembly tasks, and the results are shown in Fig. 13. It can be seen that from the fourth product, the disassembly time per product for adjusting the disassembly sequence considering human fatigue, with a slight growth trend, is shorter than that without adjusting. The area enclosed by the two curves is the

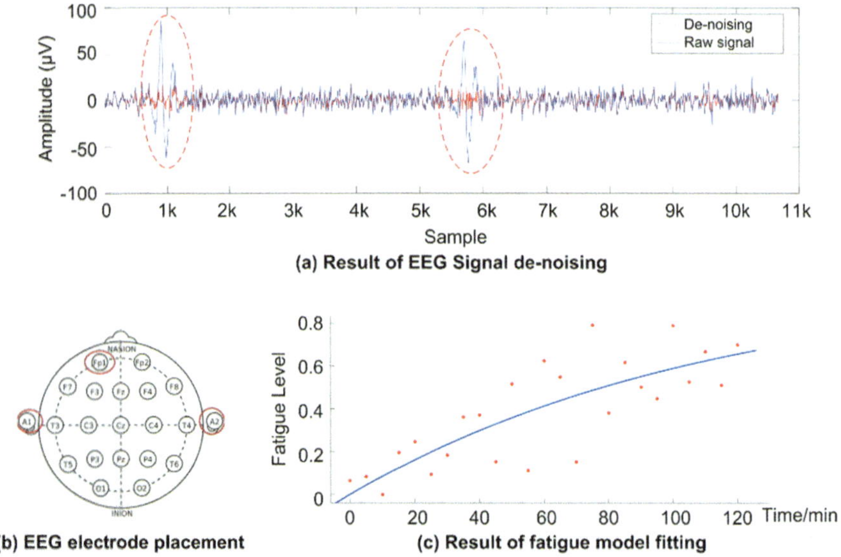

Fig. 12 EEG-based human fatigue model fitting [50]

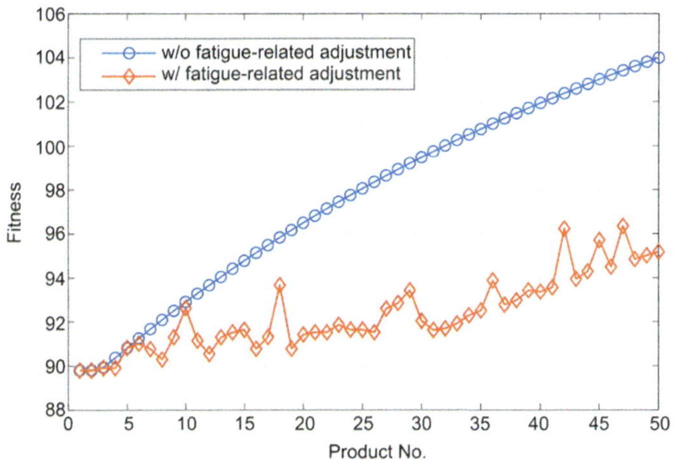

Fig. 13 The fitness of MDBA with/without the fatigue-related adjustment of disassembly sequence [50]

time saved by the disassembly sequence for batch products after considering human fatigue.

To verify the effectiveness of MDBA, we selected three baseline algorithms for comparison, namely, (1) EDBA-WMO [27], (2) IDABC [51], and (3) IGA [49]. For EDBA-WMO, $m = 20$, $n = 5$, $mb = 5$, and $nb = 5$. For IDABC, the parameter $limit = 50$, $foodNumber = 100$, the number of *Onlooker Bees* is 10, the task assignment probability $p_R = 0.6$, and $p_H = 0.4$. For IGA, the crossover variation rate $p_c = 0.5$ and the genetic variation rate $p_m = 0.2$. The four algorithms were run 10 times with different numbers of iterations, from 100 to 500 with a step size of 100 and a fixed population size of 100, and the results are shown in Fig. 14. From the results in Fig. 14a–b, the quality of the solution obtained by MDBA is better than that of the other three algorithms, while MDBA has the second longest running time. In addition, the performance of the four algorithms under different population sizes, from 50 to 100, and a fixed number of iterations, 300, were compared, and each experiment was run 10 times. The result is shown in Fig. 14c–d, where the IDBA algorithm can find the optimal solution compared to the other three algorithms.

For the disassembly of batch products, if all products adopt the same disassembly sequence as the first product, then the actual time to disassemble each product should also increase under the cumulative levels of human fatigue. Figure 14e shows the comparison of disassembly time per product for the four different algorithms in this case for batch products. As seen from the figure, the disassembly time of each product obtained by MDBA is lower than that obtained by the other three algorithms. To compare the performance of the disassembly task execution time with fatigue-related adjustment, we selected four indicators for evaluation, namely, the total time and manual disassembly time with (without) the fatigue-related adjustment. The result is shown in Fig. 14f, which shows that the adjustment considering human

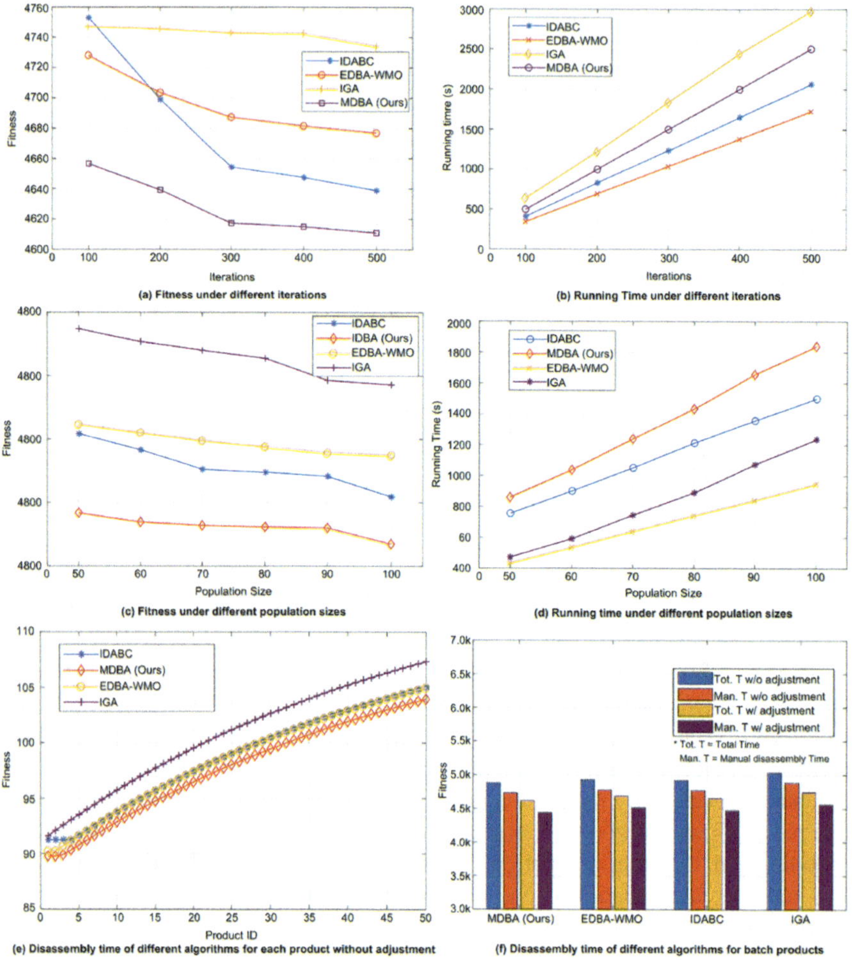

Fig. 14 Performance comparison of different algorithms with different settings [50]

fatigue can shorten the task-execution time, and the disassembly solutions obtained by MDBA in each indicator have the shortest disassembly time. Table 3 shows the solution results of four different algorithms for batch products, where the disassembly sequence, total disassembly time, manual disassembly time, and human fatigue level at the completion of the No. 1 product and the last (No. 50) product are presented. The disassembly sequence solved by MDBA comes with a human fatigue level of 48.19% at task completion, which is lower than the results obtained by the other three algorithms.

Table 3 The solutions from different algorithms for batch disassembly tasks [50]

Algorithms	Product No	Disassembly sequence	Total time (s)	Manual time (s)
MDBA (Ours)	1	3/9/10/4/2/1/6/5/8/7/11/12/16/15/14/21/13/20/17/18/19	4617.6	4443.5
		H/H/H/R/R/R/R/R/R/R/R/H/H/H/R/R/H/R/H/R/H/R		
	50	4/3/2/11/12/9/1/10/5/14/15/13/16/8/7/17/6/21/18/20/19		
		R/R/R/H/H/H/R/H/R/H/R/H/R/R/R/H/R/R/H/H/R		
EDBA-WMO	1	10/3/4/9/2/11/1/12/15/7/8/6/16/14/13/5/17/21/18/20/19	4687.7	4515.5
		R/R/R/H/R/H/R/H/H/R/R/R/H/H/H/R/H/R/H/H/R		
	50	3/4/11/9/1/10/12/15/2/7/6/8/16/14/5/21/13/17/20/18/19		
		H/H/R/R/H/R/R/R/H/R/H/R/R/R/H/H/R/R/H/H/R		
IDABC	1	11/9/10/3/2/4/8/1/12/7/6/15/16/5/13/21/14/20/17/18/19	4654.6	4474.6
		R/H/H/R/R/H/H/R/R/H/H/R/R/H/R/H/R/H/R/H/R		
	50	1/2/4/12/11/10/9/3/15/13/16/14/8/7/17/5/6/21/18/20/19		
		R/H/R/R/R/H/H/H/R/H/R/H/R/R/H/R/R/R/H/H/R		

6 Discussion

The manufacturing cell layer, production process layer, and manufacturing service layer of RC-Mfg have many optimisation problems that have multiple objectives, including RDLBP, MO-CMSAOS problems, and cross-layer optimisation problems. These optimisation problems are difficult to solve due to the complex relationship between them. The emergence of BA provides a new way to solve such optimisation problems. In particular, the improved DBA based on Pareto shows good optimisation efficiency when solving Pareto optimisation problems. With the rapid development of machine learning, the combination of deep learning and reinforcement learning with BA can be used to solve more complex optimisation problems arising from RC-Mfg.

Towards HRC-Mfg, BA provides a powerful solution to task sequence planning optimisation problems, especially when human fatigue is considered. Beyond human fatigue, many other factors need to be focused on in HRC-Mfg task planning, such as working preference and behaviour. These factors are difficult to manually model due to the high-dimensional features, where combining them with the power of machine learning methods could be a promising solution. With the modelled human factors, how to determine the optimisation objective of BA should also be considered. The dynamic and uncertain interaction between humans and robots can also disturb the plan, and how to capture these interaction states and feed them into the BA-based planner is another issue. To address the above problems, more studies should be conducted in the future. Moreover, the main application of HRC-Mfg we focused on is mainly on disassembly. More potential applications should also be explored.

7 Conclusion

In this chapter, optimisation problems in robotics-enabled collaborative manufacturing are solved by BA and its improved algorithms. In terms of RC-Mfg, the BA algorithms are utilised to solve the RDLBP, MO-CMSAOS, and cross-layer optimisation problems. In terms of HRC-Mfg, the BA algorithms are utilised to solve the task sequence planning optimisation problem. Based on two case studies, simulations are carried out to show the performance of the BA and its improved algorithms under different conditions. The performance of BA and its improved algorithms are also compared with the other optimisation algorithms. The results show that the BA and its improved algorithms can generate better solutions than the other optimisation algorithms. However, there are still many optimisation problems in robotics-enabled collaborative manufacturing, which are difficult to solve by BA. In the future, with the help of deep learning and reinforcement learning, more work should be completed to improve BA to solve the complex and uncertain optimisation problems of robotics-enabled collaborative manufacturing.

Acknowledgements This research was supported by the National Natural Science Foundation of China [grant number: 51775399] and the Hubei Provincial Natural Science Foundation of China [grant numbers: 2021CFA077 and 2021CFB044].

References

1. Hashemi-Petroodi SE, Thevenin S, Kovalev S, Dolgui A (2020) Operations management issues in design and control of hybrid human-robot collaborative manufacturing systems: a survey. Annu Rev Control 49:264–276
2. Li Z, Li S, Luo X (2021) An overview of calibration technology of industrial robots. IEEE/CAA J Autom Sinica 8(1):23–36
3. Liu J, Zhou Z, Pham DT, Xu W, Ji C, Liu Q (2020) Collaborative optimization of robotic disassembly sequence planning and robotic disassembly line balancing problem using improved discrete Bees algorithm in remanufacturing☆. Robot Comput Int Manuf 61:101829
4. Hietanen A, Pieters R, Lanz M, Latokartano J, Kämäräinen JK (2020) AR-based interaction for human-robot collaborative manufacturing. Robot Comput Int Manuf 63:101891

5. Wang L, Liu S, Liu H, Wang XV (2020) Overview of human-robot collaboration in manufacturing. In: Proceedings of 5th international conference on the industry 4.0 model for advanced manufacturing (pp 15–58). Springer, Cham
6. Pérez L, Rodríguez-Jiménez S, Rodríguez N, Usamentiaga R, García DF (2020) Digital twin and virtual reality based methodology for multi-robot manufacturing cell commissioning. Appl Sci 10(10):3633
7. Shen H, Pan L, Qian J (2019) Research on large-scale additive manufacturing based on multi-robot collaboration technology. Addit Manuf 30:100906
8. Wang L, Gao R, Váncza J, Krüger J, Wang XV, Makris S, Chryssolouris G (2019) Symbiotic human-robot collaborative assembly. CIRP annals 68(2):701–726
9. Liu H, Wang L (2020) Remote human–robot collaboration: a cyber–physical system application for hazard manufacturing environment. J Manuf Syst 54:24–34
10. Pham DT, Baronti L, Zhang B, Castellani M (2018) Optimization of engineering systems with the Bees algorithm. Int J Artif Life Res 8(1):1–15
11. Hameed NSS, Othman WAFW, Wahab AAA, Alhady SSN (2019) Optimising pid controller using Bees algorithm and firefly algorithm. Robotika 1(1):22–27
12. Xu W, Tian S, Liu Q, Xie Y, Zhou Z, Pham DT (2016) An improved discrete Bees algorithm for correlation-aware service aggregation optimization in cloud manufacturing. Int J Adv Manuf Technol 84:17–28
13. Liu J, Zhou Z, Pham DT, Xu W, Yan J, Liu A, Liu Q (2018) An improved multi-objective discrete Bees algorithm for robotic disassembly line balancing problem in remanufacturing. Int J Adv Manuf Technol 97(9):3937–3962
14. Zhang J, Liu Q, Xu W, Zhou Z, Pham DT (2018) Cross-layer optimization model toward service-oriented robotic manufacturing systems. J Manuf Sci Eng 140(4)
15. Marvel JA, Bostelman R, Falco J (2018) Multi-robot assembly strategies and metrics. ACM Comput Surv 51(1):1–32
16. Hovgard M, Lennartson B, Bengtsson K (2021) Applied energy optimization of multi-robot systems through motion parameter tuning. CIRP J Manuf Sci Technol 35:422–430
17. Kapanoglu M, Alikalfa M, Ozkan M, Parlaktuna O (2012) A pattern-based genetic algorithm for multi-robot coverage path planning minimizing completion time. J Intell Manuf 23(4):1035–1045
18. Boschetti G, Bottin M, Faccio M, Minto R (2021) Multi-robot multi-operator collaborative assembly systems: a performance evaluation model. J Intell Manuf 32(5):1455–1470
19. Liu J, Zhou Z, Pham DT, Xu W, Cui J, Yang C (2020) Service platform for robotic disassembly planning in remanufacturing. J Manuf Syst 57:338–356
20. Rahman SM (2019) Cognitive cyber-physical system (C-CPS) for human-robot collaborative manufacturing. In: 2019 14th annual conference system of systems engineering (pp 125–130). IEEE
21. Xu W, Tang Q, Liu J, Liu Z, Zhou Z, Pham DT (2020) Disassembly sequence planning using discrete Bees algorithm for human-robot collaboration in remanufacturing. Robot Comput Int Manuf 62:101860
22. Pérez L, Rodríguez-Jiménez S, Rodríguez N, Usamentiaga R, García DF, Wang L (2020) Symbiotic human–robot collaborative approach for increased productivity and enhanced safety in the aerospace manufacturing industry. Int J Adv Manuf Technol 106(3):851–863
23. Li S, Wang R, Zheng P, Wang L (2021) Towards proactive human–robot collaboration: a foreseeable cognitive manufacturing paradigm. J Manuf Syst 60:547–552
24. Liu H, Wang L (2021) Collision-free human-robot collaboration based on context awareness. Robot Comput Int Manuf 67:101997
25. Zeybek S, Pham DT, Koç E, Seçer A (2021) An improved bees algorithm for training deep recurrent networks for sentiment classification. Symmetry 13(8):1347
26. Laili Y, Tao F, Pham DT, Wang Y, Zhang L (2019) Robotic disassembly re-planning using a two-pointer detection strategy and a super-fast bees algorithm. Robot Comput Int Manuf 59:130–142

27. Liu J, Zhou Z, Pham DT, Xu W, Ji C, Liu Q (2018) Robotic disassembly sequence planning using enhanced discrete bees algorithm in remanufacturing. Int J Prod Res 56(9):3134–3151
28. Arif Şen M, Tinkir M, Kalyoncu M (2018) Optimization of a PID controller for a two-floor structure under earthquake excitation based on the bees algorithm. J Low Frequen Noise, Vibrat Active Control 37(1):107–127.
29. Seyed Aboutorabi SJ, Rezvani MH (2020) An optimized meta-heuristic bees algorithm for players' frame rate allocation problem in cloud gaming environments. Comput Games J 9(3):281-304.
30. Luo YL, Zhang L, Tao F, Zhang XS, Ren L (2012) Key technologies of manufacturing capability modeling in cloud manufacturing mode. Comput Integr Manuf Syst 18(7):1357–1367
31. Xu W, Yu J, Zhou Z, Xie Y, Pham DT, Ji C (2015) Dynamic modeling of manufacturing equipment capability using condition information in cloud manufacturing. J Manuf Sci Eng 137(4).
32. Wang B, Guan Z, Ullah S, Xu X, He Z (2017) Simultaneous order scheduling and mixed-model sequencing in assemble-to-order production environment: a multi-objective hybrid artificial bee colony algorithm. J Intell Manuf 28(2):419–436
33. Wang L, Schmidt B, Givehchi M, Adamson G (2015) Robotic assembly planning and control with enhanced adaptability through function blocks. Int J Adv Manuf Technol 77(1):705–715
34. Alavidoost MH, Zarandi MH, Tarimoradi M, Nemati Y (2017) Modified genetic algorithm for simple straight and U-shaped assembly line balancing with fuzzy processing times. J Intell Manuf 28(2):313–336
35. McGovern SM, Gupta SM (2007) A balancing method and genetic algorithm for disassembly line balancing. Eur J Oper Res 179(3):692–708
36. Tian G, Zhou M, Chu J, Liu Y (2012) Probability evaluation models of product disassembly cost subject to random removal time and different removal labor cost. IEEE Trans Autom Sci Eng 9(2):288–295
37. Guo X, Liu S, Zhou M, Tian G (2015) Disassembly sequence optimization for large-scale products with multiresource constraints using scatter search and Petri nets. IEEE Trans Cybern 46(11):2435–2446
38. Jin GQ, Li WD, Xia K (2013) Disassembly matrix for liquid crystal displays televisions. Procedia CIRP 11:357–362
39. Kalayci CB, Gupta SM (2013) Artificial bee colony algorithm for solving sequence-dependent disassembly line balancing problem. Expert Syst Appl 40(18):7231–7241
40. Deb K, Agrawal S, Pratap A, Meyarivan T (2000) A fast elitist non-dominated sorting genetic algorithm for multi-objective optimization: NSGA-II. In: International conference on parallel problem solving from nature (pp 849–858). Springer, Berlin, Heidelberg
41. Coello CC, Lechuga MS (2002) MOPSO: a proposal for multiple objective particle swarm optimization. In: Proceedings of the 2002 congress on evolutionary computation. CEC'02 (Cat. No. 02TH8600) (Vol 2, pp 1051–1056). IEEE
42. Liu SL, Liu YX, Zhang F, Tang GF, Jing N (2007) Dynamic web services selection algorithm with QoS global optimal in web services composition. J Softw 18(3):646–656
43. Baizid K, Ćuković S, Iqbal J, Yousnadj A, Chellali R, Meddahi A, Ghionea I (2016) IRoSim: industrial robotics simulation design planning and optimization platform based on CAD and knowledgeware technologies. Robot Comput Int Manuf 42:121–134
44. Ajoudani A, Zanchettin AM, Ivaldi S, Albu-Schäffer A, Kosuge K, Khatib O (2018) Progress and prospects of the human–robot collaboration. Auton Robot 42(5):957–975
45. Villani V, Pini F, Leali F, Secchi C (2018) Survey on human–robot collaboration in industrial settings: safety, intuitive interfaces and applications. Mechatronics 55:248–266
46. Liu Z, Liu Q, Wang L, Xu W, Zhou Z (2021) Task-level decision-making for dynamic and stochastic human-robot collaboration based on dual agents deep reinforcement learning. Int J Adv Manuf Technol 115(11):3533–3552
47. Glock CH, Grosse EH, Kim T, Neumann WP, Sobhani A (2019) An integrated cost and worker fatigue evaluation model of a packaging process. Int J Prod Econ 207:107–124

48. Ferjani A, Ammar A, Pierreval H, Elkosantini S (2017) A simulation-optimization based heuristic for the online assignment of multi-skilled workers subjected to fatigue in manufacturing systems. Comput Ind Eng 112:663–674
49. Tian Y, Zhang X, Liu Z, Jiang X, Xue J (2019) Product cooperative disassembly sequence and task planning based on genetic algorithm. Int J Adv Manuf Technol 105(5):2103–2120
50. Li K (2020) Research on sequence planning for human-robot collaboration in disassembly based on human fatigue (Master thesis, Wuhan University of Technology)
51. Ren Y, Tian G, Zhao F, Yu D, Zhang C (2017) Selective cooperative disassembly planning based on multi-objective discrete artificial bee colony algorithm. Eng Appl Artif Intell 64:415–431

Bees Algorithm for Hyperparameter Search with Deep Learning to Estimate the Remaining Useful Life of Ball Bearings

Anurakt Kumar, Satyam Kumar, Neha Gupta, Nathinee Theinnoi, and D. T. Pham

Abstract Hyperparameter searching is one of the significant challenges in training deep learning models. To solve this challenge, the Bees Algorithm (BA), which simulates the foraging behaviour of honey bees, is used for hyperparameter searching and finding the best set of hyperparameters for a given deep learning model. This study applies a two-parameter version of the Bees Algorithm (BA_2) to search for the best set of hyperparameters for a Convolutional Neural Network (CNN) combined with a Long Short-Term Memory (LSTM) model. Then, the model is used to predict the remaining useful life (RUL) of ball bearings. BA_2 uses the traplining foraging technique of bees to integrate explorative and exploitative search mechanisms, reduce the number of parameters to only two and improve the remaining useful life (RUL) prediction. The algorithm can find a set of hyperparameters that makes the deep learning model perform better than the IEEE PHM 2012 Prognostic challenge winner by 38.97%.

A. Kumar
Department of Electronics and Electrical Communication Engineering, Indian Institute of Technology Kharagpur, Kharagpur, India
e-mail: anurakt@kgpian.iitkgp.ac.in

S. Kumar
Department of Civil Engineering, Indian Institute of Technology Kharagpur, Kharagpur, India
e-mail: satyamk1102@kgpian.iitkgp.ac.in

N. Gupta
Department of Chemical Engineering, Indian Institute of Technology Kharagpur, Kharagpur, India
e-mail: neha_gupta@kgpian.iitkgp.ac.in

N. Theinnoi (✉) · D. T. Pham
Department of Mechanical Engineering, College of Engineering and Physical Sciences, The University of Birmingham, Birmingham B15 2TT, UK
e-mail: nxt794@alumni.bham.ac.uk

D. T. Pham
e-mail: D.T.Pham@bham.ac.uk

D. T. Pham and N. Hartono (eds.), *Intelligent Engineering Optimisation with the Bees Algorithm*, Springer Series in Advanced Manufacturing,
https://doi.org/10.1007/978-3-031-64936-3_11

Keywords Bees algorithm · Convolutional neural network (CNN) · Long short-term memory (LSTM) · Hyperparameter optimisation · Remaining useful life (RUL) prediction

1 Introduction

High competition to capture a significant market share plays an essential role in all manufacturing. The best quality products with the lowest production costs and less waste are desired. Production failures with unexpected consequences, such as downtime or machine breakdowns, are obstacles to production flow. As a result, it is imperative to maintain equipment, machinery, and infrastructure to ensure their optimal performance, safety and longevity.

There are several types of maintenance based on their purpose, timing, and approach. According to the BSI standards [1], maintenance techniques are categorised based on the causes of issues. For instance, preventive maintenance (PM), condition-based -based maintenance (CBM) and predictive maintenance (PdM) are a group of maintenance strategies to prevent machine breakdowns. On the other hand, following the occurrence of failures, a maintenance approach known as corrective maintenance is executed to repair or service malfunctioning machinery. Nevertheless, the expenses and downtime associated with corrective maintenance are frequently higher than those with preventative and predictive maintenance techniques.

To reduce total maintenance expenses and operational downtime, PdM provides the most efficient solution for optimising maintenance, particularly preceding equipment failures [2]. In concurrence with [3], the authors confirm that CBM and PdM constitute successful, cost-effective maintenance approaches. These strategies aim to analyse data procured from sensors and monitoring systems to evaluate the equipment's condition and predict when maintenance is needed. Furthermore, this technique progressively demonstrates enhanced reliability and precision and may eventually replace schedule-based maintenance [4].

To facilitate effective maintenance, monitoring essential machine parts can reduce unforeseen problems that are difficult to manage in the maintenance plan. In the majority of machines, bearings are one of the most critical components of mechanical systems [2, 4, 5]. Bearings function to reduce friction when another component rotates or slides upon them [5]. Deficiencies in bearings can precipitate machinery malfunctions, contributing to up to 40% of machine failures [4]. Therefore, observing and preparing for maintenance can prolong bearing longevity and decrease machine downtime.

Based on PdM, a fundamental notion in maintenance planning is the Remaining Useful Life (RUL) of bearings, ensuring the secure operation of the machinery before failure. RUL aims to estimate the residual operational duration until bearings reach their end of life. Owing to RUL prediction, predictive maintenance can be direct, thereby augmenting system availability, reliability, and safety [6]. Nonetheless, challenges persist in RUL prediction due to uncertainties, including noise and fluctuating working conditions [7]. To detect the changing bearing conditions, vibration signals derived from sensor data are collected to identify these localised defects [4].

Due to the growth of the industrial Internet of Things (IoT) and big data, a wealth of information is readily accessible from sensors mounted on prognosis and health management (PHM) systems. The objective of PHM entails monitoring health conditions using sensory signals to identify anomalies, diagnose faults and forecast RUL [8]. Intelligent algorithms, particularly deep learning algorithms (DL), are now capable of analysing massive amounts of collected data [7, 9, 10], extracting information with high-dimensional features [8, 10, 11], comprehending functional and temporal correlations between and within condition monitoring time series signals and transferring knowledge across diverse operational contexts and units [10].

As mentioned above, DL has attained significant success across various fields, such as image processing [12–16], speech recognition [17–21], fault detection [22–27] and RUL prediction [7, 9, 11, 28–32]. Although DL is widely used for many problem domains, the challenge in implementing DL across different datasets lies in hyperparameter tuning, attributable to the multiple hyperparameters that dictate various aspects of DL functionality. For this reason, hyperparameters influence the execution time and memory requirements, impact model performance, and affect results when applied to unseen or new data inputs [33].

Hyperparameter settings can be classified into two categories: manual and automatic. Manually tuning hyperparameters through trial and error is a time-consuming process requiring extensive knowledge to identify the optimal hyperparameters for high performance. On the other hand, automatic hyperparameter selection demands minimal understanding to search for optimal hyperparameters; however, it is typically more computationally intensive than manual approaches [33]. Although automatic hyperparameter selection algorithms consume computation time, several techniques, such as grid search, random search, and model-based hyperparameter optimisation [33–38], are commonly employed to optimise DL hyperparameters.

Metaheuristic algorithms, especially particle swarm optimisation (PSO) [39–43], represent alternative techniques generally applied for hyperparameter optimisation. Presently, metaheuristic algorithms inspired by animal behaviour have gained popularity for hyperparameter optimisation. Examples of such metaheuristic algorithms include Artificial Bee Colony (ABC) [44–47], Firefly Algorithm (FA) [44, 48], Bat Algorithm (BA) [44], Grey Wolf Optimisation (GWO) [49–51], Ant Colony [52, 53] and Cuckoo Search (CS) [54], all of which have been utilised to tune hyperparameters.

To improve the performance of the RUL prediction, this chapter proposes applying the two-parameter Bees Algorithm (BA_2), adapted from the Bees Algorithm (BA), to search for optimal hyperparameters of the two deep learning combinations and estimate the RUL of ball bearings. The best set of hyperparameters for predicting the RUL of ball bearings is assigned, and the prediction performance is compared to the IEEE PHM 2012 Prognostic challenge [55] and the winner of this challenge [56].

2 Related Work

Multiple methodologies can be followed to estimate the remaining useful life of ball bearings under experimental loading conditions. RUL estimation methods can be coarsely classified as model-based (physics-based), data-driven-based and hybrid based [57–60]. Model-based methods depend on building mathematical models to reflect the degradation trend of machines after the fault mechanism is fully understood. For data-driven methodologies, historical information and monitoring data are utilised to compute RUL based on statistical and probabilistic methods. The latter is the hybrid-based approach, which takes advantage of the benefits of physics-based and data-driven approaches by eliminating their drawbacks. However, hybrid approaches have the disadvantage of combining the weaknesses of each approach with increased complexity in the solution-finding process [57].

Due to increasing data availability, data-driven models have recently offered a new approach to predictive maintenance and RUL prediction [58, 61–65]. A prognostic model generates a model based on trends and states obtained from monitoring sensors, and this model is then used to predict RULs [66]. Although the prediction performance of data-driven approaches is not better than that of model-based approaches, data-driven prognostics spread much faster than physics-based prognostics in the RUL prediction field because of quick implementation and deployment based on AI tools [57, 61]. Most techniques can be applied directly or modified slightly. For these reasons, a data-driven method is widely used to predict RUL.

A wide range of AI techniques, especially DL techniques, are becoming increasingly attractive. Due to the development of DL technology, the ability to analyse complex data has greatly improved, and the DL method is increasingly used in prediction problems. DL networks are used as a solution, which digs a deeper level of information and has shown good progress in recent years in different fields. There have been various deep learning models used for the prediction of the RUL, such as the recurrent neural network (RNN)-based approach of [67], the long short-term memory (LSTM)-based approach of [29, 39], and the CNN-based approach of [11, 28, 68].

As mentioned above, one of the popular DL techniques for RUL prediction is a variant of RNN called LSTM. It has a greater ability to learn long-term dependencies than simple recurrent architectures [33]. Supercapacitors and rolling bearings are the case study for [29, 39], respectively. The experimental results are compared with the error analysis of RNN and gated recurrent unit (GRU). It has been demonstrated that LSTM performance is accurate and robust in predictions [39]. As with [39], this model also yields the lowest root mean square error (RMSE) when comparing the results with the BP neural network and support vector regression [29].

In addition to LSTM, CNN is becoming a popular technique to apply for RUL prediction. The first proposed use of CNNs is for image processing, and several research studies have been conducted on computer vision or speech processing [28]. Although CNN structures have demonstrated excellent feature extraction capabilities, there is a lack of studies on their application in predicting the remaining useful life of machines. Li et al. [28] introduced a CNN to extract fault conditions before predicting the RUL and found that the RUL can be predicted with reasonable accuracy during the engine's lifetime, especially in the near-failure period. On the other hand, CNNs work in hybrids with another feature extraction technique, the spectrum-principal-energy-vector, to improve the prediction results [11]. This study significantly improves the predictability of the results. Researchers recommend combining CNNs with other DLs, such as RNNs, for time-sequential data to improve accuracy.

The idea of [11] is presented by [68]. It is proposed that machine RUL predictions can be achieved by using a recurrent convolutional neural network (RCNN). Time-series sensor data are processed through a recurrent convolutional layer to determine the temporal dependencies, allowing the prognostics network to remember degradation information over time and enhance its representation. The inherent limitations of the CNN are overcome by variational inference to obtain a probabilistic RUL prediction. As a result, RCNN offers a clear advantage in accuracy and convergence over machine learning models and some CNN-based prognostic approaches.

Currently, DL techniques remain widely used for RUL prediction. A combination of two techniques, CNN with LSTM, is demonstrated in [7, 69–71]. According to [7], CNNs can be used to detect bearing faults with good representation ability. Then, the RUL prediction model is constructed by feeding features into LSTM. Unlike [69–71], CNN and LSTM operate as parallel models and are used for feature extraction. Temporal and spatial features are extracted using the LSTM and CNN functions, respectively. Next, these features are merged to provide a RUL prediction [69–71]. There are a few differences compared to [70]. Processed data are split into two paths. One is introduced into the CNN, and the other is for the first LSTM. The outputs of both paths are summed and entered into the second LSTM for prediction. Based on all results, the prediction performance improves significantly better than that of previously published methods.

Due to several hyperparameters of deep learning algorithms, it is possible to manually tune hyperparameters when having a good starting point or by becoming familiar with hyperparameter values [33]. However, it usually takes a long time to achieve the best deal. Using hyperparameter optimisation such as grid search, random search or model-based methods, good hyperparameter values can be determined [33, 34]. The different examples of hyperparameter tuning are presented in [35–38]. Kaur et al. [35] applied grid search optimisation to fine-tune the number of layers, neurons per layer, type of activation function per layer, learning rates, batch sizes and the number of epochs to distinguish Parkinson's disease patients from healthy controls based on voice samples. The success of applying this algorithm is an average accuracy of 91.69%.

The most popular hyperparameter optimisation is Bayesian optimisation (BO), which is model-based hyperparameter optimisation [33]. Wu et al. [36] claimed that BO is very effective in solving the optimisation problem. It was applied to tune hyperparameters based on the Gaussian method for three machine learning models: random forest, neural network and multi-grained cascade forest. For the neural network, the learning rate and the number of samples in each batch of CNN and RNN are tuned. The model performance reaches 99.14 and 97.13% prediction accuracy, respectively, after hyperparameter adjustment.

The following research predicts the RUL of lithium-ion batteries using a combination of a deep convolution neural network (DCNN) with double-layer LSTM and improves the performance of the BO network for hyperparameter optimisation [37]. Five hyperparameters of each deep neural network were tuned. As a result, this optimiser shows a great impact on network performance, represented by a low root mean square error, which is 0.0627 for the RUL prediction. The other case is [38]. This study has four tuned hyperparameters: the optimiser, the learning rate, the activation function, and the number of neurons. Optimal values for deep CNN hyperparameters are chosen through BO. In comparison with the default values, the optimal values for hyperparameters showed a significant performance improvement. For all examples, it can be seen that BO is a wild implementation for fine-tuning hyperparameter optimisation.

A metaheuristic algorithm effectively solves optimisation problems and is usually applied to automated hyperparameter selection. Using this method, the algorithm explores and exploits the search space to find the global optima solution, which is the main advantage of this method [49]. The first example is PSO. The success of this technique is presented in [39–41]. This technique confirms that it efficiently uses computational resources and automates the selection of deep neural network (DNN) hyperparameters [40]. As neural networks become more complex, more hyperparameters need to be adjusted. In [41], the PSO determines the optimal CNN-LSTM hyperparameter structure during particle movement to predict energy consumption. CNN kernels, LSTM hidden units, and CNN-LSTM fully connected units are encoded. This PSO-based CNN-LSTM method achieves nearly perfect prediction performance for energy prediction by exploring the optimal prediction structure.

The various metaheuristic algorithms inspired by animal behaviour such as the ABC [45–47] and GWO [49] algorithms, have also been adapted for hyperparameter optimisation. ABC is implemented to optimise LSTM hyperparameters to predict Bitcoin prices and traded volume [45] and use it for large datasets [47]. Additionally, an optimal deep CNN (ODC) classifier based on the ABC algorithm is proposed to optimise hyperparameters of deep CNNs and improve plant species classification [46]. With optimisation, the model performs better than without optimisation [45] and achieves high accuracy when benchmarked with genetic, particle swarm and firefly algorithms [46].

In addition to ABC, GWO will find a configuration of the best combination of hyperparameters to generate LSTM models with good performance in language modelling [49]. Moreover, the new balance grey wolf optimiser (NB-GWO), the improved version of GWO, was successfully developed for the first time [50]. Compared to the original GWO, it achieves a better balance between exploration and exploitation. From the results obtained in this study, it is evident that NB-GWO has proven to be a very effective and reliable optimisation tool to optimise DNN hyperparameters.

Referring to all related works, it can be noted that DL techniques are accepted to be applied for RUL prediction and that metaheuristic algorithms play a crucial role in tuning DL hyperparameters. Therefore, the idea of this study is to implement the BA_2 algorithm to perform a hyperparameter search on CNN in combination with LSTM, which increases RUL prediction accuracy by finding an optimal set of hyperparameters.

3 Prediction of Ball Bearings' Remaining Useful Life

3.1 IEEE Challenge

The IEEE challenge [55] focused on estimating the RUL of bearings, a significant issue since the majority of rotating machinery failures are related to these components. Under this challenge, three different operating conditions were examined as shown in Table 1.

Experiments were performed on an experimental laboratory setup allowing rapid deterioration of ball bearings and dedicated to evaluating and verifying methods for predicting and diagnosing ball bearing failures under constant or varying conditions. Vibration and temperature signals were collected, consisting of seventeen datasets from three conditions with different bearing lives.

Due to the different bearing lifetimes, it is more challenging to perform reasonable estimates. In this study, vibration signals are selected as input data for the DL model. Six training sets (two sets of each condition) were obtained from run-to-failure experiments, and eleven test sets (five sets of conditions one and two and one set of condition three) were used to develop prognostic models and precise RUL prediction. This study follows the concept of sequence-to-sequence classification. The vibration signals will be classified into two groups: normal and abnormal signals at a threshold point, which is the initial point of failure. From this point until the end of the time step, it represents a remaining useful life. The RUL prediction procedure is shown in Fig. 1.

The first step is data preparation. Due to a few training dataset samples of each condition, more training samples are needed for training the model. A sampling technique has been implemented to create more training samples. Raw vibration signals are sampled with the same time step size until the end of the sequence. New sequences or samples are designed with the same training sequence length in each condition, while the recording period of the original experiment is not changed. Then, new RULs are calculated using the rule of three from the actual RUL for the new time step, and the threshold point is marked for training. Next, the processing datasets are trained and tested with the model. The first time step, which divides the normal

Table 1 The experimental operating conditions

Condition	Loads
1	4000 N, 1800 rpm
2	4200 N, 1650 rpm
3	5000 N, 1500 rpm

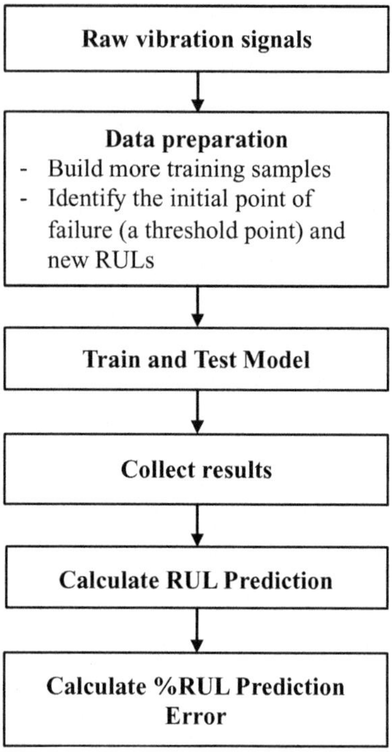

Fig. 1 RUL prediction procedure

and abnormal signals of the prediction graph, is selected to calculate the RUL and compare it to the actual results of the competition.

3.2 RUL Prediction Model

In this study, the model uses a combined CNN and LSTM architecture. The model uses stack layers, including an input layer, a convolutional layer, a batch normalisation, an activation function, a pooling layer, a dropout, two fully connected layers, an LSTM layer, a softmax layer and an output layer. An illustration of the detailed structure can be found in Fig. 2.

The 1-D time series sequence is an input passed through the first layer that is part of the CNN. This structure consists of layers: convolutional, pooling and fully connected. In addition to these three layers, as mentioned above, the dropout layer and activation function are two other essential factors in this structure. The dropout layer is applied for regularisation to prevent overfitting. The model's size is reduced by removing a small number of neurons during training. The activation function is used to help the neural network select the most valuable data while discarding irrelevant information. In general, there are two types of activation functions: linear

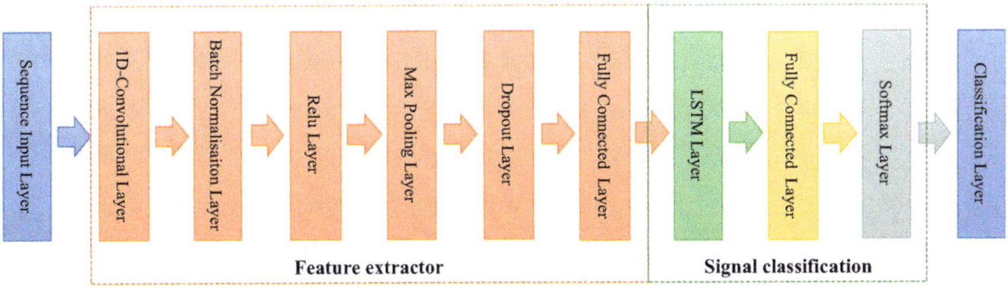

Fig. 2 A structure of CNN in combination with LSTM

and nonlinear activation functions. For our case, a nonlinearity called the rectified linear unit (ReLU) can be introduced into the proposed model.

The input features can extract the different characteristics by the convolutional layer. A batch normalisation layer between the convolutional layer and activation function layer can speed up the CNN training process and decrease the sensitivity to network initialisation. Next to the convolutional layer, the max pooling layer downsamples the input by dividing it into 1-D pooling regions and calculating the maximum for each region. As a result, the convolved feature map is scaled down before moving to the next layer.

In a CNN architecture, a fully connected layer represents the last few levels before the output layer. The input is multiplied by a weight matrix, followed by a bias vector, which is the input of the fully connected layer. Consequently, every input influences every output, thus allowing all layer-to-layer connections to exist. At the end of this process, the extracted signals are sent to the signal classification stage. Finally, 1-D signal sequences are fed into the LSTM layer, which can be used to learn the order dependence of sequences.

The structure of an LSTM cell is shown in Fig. 3. The LSTM cell contains four gates: the cell state gate, the forget gate, the input gate, and the output gate [72]. Additionally, an LSTM block contains three inputs, namely, the cell state c^{t-1}, the previously hidden state h^{t-1}, and the current input x^t, and three outputs, namely, the cell state c^t, the hidden state h^t, and the current output y^t, while b, b_f, b_i and b_o are the biases for the cell state gate, the forget gate, the input gate, and the output gate, respectively. The purpose of each gate and the mathematical formula that governs LSTM can be described in Table 2.

The LSTM layer classified the signals into two groups: normal and abnormal signals. The first time step, dividing the normal and abnormal signals, is a starting point for diagnosis. From this point until the end of the time steps, the time span is converted into a time unit, and the RUL prediction is calculated.

Fig. 3 LSTM cell [72]

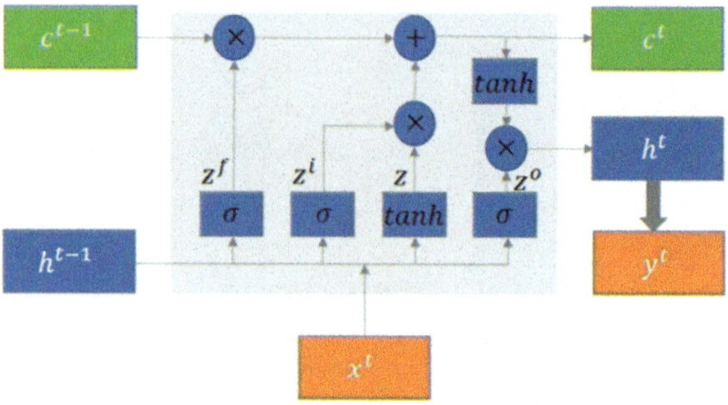

Table 2 A description of an LSTM cell

Component	Purpose	Formula
Cell state gate (z)	Track information over time	$z = \tan h(w[h_{t-1}, x_t] + b)$
Forget gate (z^f)	Control how much value is kept in a cell	$z^f = \sigma(w_f[h_{t-1}, x_t] + b_f)$
Input gate (z^i)	Control the amount of value flowing through a cell	$z^i = \sigma(w_i[h_{t-1}, x_t] + b_i)$
Output gate (z^o)	Control the amount of value extracted from the cell to calculate the output	$z^o = \sigma(w_o[h_{t-1}, x_t] + b_O)$
Cell state at time step $t(c^t)$		$c^t = z^f * c^{t-1} + z^i * z$
Hidden state at time step $t(h^t)$		$h^t = z^o * \tan h(c^t)$

4 Hyperparameter Search for the RUL of Ball Bearings

4.1 *The BA_2 Algorithm*

The two-parameter Bees Algorithm (BA_2) is an advanced version of the basic Bees Algorithm (BA) [73] that uses the traplining metaphor to combine explorative and exploitative search strategies. The algorithm runs using only two parameters, the number of scout bees (n) and worker bees on the elite patches (nep) [74]. The recruitment and search mechanisms distinguish BA_2 from BA. The recruited bees are generated once by analysing the rank of a patch compared to other patches. A linear function determines worker bee populations in each patch as in Eq. (1).

$$w_k = w_{\max} + (k - 1)\left(\frac{w_{\min} - w_{\max}}{n - 1}\right) \tag{1}$$

In the above formula, w_k is the number of worker bees in the k'th patch, $w_{\max}$ and $w_{\min}$ are the maximum and minimum number of workers, which are equal to nep and 1, respectively, and k is the rank of the patch.

As mentioned above, neither exploration nor exploitation is separated in BA_2's foraging mechanism. Bees swarm densely in areas with good chances of success in finding plentiful amounts of nectar, and in poorer areas, they are likely to move further afield. Each site has a unique size of the search neighbourhood (ngh_k). The radius of the neighbourhood does not represent the maximum radius but rather the most densely populated area. It is presented in Eq. (2).

$$for\,aging\,point_{k,w_k} = \begin{cases} patch\,point \pm T\left[0, ngh_k, 1\right] * (x_{\max} - x_{\min}) \\ x_{\min;} for\,aging\,point_{k,w_k} < x_{\min} \\ x_{\max;} for\,aging\,point_{k,w_k} > x_{\max} \end{cases} \quad (2)$$

Based on the value of ngh_k, a probability distribution ($T\left[0, ngh_k, 1\right]$) controls the ratio of exploitation and exploration, where $x_{\max}$ and $x_{\min}$ are the upper and lower bounds of the search space. The finest patch would have a maximum number of worker bees and a small swarming radius ($ngh_k = 0$). On the other hand, a patch of less desirable quality will have fewer worker bees and a larger swarming radius ($ngh_k = 1$). A flowchart showing the BA_2 algorithm is shown in Fig. 4.

4.2 Experimental Setup

According to this study, BA_2 was used for the CNN in combination with the LSTM hyperparameter search. The initial parameters of BA_2, namely, n and nep, are set, and then the stopping criterion is defined. The hyperparameter range (maximum and minimum values) is defined for six hyperparameters: (1) number of filters, (2) filter sizes, (3) number of hidden units, (4) dropout probability, (5) initial learning rate, and (6) mini-batch size. After that, the hyperparameters are automatically decided and tuned to achieve the maximum accuracy they can achieve. The algorithm is followed by the following steps:

- Setting the initial BA_2 parameter and the maximum number of iterations for the stopping criterion. In our implementation, the values were $n = 5$, $nep = 6$ and maximum epochs = 100
- Setting the search range for the algorithm. The upper and lower bounds are different in various hyperparameters and can be adjusted to search in any arbitrary range. In our implementation, the search spaces are shown in Table 3.
- A set of n models is then generated with hyperparameters within the defined range, and each of the models is trained for a set number of epochs.

For each site, bees are then assigned to conduct an explorative and exploitative search depending on the fitness of the patch. These sites are then sorted in ascending order, with the best site producing the minimum loss. The best sites are then searched

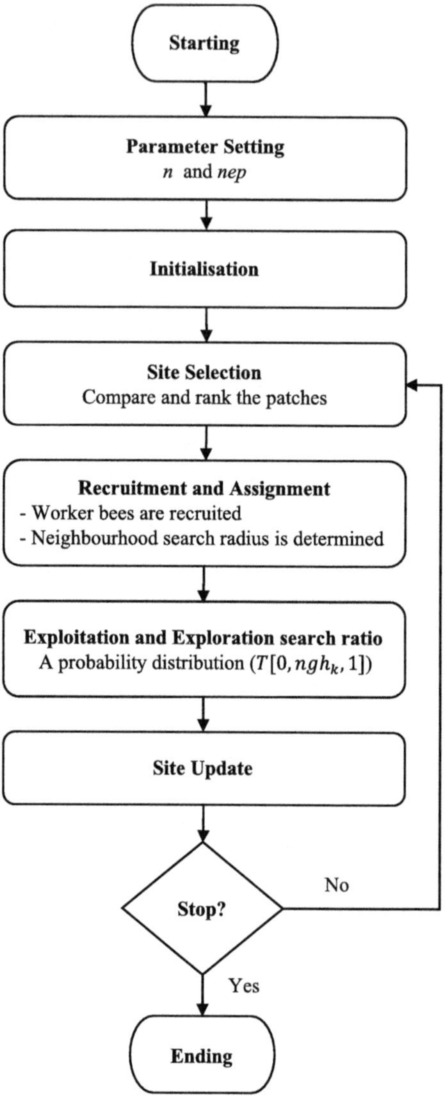

Fig. 4 Flowchart for the BA_2 algorithm

Table 3 The search space of the hyperparameters

Hyperparameter	Lower limit	Upper limit
(1) Number of filters	4	256
(2) Filter sizes	2	50
(3) Number of hidden units	10	200
(4) Dropout probability	0.1	0.9
(5) Learning rate	0.00001	0.01
(6) Mini-batch size	128	2048

more exploitatively, whereas the sites that had high losses are searched more exploratively. These steps are repeated for a predefined number of iterations, and then the best bee is saved in the best bee variable of each iteration.

5 Results

The algorithm is able to search for the best set of hyperparameters. The loss of the model decreased with each iteration of BA_2 and remained constant as it searched for better and better hyperparameter combinations. An example of the model training and loss is plotted for each iteration of BA_2, as shown in Fig. 5.

The best hyperparameters were determined for the CNN + LSTM model to train and test on the IEEE PHM 2012 Prognostic challenge dataset [55]. The % prediction error on test set i can be calculated using Eq. (3), where $ActRUL_i$ is an RUL prediction from our model and RUL_i is the actual RUL from the IEEE PHM 2012 challenge. Additionally, early predictions ($\%Er_i > 0$) score is not treated the same as late predictions ($\%Er_i \leq 0$). The score calculation followed Eqs. (4) and (5). All the results of the experiment are summarised in Table 4.

$$\%Er_i = 100 \times \frac{ActRUL_i - RUL_i}{ActRUL_i} \quad (3)$$

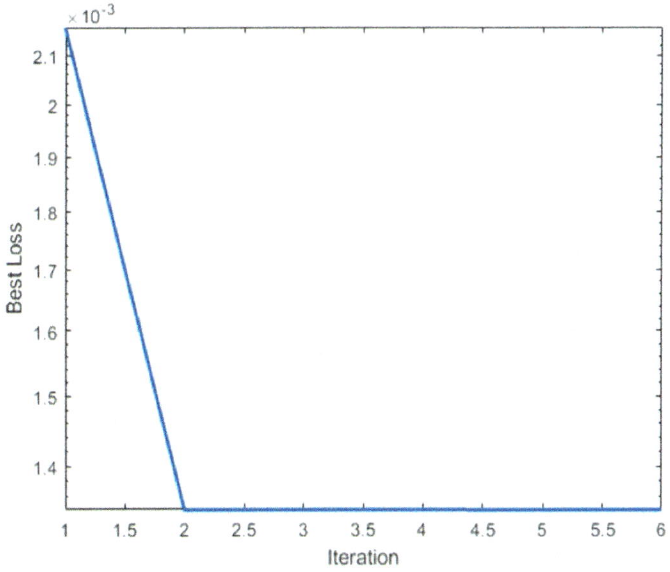

Fig. 5 The best loss vs iterations for the model

Table 4 RUL prediction and model score

Bearing test set	Actual RUL (s) [55]	Predicted RUL (s)	ER_i of the winner [56]	ER_i of our model
B13	5730	4520	37%	– 26.77%
B14	339	590	80%	42.54%
B15	1610	1420	9%	– 13.38%
B16	1460	1420	– 5%	– 2.82%
B17	7570	6920	– 2%	– 9.39%
B23	7530	7670	64%	1.83%
B24	1390	1770	10%	21.47%
B25	3090	2540	– 440%	– 21.65%
B26	1290	1370	49%	5.84%
B27	580	633	– 317%	8.30%
B33	820	1260	90%	34.92%
SCORE			0.3066	0.4259

$$A_i = \begin{cases} \exp^{-ln(0.5).\left(\frac{Er_i}{5}\right)} & if\ Er_i \leq 0 \\ \exp^{+ln(0.5).\left(\frac{Er_i}{20}\right)} & if\ Er_i > 0 \end{cases} \tag{4}$$

$$Score = \frac{1}{11}\sum_{i=1}^{11}(A_i) \tag{5}$$

Compared to the challenge winner, eight of eleven test sets, except for B15, B17 and B24, provide better accuracy than the winner's results, accounting for 27.27% of the total. Overall, the model outperforms the challenge winning results by 38.97% of the score. The study model can produce very accurate results for the remaining useful life of ball bearings.

6 Conclusion

This chapter proposes to apply BA_2, adapted from BA, to search for optimal hyperparameters of the two deep learning combinations and estimate the RUL of ball bearings. Thus, we created a model that searches for hyperparameters in a CNN + LSTM deep learning model. The best set of hyperparameters in each condition of the ball bearing is selected for RUL prediction. This study shows that hyperparameter searching can be significantly optimised and applied to a prediction problem. The results showed massive improvements. It can be confirmed that this approach is general enough to search for hyperparameters for any deep learning model efficiently. Further research should focus on using BA_2 with parameters in combination with hyperparameter search to increase the prediction efficiency of deep learning models.

Acknowledgements This research was funded by the Royal Thai Army scholarship, which is associated with Nathinee's research study. The authors would like to thank Dr A.H. Ismail for providing BA_2 MATLAB code and the University of Birmingham for offering access to their high-performance computing resources.

References

1. BS EN 13306: 2017 BSI Standards Publication Maintenance—Maintenance Terminology
2. Nectoux P et al (2012) PRONOSTIA: an experimental platform for bearings accelerated degradation tests [Online]. Available: https://www.researchgate.net/publication/258028751
3. Nata C, Laurence, Hartono N, Cahyadi L (2021) Implementation of condition-based and predictive-based maintenance using vibration analysis. In: 2021 4th International conference of computer and informatics engineering (IC2IE), pp 90–95. https://doi.org/10.1109/IC2IE53219.2021.9649400
4. Kahr M, Kovács G, Loinig M, Brückl H (2022) Condition monitoring of ball bearings based on machine learning with synthetically generated data. Sensors 22(7):1–17. https://doi.org/10.3390/s22072490
5. Abu-Zeid MA, Abdel-Rahman SM (2013) Bearing problems' effects on the dynamic performance of pumping stations. Alex Eng J 52(3):241–248. https://doi.org/10.1016/j.aej.2013.02.002
6. Chen J, Chen D, Liu G (2021) Using temporal convolution network for remaining useful lifetime prediction. Eng Rep 3(3):1–17. https://doi.org/10.1002/eng2.12305
7. Mao W, He J, Tang J, Li Y (2018) Predicting remaining useful life of rolling bearings based on deep feature representation and long short-term memory neural network. Adv Mech Eng 10(12):1–18. https://doi.org/10.1177/1687814018817184
8. Rezaeianjouybari B, Shang Y (2020) Deep learning for prognostics and health management: state of the art, challenges, and opportunities. Measurement 163:107929. https://doi.org/10.1016/j.measurement.2020.107929
9. Yang J, Peng Y, Xie J, Wang P (2022) Remaining useful life prediction method for bearings based on lstm with uncertainty quantification. Sensors 22(12). https://doi.org/10.3390/s22124549
10. Fink O, Wang Q, Svensén M, Dersin P, Lee W-J, Ducoffe M (2020) Potential, challenges and future directions for deep learning in prognostics and health management applications. Eng Appl Artif Intell 92:103678. https://doi.org/10.1016/j.engappai.2020.103678
11. Ren L, Sun Y, Wang H, Zhang L (2018) Prediction of bearing remaining useful life with deep convolution neural network. IEEE Access 6:13041–13049. https://doi.org/10.1109/ACCESS.2018.2804930
12. Razzak MI, Naz S, Zaib A (2018) Deep learning for medical image processing: overview, challenges and the future. In: Dey N, Ashour AS, Borra S (eds) Classification in bioapps: automation of decision making. Springer International Publishing, Cham, pp 323–350. https://doi.org/10.1007/978-3-319-65981-7_12
13. Jiao L, Zhao J (2019) A survey on the new generation of deep learning in image processing. IEEE Access 7:172231–172263. https://doi.org/10.1109/ACCESS.2019.2956508
14. Maier A, Syben C, Lasser T, Riess C (2019) A gentle introduction to deep learning in medical image processing. Z Med Phys 29(2):86–101. https://doi.org/10.1016/j.zemedi.2018.12.003
15. Monga V, Li Y, Eldar YC (2021) Algorithm unrolling: interpretable, efficient deep learning for signal and image processing. IEEE Signal Process Mag 38(2):18–44. https://doi.org/10.1109/MSP.2020.3016905

16. Wang Z, Chen J, Hoi SCH (2021) Deep learning for image super-resolution: a survey. IEEE Trans Pattern Anal Mach Intell 43(10):3365–3387. https://doi.org/10.1109/TPAMI.2020.2982166
17. Deng L, Hinton G, Kingsbury B (2013) New types of deep neural network learning for speech recognition and related applications: an overview. In: 2013 IEEE international conference on acoustics, speech and signal processing, pp 8599–8603. https://doi.org/10.1109/ICASSP.2013.6639344
18. Deng L et al (2013) Recent advances in deep learning for speech research at microsoft. In: 2013 IEEE International conference on acoustics, speech and signal processing, pp 8604–8608. https://doi.org/10.1109/ICASSP.2013.6639345
19. Noda K, Yamaguchi Y, Nakadai K, Okuno HG, Ogata T (2015) Audio-visual speech recognition using deep learning. Appl Intell 42(4):722–737. https://doi.org/10.1007/s10489-014-0629-7
20. Amodei D et al (2016) Deep speech 2 : end-to-end speech recognition in english and mandarin. In: Balcan MF, Weinberger KQ (eds) Proceedings of the 33rd international conference on machine learning, Proceedings of machine learning research, vol 48. New York, USA. PMLR, pp 173–182 [Online]. Available: https://proceedings.mlr.press/v48/amodei16.html
21. Nassif AB, Shahin I, Attili I, Azzeh M, Shaalan K (2019) Speech recognition using deep neural networks: a systematic review. IEEE Access 7:19143–19165. https://doi.org/10.1109/ACCESS.2019.2896880
22. Hoang DT, Kang HJ (2019) A survey on deep learning based bearing fault diagnosis. Neurocomputing 335:327–335. https://doi.org/10.1016/j.neucom.2018.06.078
23. Zhang S, Zhang S, Wang B, Habetler TG (2020) Deep learning algorithms for bearing fault diagnostics—a comprehensive review. IEEE Access 8:29857–29881. https://doi.org/10.1109/ACCESS.2020.2972859
24. Neupane D, Seok J (2020) Bearing fault detection and diagnosis using case western reserve university dataset with deep learning approaches: a review. IEEE Access 8:93155–93178. https://doi.org/10.1109/ACCESS.2020.2990528
25. Khorram A, Khalooei M, Rezghi M (2021) End-to-end CNN + LSTM deep learning approach for bearing fault diagnosis. Appl Intell 51(2):736–751. https://doi.org/10.1007/s10489-020-01859-1
26. Sanakkayala DC et al (2022) Explainable AI for bearing fault prognosis using deep learning techniques. Micromachines (Basel) 13(9). https://doi.org/10.3390/mi13091471
27. Hakim M, Omran AAB, Ahmed AN, Al-Waily M, Abdellatif A (2023) A systematic review of rolling bearing fault diagnoses based on deep learning and transfer learning: taxonomy, overview, application, open challenges, weaknesses and recommendations. Ain Shams Eng J 14(4):101945. https://doi.org/10.1016/j.asej.2022.101945
28. Li X, Ding Q, Sun JQ (2018) Remaining useful life estimation in prognostics using deep convolution neural networks. Reliab Eng Syst Saf 172:1–11. https://doi.org/10.1016/j.ress.2017.11.021
29. Wang F, Liu X, Deng G, Yu X, Li H, Han Q (2019) Remaining life prediction method for rolling bearing based on the long short-term memory network. Neural Process Lett 50(3):2437–2454. https://doi.org/10.1007/s11063-019-10016-w
30. Zhang B, Zhang S, Li W (2019) Bearing performance degradation assessment using long short-term memory recurrent network. Comput Ind 106:14–29. https://doi.org/10.1016/j.compind.2018.12.016
31. Yao D, Li B, Liu H, Yang J, Jia L (2021) Remaining useful life prediction of roller bearings based on improved 1D-CNN and simple recurrent unit. Measurement (Lond) 175:109166. https://doi.org/10.1016/j.measurement.2021.109166
32. Liu ZH et al (2021) A regularized LSTM method for predicting remaining useful life of rolling bearings. Int J Autom Comput 18(4): 581–593. https://doi.org/10.1007/s11633-020-1276-6
33. Goodfellow I, Bengio Y, Courville A (2017) Deep learning book [Online]. Available: http://www.deeplearningbook.org/
34. Yu T, Zhu H (2020) Hyper-parameter optimization: a review of algorithms and applications, pp 1–56 [Online]. Available: http://arxiv.org/abs/2003.05689

35. Kaur S, Aggarwal H, Rani R (2020) Hyper-parameter optimization of deep learning model for prediction of Parkinson's disease. Mach Vis Appl 31(5):1–15. https://doi.org/10.1007/s00138-020-01078-1
36. Wu J, Chen XY, Zhang H, Xiong LD, Lei H, Deng SH (2019) Hyperparameter optimization for machine learning models based on Bayesian optimization. J Electron Sci Technol 17(1):26–40. https://doi.org/10.11989/JEST.1674-862X.80904120
37. Kong D, Wang S, Ping P (2022) State-of-health estimation and remaining useful life for lithium-ion battery based on deep learning with Bayesian hyperparameter optimization. Int J Energy Res 46(5):6081–6098. https://doi.org/10.1002/er.7548
38. Subramanian M, Shanmugavadivel K, Nandhini PS (2022) On fine-tuning deep learning models using transfer learning and hyper-parameters optimization for disease identification in maize leaves. Neural Comput Appl 34(16):13951–13968. https://doi.org/10.1007/s00521-022-07246-w
39. Zhou Y, Huang Y, Pang J, Wang K (2019) Remaining useful life prediction for supercapacitor based on long short-term memory neural network. J Power Sources 440:227149. https://doi.org/10.1016/j.jpowsour.2019.227149
40. Lorenzo PR, Nalepa J, Kawulok M, Ramos LS, Pastor JR (2017) Particle swarm optimization for hyper-parameter selection in deep neural networks. In: GECCO 2017—proceedings of the 2017 genetic and evolutionary computation conference, pp 481–488. https://doi.org/10.1145/3071178.3071208
41. Kim TY, Cho SB (2019) Particle swarm optimization-based CNN-LSTM networks for forecasting energy consumption. In: 2019 IEEE congress on evolutionary computation, CEC 2019—proceedings, pp 1510–1516. https://doi.org/10.1109/CEC.2019.8789968
42. İnik Ö (2023) CNN hyper-parameter optimization for environmental sound classification. Appl Acoust 202:109168. https://doi.org/10.1016/j.apacoust.2022.109168
43. Bouktif S, Fiaz A, Ouni A, Serhani MA (2020) Multi-sequence LSTM-RNN deep learning and metaheuristics for electric load forecasting. Energies (Basel) 13(2). https://doi.org/10.3390/en13020391
44. Bacanin N, Stoean C, Zivkovic M, Rakic M, Strulak-Wójcikiewicz R, Stoean R (2023) On the benefits of using metaheuristics in the hyperparameter tuning of deep learning models for energy load forecasting. Energies (Basel) 16(3). https://doi.org/10.3390/en16031434
45. Yuliyono AD, Girsang AS (2019) Artificial bee colony-optimized LSTM for bitcoin price prediction. Adv Sci Technol Eng Syst 4(5):375–383. https://doi.org/10.25046/aj040549
46. Erkan U, Toktas A, Ustun D (2022) Hyperparameter optimization of deep CNN classifier for plant species identification using artificial bee colony algorithm. J Ambient Intell Humaniz Comput Dudani 1976. https://doi.org/10.1007/s12652-021-03631-w
47. Mittal S, Sangwan OP (2020) ABC LSTM optimizing parameters of deep LSTM using ABC algorithm for big datasets. Int J Eng Adv Technol 9(5):221–226. https://doi.org/10.35940/ijeat.d7649.069520
48. Strumberger I, Tuba E, Bavcanin N, Zivkovic M, Beko M, Tuba M (2019) Designing convolutional neural network architecture by the firefly algorithm. In: 2019 International young engineers forum (YEF-ECE), pp 59–65
49. Aufa BZ, Suyanto S, Arifianto A (2020) Hyperparameter setting of LSTM-based language model using grey wolf optimizer. In: 2020 International conference on data science and its applications, ICoDSA 2020. https://doi.org/10.1109/ICoDSA50139.2020.9213031
50. Cuong-Le T, Minh H-L, Sang-To T, Khatir S, Mirjalili S, Abdel Wahab M (2022) A novel version of grey wolf optimizer based on a balance function and its application for hyperparameters optimization in deep neural network (DNN) for structural damage identification. Eng Fail Anal 142:106829. https://doi.org/10.1016/j.engfailanal.2022.106829
51. Nematzadeh S, Kiani F, Torkamanian-Afshar M, Aydin N (2022) Tuning hyperparameters of machine learning algorithms and deep neural networks using metaheuristics: a bioinformatics study on biomedical and biological cases. Comput Biol Chem 97:107619. https://doi.org/10.1016/j.compbiolchem.2021.107619

52. Lohvithee M, Sun W, Chretien S, Soleimani M (2021) Ant colony-based hyperparameter optimisation in total variation reconstruction in X-ray computed tomography. Sensors 21(2). https://doi.org/10.3390/s21020591
53. Trajkovski A, Madjarov G (2022) Model hyper parameter tuning using ant colony optimization, pp 37–41
54. Tong Y, Yu B (2022) Research on hyper-parameter optimization of activity recognition algorithm based on improved cuckoo search. Entropy 24(6). https://doi.org/10.3390/e24060845
55. IEEE (2012) IEEE PHM 2012 prognostic challenge—outline, experiments, scoring of results, winners. IEEE PHM, pp 1–11
56. Sutrisno E, Oh H, Vasan ASS, Pecht M (2012) Estimation of remaining useful life of ball bearings using data driven methodologies. In: PHM 2012—2012 IEEE international conference on prognostics and health management: enhancing safety, efficiency, availability, and effectiveness of systems through PHM technology and application, conference program, vol 2, pp 1–7. https://doi.org/10.1109/ICPHM.2012.6299548
57. Wang Y, Zhao Y, Addepalli S (2020) Remaining useful life prediction using deep learning approaches: a review. Procedia Manuf 49(2019):81–88. https://doi.org/10.1016/j.promfg.2020.06.015
58. Lei Y, Li N, Guo L, Li N, Yan T, Lin J (2018) Machinery health prognostics: a systematic review from data acquisition to RUL prediction. Mech Syst Signal Process 104:799–834. https://doi.org/10.1016/j.ymssp.2017.11.016
59. Son S, Oh K-Y (2022) Integrated framework for estimating remaining useful lifetime through a deep neural network. Appl Soft Comput 122:108879. https://doi.org/10.1016/j.asoc.2022.108879
60. Ferreira C, Gonçalves G (2022) Remaining useful life prediction and challenges: a literature review on the use of machine learning methods. J Manuf Syst 63:550–562. https://doi.org/10.1016/j.jmsy.2022.05.010
61. Yan H, Wan J, Zhang C, Tang S, Hua Q, Wang Z (2018) Industrial big data analytics for prediction of remaining useful life based on deep learning. IEEE Access 6:17190–17197. https://doi.org/10.1109/ACCESS.2018.2809681
62. Ozkat EC, Bektas O, Nielsen MJ, la Cour-Harbo A (2023) A data-driven predictive maintenance model to estimate RUL in a multi-rotor UAS. Inter J Micro Air Veh 15. https://doi.org/10.1177/17568293221150171
63. Mitici M, de Pater I, Barros A, Zeng Z (2023) Dynamic predictive maintenance for multiple components using data-driven probabilistic RUL prognostics: the case of turbofan engines. Reliab Eng Syst Saf 234:109199. https://doi.org/10.1016/j.ress.2023.109199
64. Shaheen B, Kocsis Á, Németh I (2023) Data-driven failure prediction and RUL estimation of mechanical components using accumulative artificial neural networks. Eng Appl Artif Intell 119:105749. https://doi.org/10.1016/j.engappai.2022.105749
65. Gawde S, Patil S, Kumar S, Kamat P, Kotecha K, Abraham A (2023) Multi-fault diagnosis of industrial rotating machines using data-driven approach: a review of two decades of research. Eng Appl Artif Intell 123:106139. https://doi.org/10.1016/j.engappai.2023.106139
66. Medjaher M, Zerhouni N, Baklouti J (2013) Data-driven prognostics based on health indicator construction: application to PRONOSTIA's data. In: 2013 European control conference (ECC), pp 1451–1456. https://doi.org/10.23919/ecc.2013.6669223
67. Guo L, Li N, Jia F, Lei Y, Lin J (2017) A recurrent neural network based health indicator for remaining useful life prediction of bearings. Neurocomputing 240:98–109. https://doi.org/10.1016/j.neucom.2017.02.045
68. Wang B, Lei Y, Yan T, Li N, Guo L (2020) Recurrent convolutional neural network: a new framework for remaining useful life prediction of machinery. Neurocomputing 379:117–129. https://doi.org/10.1016/j.neucom.2019.10.064
69. Al-Dulaimi A, Zabihi S, Asif A, Mohammadi A (2019) A multimodal and hybrid deep neural network model for remaining useful life estimation. Comput Ind 108:186–196. https://doi.org/10.1016/j.compind.2019.02.004

70. Li J, Li X, He D (2019) A directed acyclic graph network combined with CNN and LSTM for remaining useful life prediction. IEEE Access 7:75464–75475. https://doi.org/10.1109/ACCESS.2019.2919566
71. Abdelli K, Grießer H, Pachnicke S (2021) A hybrid CNN-LSTM approach for laser remaining useful life prediction. In: Optics infobase conference papers, no June, pp 1–4. https://doi.org/10.1364/oecc.2021.s3d.3
72. Xu G, Ren T, Chen Y, Che W (2020) A one-dimensional CNN-LSTM model for epileptic seizure recognition using EEG signal analysis. Front Neurosci 14:1–9. https://doi.org/10.3389/fnins.2020.578126
73. Pham DT, Ghanbarzadeh A, Koç E, Otri S, Rahim S, Zaidi M (2006) The bees algorithm—a novel tool for complex optimisation problems. In: Intelligent production machines and systems—2nd I*PROMS virtual international conference 3–14 July 2006, no December, pp 454–459. https://doi.org/10.1016/B978-008045157-2/50081-X
74. Ismail AH, Pham DT (2023) Bees traplining metaphors for the vehicle routing problem using a decomposition approach. In: Pham BT, Hartono N (eds) Intelligent production and manufacturing optimisation—the bees algorithm approach. Springer International Publishing, Cham, pp 261–287. https://doi.org/10.1007/978-3-031-14537-7_16

Bees Local Phase Quantisation Feature Selection for RGB-D Facial Expression Recognition

Seyed Muhammad Hossein Mousavi and Atiye Ilanloo

Abstract Feature selection can be defined as an optimisation problem and solved by bioinspired algorithms. The Bees Algorithm (BA) returns great performance in the feature selection optimisation task. On the other hand, local phase quantisation (LPQ) is a frequency domain feature that has excellent performance on depth images. Here, after extracting LPQ features from RGB (colour) and depth images from the Iranian Kinect Face Database (IKFDB), the Bees feature selection algorithm is applied to select the desired number of features for final classification tasks. IKFDB is recorded with Kinect sensor V.2 and contains colour and depth images for facial and facial microexpression recognition purposes. Here, five facial expressions, Anger, Joy, Surprise, Disgust and Fear, are used for final validation. The proposed Bees LPQ method is compared with Particle Swarm Optimisation (PSO) LPQ, PCA LPQ, Lasso LPQ, and just LPQ features for classification tasks with Support Vector Machines (SVM), K-Nearest Neighbourhood (KNN), Shallow Neural Network and Ensemble Subspace KNN. The returned results show a promising performance of the proposed algorithm (99% accuracy) in comparison with others.

Keywords Bees algorithm · Feature selection · Local phase quantisation · Optimisation · Kinect · Facial expressions · Depth images

S. M. H. Mousavi (✉)
Tehran, Iran
e-mail: mosavi.a.i.buali@gmail.com

A. Ilanloo
Faculty of Humanities-Psychology, Islamic Azad University, Rasht, Gilan, Iran

D. T. Pham and N. Hartono (eds.), *Intelligent Engineering Optimisation with the Bees Algorithm*, Springer Series in Advanced Manufacturing,
https://doi.org/10.1007/978-3-031-64936-3_12

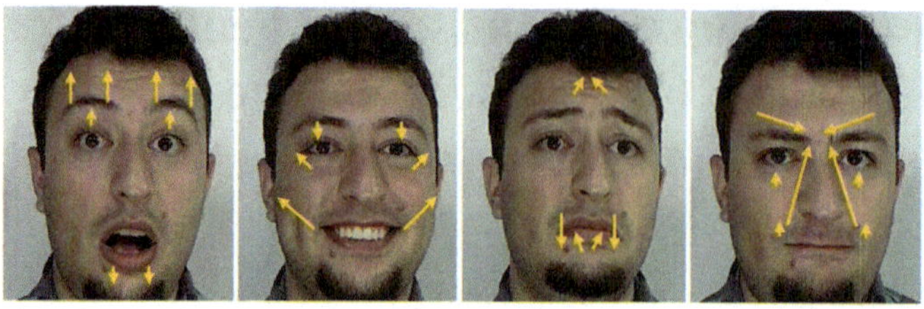

Fig. 1 Four main facial expressions—these images were adapted from IKFDB [2] by adding arrows to represent Action Units (AUs) (Licence: https://creativecommons.org/licenses/by/4.0/)

1 Introduction

1.1 Facial Expressions Recognition

After face detection [1] and face recognition [2], analysing facial muscles' movements comes to mind. Recognising these movements and micromovements is called facial expression recognition (FER) [2] and facial microexpression recognition (FMER) [2], which have high importance in image processing and psychology. Practical applications are human-computer interaction (HCI), such as face image processing, facial video surveillance, and facial animation. Recognising facial expressions is harder than recognising faces, as they could have micromovements and differ from one person to another, which is the main problem in FER. Additionally, each expression could be misunderstood in different races, as skin wrinkles affect the recognition direction. Facial expressions are made and calculated by Facial Action Coding Systems (FACS) [3], and they are categorised into seven primary facial expressions of joy, anger, disgust, sadness, fear, surprise and neutral. For example, for Joy expression, Action Units (AU) of + and 12 are involved. FMER is FER with more precise calculations, as few changes appear on the face. Some of these expressions are illustrated in Fig. 1 alongside their corresponding action units.

1.2 Bioinspired Algorithms and Bees Algorithm

Bioinspired or nature-inspired algorithms [4, 5] are normally mathematical models of animal and insect social behaviour in a manner that leads to problems with an optimal solution during iterations by a specific number of populations in each generation. These algorithms could be employed in multiple optimisation tasks, such as regression [6], clustering [7, 8], feature selection [9], minimum spanning tree (MST), and hub location allocation (HLA). One of these bioinspired algorithms that has high efficiency is called the Bees Algorithm (BA) [10–12]. The Bees Algorithm has many applications in which the Bees Algorithm simply implements the social behaviour of honey bees in searching for food in flowers. Agents, scouts and Forager bees are involved in global and local searches to reach the best solution. Waggle dancing is performed by scout bees that found the best sites. Those that landed on elite sites recruit more new members. The list of best bees based on local and global information goes to the next generation and the cycle stops when termination criteria conditions are met.

1.3 Kinect Sensor and Depth Data

There are many infrared sensors, of which the Kinect sensor [13] is one of the most efficient and cheapest for educational purposes. Here, data recorded with the Kinect Version 2 sensor are used in the validation section. Kinect V.2 is capable of recording colour data with 1920 * 1080 resolutions and depth data with 512 * 424 resolutions with 30 frames per second (FPS). Colour data are based on red, green and blue (RGB) channels, and depth data are grey-like images, which are sometimes called 2.5-dimensional (2.5-D) images. Each pixel of the depth image in the Kinect sensor represents the distance between the sensor and the subject in millimetres. The range is 0.8–5 m, and for instance, a value of 2000 pixels means that the object is 2 m from the sensor. Depth images could be transformed into 3-D images and aid colour images to increase recognition accuracy. As depth images are generated from infrared particles, the Kinect sensor could operate in absolute darkness.

1.4 Feature Selection

In dealing with big data [14] or a massive number of samples (especially image samples), and after the feature extraction [15] step, selecting the most impactful features out of data is essential. Feature selection [16] or dimensionality reduction helps to decrease computational time and volume plus eliminating outliers in the dataset. Outliers lead the classification [17] task into misclassification, and removing them is vital. In this paper, the Bees Algorithm is employed as a feature selection tool. Here, the Local Phase Quantisation (LPQ) [18] feature is extracted, which provides 256 features and Bees feature selection and selects fewer than 128 features without a classification accuracy drop. The paper consists of five main sections: introduction, prior related research, proposed method, validation and results and conclusion.

2 Prior Related Research

As this research is about feature selection, famous and benchmark feature selection or dimensionality reduction techniques belonging to traditional and bioinspired categories will be explained. These algorithms could be applied directly to extracted images or signal features or could be applied to unfolded versions of images, signals or any type of numerical matrix to eliminate outliers and reduce computational time.

Principal component analysis (PCA) [19] generates new matrixes, named principal components (PAs). Each PA is a linear mixture of the original matrixes. All the PAs are orthogonal to one another, so there is no extra information. By selecting the best principles components, the best features will be chosen.

On the other hand, Lasso [20] is a regularisation [20] method for estimating generalised linear models. Lasso includes a red line that limits the size of the approximated coefficients. Therefore, it is similar to ridge regression [21]. Lasso is a shrinkage approximator: it makes coefficient approximations that are biased to be tiny. By shrinking features, the best features remain.

The chi-square test [22] is a statistical test employed to contrast observed data and expected data. The objective is to specify if a change between observed data and expected data is because of chance or because of a relationship between the variables that are being studied. Having two variables, it is possible to obtain the observed count and expected count. Chi-square measures how the desired count and observed count deviate from each other. If the target variable is independent of the feature variable, it is possible to ignore that specific feature. If they are dependent, the feature is very significant.

Selecting the best features based on the highest Data Envelopment Analysis (DEA) [23] for Data Management Units (DMUs) [23] is a management-related feature selection technique. In that case, those features or DMUs with the best efficiency will be selected, and the rest will be considered leftovers or outliers [24].

Among bioinspired feature selection techniques, few have proper performance, including PSO feature selection [25], firefly feature selection [26] and bee feature selection [27].

3 Proposed Method

The proposed method consists of multiple parts: data acquisition, preprocessing, LPQ feature extraction, Bee feature selection, labelling, classification and comparison. These operations are illustrated in Fig. 2. For validation, the Iranian Kinect Face Data Base (IKFDB) [2] is employed, which consists of 40 subjects. This dataset contains facial images for facial recognition, age estimation facial expressions and facial micro-expression recognition tasks. Additionally, data are recorded with Kinect sensor V.2 in colour and depth formats. All seven main facial expressions along the side with pitch, roll and yaw as video frames are available in this database.

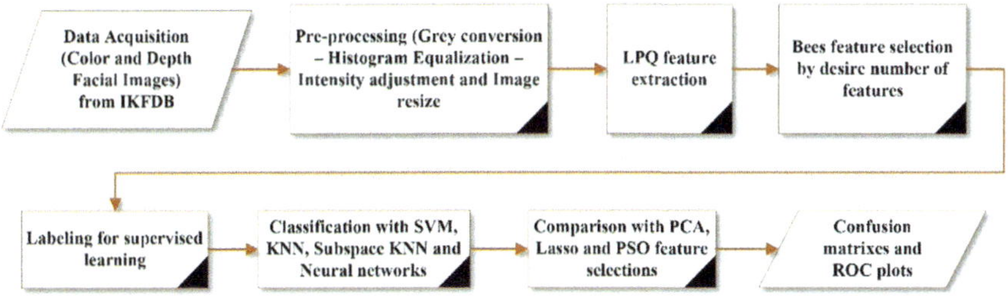

Fig. 2 Proposed method flowchart

Here, five expressions of joy, anger, fear, disgust and surprise in colour and depth formats are used for the main experiment. After reading the data images from the input, the preprocessing stage, including grey level conversion, intensity adjustment, histogram equalisation and image resizing, takes place. The next step is extracting the LPQ feature from the frequency domain from all colour and depth samples. For each image sample, 256 LPQ features will be extracted. The fourth step consists of feature selection by the Bees Algorithm in the desired number of features. Here, the range of 2–255 features could be selected. As the classification task is a supervised learning method, labelling is needed for the fifth step. Classification algorithms are Support Vector Machines (SVM) [5], K-Nearest Neighborhood (KNN) [28], Shallow Neural Network [2] and Ensemble Subspace KNN [29]. Next, comparison by PCA, Lasso, PSO feature selection and LPQ using only the mentioned classification algorithms takes place. Finally, confusion matrixes and receiver operating characteristic (ROC) [30] curve plots were generated.

3.1 *Local Phase Quantisation*

LPQ [18] is a frequency neighbourhood-based feature based on Fourier transform [31]. It manipulated the blurring effect in magnitude and phase channels. The phase channel is capable of deactivating low-pass filters that exist in some images. LPQ features are perfect for use on depth data in the frequency domain. Figure 3 illustrates the LPQ algorithm process. Furthermore, this feature has perfect performance in dealing with low-pass Gaussian filters.

3.2 *Bee Feature Selection*

In feature selection, we address the number of features (NF), the weight of features (w) and the mean square error (MSE) [32], which should be minimised to select the feature. Additionally, if x_i are values of NF, then $x_i^{\wedge}$ would be selected features

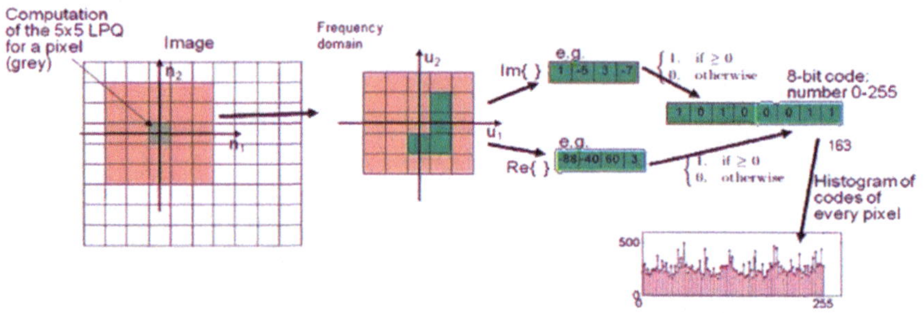

Fig. 3 LPQ algorithm workflow

out of NF. Therefore, considering the number of features entering the system, "y' would be the output and "t" would be the target. To calculate the final error (objective function), e_i must be calculated, which is $t_i - y_i$. Thus, the final error is:

$$\text{Objective Function} = \text{Min}\, MSE = \frac{1}{n}\sum_{i=1}^{n} e_i^2 + w * NF \quad (1)$$

This goes for all features, and finally, those features with the lowest MSE will be selected. In the combination of Bees and feature selection, each feature vector is considered a bee with a different fitness function. Those bees that could fit into the final iteration would be selected alongside their related features with lower error, as mentioned. The pseudocode of Bees feature selection is presented below.

Start
Load LPQ features
Generating the initial population (features of F)
Define FN (number of features) and w (wight of features)
Evaluating the population based on the fitness function
Sorting
While *max iteration is not satisfied*
Select elite patches and non-elite best patches for local search
Recruit forager bees for elite patches and non-elite best patches
Evaluate the fitness value of each patch
Sorting (select NF of F)
Allocate the rest of the bees for global search
Evaluate the fitness value of non-best patches
Sorting (Select NF of F)
End While
Select best first NF's
End

4 Validation and Results

As mentioned earlier for validation, IKFDB [2] data are used, which consist of colour and depth frames of seven main facial expressions. Here, 1000 colour and depth samples of five expressions (each 200 samples) are used in this experiment.

Some samples of IKFDB in different expressions and in colour-depth form are presented in Fig. 4. Additionally, the proposed Bees LPQ, PSO LPQ [25], PCA LPQ [19], Lasso LPQ [20] and solo LPQ using SVM [5], KNN [28], Shallow Neural Network [2] and Ensembles Subspace KNN [29] classification algorithms are compared. Table 1 presents the PSO and Bees parameters that are used in the experiment. Parameters are determined based on the size of the dataset, the number of samples, and many experiments. Increasing the number of iterations and population was not necessary, as it increased the runtime and did not improve the final accuracy. Table 2 presents the classification comparison results for different feature selection algorithms with half of the features (128), a quarter of the features (64) and an octant of features (32). Obviously, all features (256) are considered for solo LPQ in this table.

Figure 5 represents confusion matrixes and ROC curves for solo LPQ with all 256 features, the Lasso algorithm with 64 features and the Bees Algorithm with 64 features for the SVM classifier. Figure 6 illustrates the same but for the KNN classifier. Figure 7 presents the performance of the Bees Algorithm for the feature selection task over 100 iterations.

Looking at Table 2 (results), the best results belong to Bees feature selection, PSO feature selection and PCA feature selection for different numbers of features. Additionally, Lasso feature selections are placed at the end of the ranking according to its weak performance. Additionally, solo LPQ with all 256 feature places is in the middle of the ranking table.

PSO and Bees compete with each other with different classifiers, and the best result ever belongs to the Bees feature selection algorithm with 128 features and by the subspace KNN classifier (99.8% accuracy). However, the weakest performance goes to Lasso with 32 features and with the SVM classifier (84.5% accuracy). Clearly, accuracy is the number of correctly classified samples divided by the number of all samples multiplied by 100. Here, facial expressions are labelled numerically as 1:

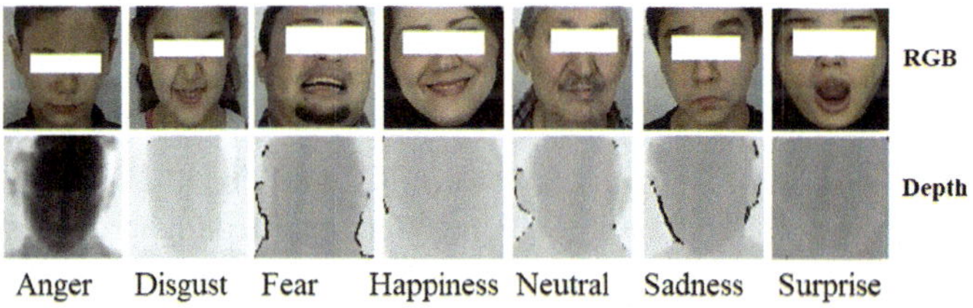

Fig. 4 Sample images taken from IKFDB [2] and changed by covering the subject's eyes (Licence: https://creativecommons.org/licenses/by/4.0/)

Table 1 Bees and PSO parameter

Algorithm	Bees	PSO
Iteration	100	100
Population	10 Bees	20 Particles
Decision variable	20	20
Decision variable size	[1, 20]	[1, 20]
Lower bound (LB)	–10	– 10
Upper bound (UB)	10	10
Mutation rate	0.2	0.2
Inertia weight	–	1
Inertia weight damping ratio	–	0.99
Personal learning coefficient	–	1.5
Global learning coefficient	–	2
Selected sites (SS)	Bees * 0.5	–
Select elite sites	SS * 0.4	–
Recruited bees for selected sites	Bees * 0.5	–
Recruited bees for elite sites	SS * 2	–
Neighbourhood radius	0.1→ * (UB–LB)	–
Neighbourhood radius damp rate	0.2→0.96	–

Table 2 Comparison classification results with different numbers of features (256, 128, 64 and 32)

	LPQ (%)	PCA (%)	Lasso (%)	PSO (%)	Bees (%)
SVM	256: 98.8	128: 98.4 64: 98.0 32: 90.8	128: 90.6 64: 89.3 32: 84.5	128: 99.1 64: 98.6 32: 97.4	128: 99.6 64: 98.9 32: 97.3
KNN	256: 98.0	128: 98.1 64: 97.9 32: 97.5	128: 93.0 64: 93.9 32: 92.8	128: 98.8 64: 98.2 32: 97.6	128: 99.2 64: 98.1 32: 97.7
Shallow NN	256: 97.6	128: 97.3 64: 96.6 32: 95.8	128: 96.7 64: 90.7 32: 91.2	128: 99.1 64: 97.8 32: 96.9	128: 99.4 64: 98.7 32: 97.9
Ensemble subspace KNN	256: 98.3	128: 97.9 64: 97.7 32: 98.1	128: 95.6 64: 93.1 32: 89.1	128: 99.5 64: 98.3 32: 98.0	128: 99.8 64: 98.9 32: 98.5

Anger, 2: Joy, 3: Fear, 4: Disgust and 5: Surprise. Additionally, the area under the ROC curve for Bees features and for all classifiers returned higher values (closer to 1) by having higher true positive rates and lower false positive rates. Additionally, Fig. 7 depicts fast convergence for BA feature selection, which occurred in the initial iterations (iterations 20–25). This means that BA feature selection could obtain

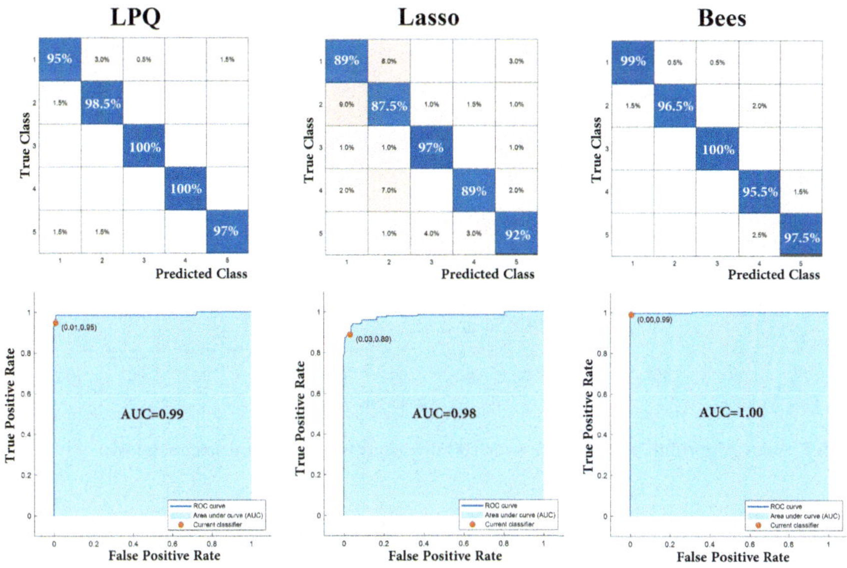

Fig. 5 Confusion matrixes and ROC curves for LPQ, Lasso and Bees algorithms with 256, 64 and 64 features with SVM classifier

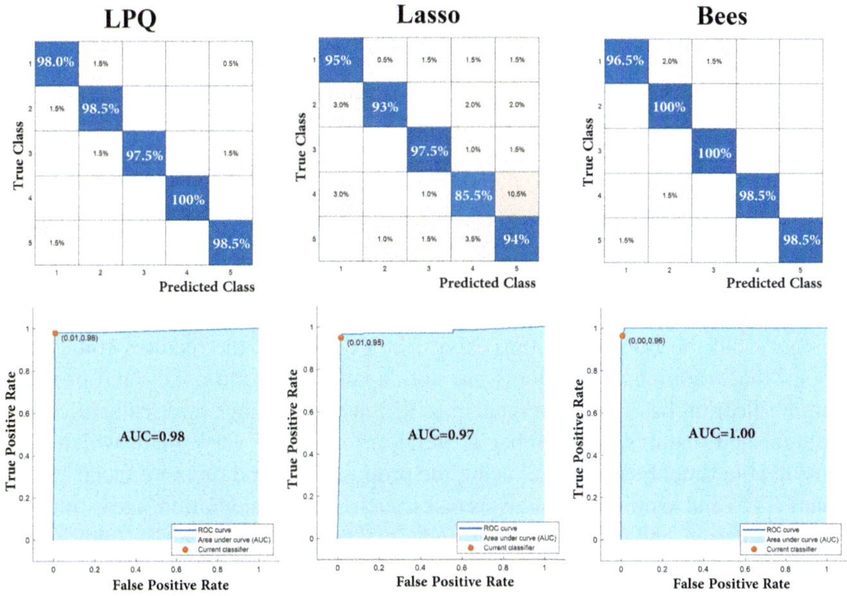

Fig. 6 Confusion matrixes and ROC curves for the LPQ, Lasso and Bees algorithms with 256, 64 and 64 features with the KNN classifier

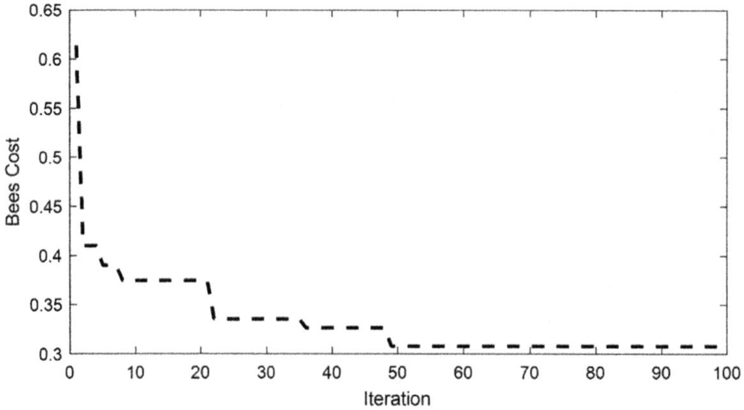

Fig. 7 Bees Algorithm performance over 100 iterations for the feature selection task

proper features very quickly. However, more iterations mean having the best features possible.

5 Conclusion and Suggestions

Employing bioinspired or nature-inspired algorithms for feature selection tasks could outperform traditional feature selection algorithms in face analysis and especially facial expression recognition tasks. The capability of extracting facial impactful features is a very difficult task. BA feature selection was able to perform this task very well, which was significant. Defining the feature selection cost function for the Bees Algorithm returned successful results for all four classification algorithms. Therefore, it can be concluded that the Bees Algorithm has a great impact on feature selection optimisation after feature extraction by LPQ features. However, the run time increases, but the Bees Algorithm using just a quarter of the features could reach a point that traditional algorithms are not capable of. Additionally, the proposed feature selection has fast convergence; in the case of having fewer categories with distinguished features, the number of iterations could be very small, which saves considerable time. Furthermore, using the proposed method on more facial expressions (7–15) and in other face analysis tasks, such as face recognition, age estimation, gender detection, and face recognition under makeup, are future works. Additionally, it is suggested to use the proposed method with more classification algorithms, such as tree and naive Bayes classifiers. In addition, comparing the proposed method with more traditional and bioinspired feature selection techniques, such as the chi-square test and Genetic Algorithm (GA), is another suggestion.

References

1. Mousavi SMH (2018) A new way to age estimation for RGB-D images, based on a new face detection and extraction method for depth images. Int J Image Graph Sig Process 10:10
2. Mousavi SHM, Mirinezhad SY (2021) Iranian kinect face database (IKFDB): a color-depth based face database collected by kinect v.2 sensor. SN Appl Sci 3(1):1–7
3. Ekman P, Friesen WV (1978) Facial action coding system. Environ Psychol Nonverbal Behav
4. Singh A, Sharma S, Singh J (2021) Nature-inspired algorithms for wireless sensor networks: a comprehensive survey. Comput Sci Rev 39:100342
5. Mousavi SM, Hossein VC, Gherman T (2020) An evolutionary pentagon support vector finder method. Expert Syst Appl 150:113284
6. Singh A et al (2021) A Gaussian process regression approach to predict the k-barrier coverage probability for intrusion detection in wireless sensor networks. Expert Syst Appl 172:114603
7. Mousavi SMH (2019) A new clustering method using evolutionary algorithms for determining initial states, and diverse pairwise distances for clustering. Int J Mechatron Electr Comput Technol (IJMEC) 9(31):4098–110
8. Pham DT et al (2007) Data clustering using the Bees algorithm
9. Ji B et al (2020) Bioinspired feature selection: an improved binary particle swarm optimization approach. IEEE Access 8:85989–86002
10. Pham DT et al (2006) The Bees algorithm—a novel tool for complex optimization problems. Intelligent production machines and systems. Elsevier Science Ltd., pp 454–59
11. Pham DT et al (2005) The Bees algorithm. Technical note, Manufacturing Engineering Centre, Cardiff University, UK, pp 44–48
12. Ismail AH et al (2021) Combinatorial Bees algorithm for vehicle routing problem. Macromol Symp 396(1)
13. Gonzalez-Jorge H et al (2015) Metrological comparison between Kinect I and Kinect II sensors. Measurement 70:21–26
14. Charles V, Gherman T (2013) Achieving competitive advantage through big data. Strategic implications. Middle-East J Sci Res 16(8):1069–1074
15. Mousavi SMH, Prasath VS, Mousavi SMH (2019) Persian classical music instrument recognition (PCMIR) using a novel Persian music database. In: 2019 9th International conference on computer and knowledge engineering (ICCKE). IEEE
16. Mousavi SMH, MiriNezhad SY, Mirmoini A (2017) A new support vector finder method, based on triangular calculations and K-means clustering. In: 2017 9th International conference on information and knowledge technology (IKT). IEEE
17. Dezfoulian, Mir Hossein et al (2016) Optimization of the Ho-Kashyap classification algorithm using appropriate learning samples. In: 2016 8th International conference on information and knowledge technology (IKT). IEEE
18. Yuan B, Cao H, Chu J (2012) Combining local binary pattern and local phase quantization for face recognition. In: 2012 International symposium on biometrics and security technologies. IEEE
19. Abdi H, Williams LJ (2010) Principal component analysis. Wiley Interdisc Rev Comput Stat 2(4):433–459
20. Tibshirani R (1996) Regression shrinkage and selection via the Lasso. J Roy Stat Soc: Ser B (Methodol) 58(1):267–288
21. Kibria BM, Banik S (2020) Some ridge regression estimators and their performances
22. Jin X et al (2006) Machine learning techniques and chi-square feature selection for cancer classification using SAGE gene expression profiles. In: International workshop on data mining for biomedical applications. Springer, Berlin, Heidelberg
23. Shiraz R, Charles V, Jalalzadeh L (2014) Fuzzy rough DEA model: a possibility and expected value approaches. Expert Syst Appl 41(2):434–444
24. Zhang Y et al (2014) Feature selection using data envelopment analysis. Knowl-Based Syst 64:70–80

25. Mistry K et al (2016) A micro-GA embedded PSO feature selection approach to intelligent facial emotion recognition. IEEE Trans Cybern 47(6):1496–1509
26. Emary E et al (2015) Firefly optimization algorithm for feature selection. In: Proceedings of the 7th Balkan conference on informatics conference
27. Alomari O, Othman ZA (2012) Bees algorithm for feature selection in network anomaly detection. J Appl Sci Res 8(3):1748–1756
28. Ding H et al (2022) Imbalanced data classification: a KNN and generative adversarial networks-based hybrid approach for intrusion detection. Future Gener Comput Syst 131:240–254
29. Ramasamy K, Balakrishnan K, Velusamy D (2022) Detection of cardiac arrhythmias from ECG signals using FBSE and Jaya optimized ensemble random subspace K-nearest neighbor algorithm. Biomed Signal Process Control 76:103654
30. Mahdizadeh M, Zamanzade E (2022) On estimating the area under the ROC curve in ranked set sampling. Stat Methods Med Res. 09622802221097211
31. Gupta V et al (2022) Detection of R-peaks using fractional Fourier transform and principal component analysis. J Ambient Intell Humanized Comput 13(2):961–972
32. Fraś M, Witkowski M, Kowalczyk K (2022) Convolutional weighted minimum mean square error filter for joint source separation and dereverberation. In: IEEE international conference on acoustics, speech and signal processing (ICASSP). IEEE

Optimisation of Convolutional Neural Network Parameters Using the Bees Algorithm

Michael S. Packianather and Nawaf Mohammad H. Alamri

Abstract The convolutional neural network (CNN) is one of the most popular deep learning algorithms that deals mainly with image data. In this study, the application of the Bees Algorithm (BA), which behaves like honeybees, was used along with the Bayesian Optimisation (BO) approach to improve CNN performance (BA-BO-CNN). Applying the hybrid algorithm on Cifar10DataDir images increased the accuracy on the validation set from 80.72% for the existing BO-CNN to 82.22% for the hybrid algorithm. Applying BA-BO-CNN on digit datasets showed the same accuracy as the existing hybrid BO-CNN, but with a computational time shortened by 3 min and 12 s. Finally, concrete crack benchmark image data yielded almost similar results to existing algorithms. Similarly, a new hybrid Bees Convolutional Neural Network (BA-CNN) algorithm was proposed that uses BA to design better CNN topology by optimising its hyperparameters, which increases the network accuracy. Applying the hybrid algorithm to the Cifar10DataDir dataset yielded the same accuracy as existing CNNs and BO-CNNs. Applying it to the digits dataset produced the lowest computational time with 4 min and 14 s reductions compared to BO-CNN, so it is the best algorithm in terms of cost-effectiveness. Finally, applying it to concrete crack images produced similar results to the existing algorithms.

Keywords Convolutional neural network (CNN) · Bees algorithm (BA) · Bees convolutional neural network (BA-CNN) · Bees bayesian convolutional neural network (BA-BO CNN)

M. S. Packianather (✉)
Department of Mechanical and Medical Engineering, School of Engineering, College of Physical Sciences and Engineering, Cardiff University, Queen's Buildings, The Parade, Cardiff CF24 3AA, UK
e-mail: packianatherms@cardiff.ac.uk

N. M. H. Alamri
King Abdulaziz University, Jeddah, Saudi Arabia
e-mail: nmhalamri@kau.edu.sa

D. T. Pham and N. Hartono (eds.), *Intelligent Engineering Optimisation with the Bees Algorithm*, Springer Series in Advanced Manufacturing,
https://doi.org/10.1007/978-3-031-64936-3_13

1 Introduction

Artificial intelligence (AI) is an important concept that allows machines to mimic human behaviour, facilitating intelligent system development [1] and improving the efficiency and effectiveness of processes such as manufacturing machines. An artificial neural network (ANN) is a popular machine learning (ML) algorithm that mimics the biological human brain; it can analyse big data and extract complex meaningful patterns for predicting the future of complex systems [2].

Deep learning (DL) algorithms are established as an extension to ANN capabilities, as DL has the advantage of automatic feature extraction by learning large numbers of nonlinear filters, as stated by Singh [3]. The convolutional neural network (CNN) is one of the most popular DL networks that deals mainly with image data, and it can be integrated with other intelligent swarm optimisation algorithms, such as the Bees Algorithm (BA), to design better CNN topology by optimising its parameters to improve network performance.

The chapter is structured as follows: Sect. 2 presents a literature review of CNNs with the gaps that may need to be addressed using BA to improve CNN performance followed by the impact of integrating nature-inspired algorithms with DL networks in Sect. 3. Section 4 proposes a hybrid Bees Bayesian Convolutional Neural Network (BA-BO-CNN) and Bees Convolutional Neural Network (BA-CNN) algorithm, and Sect. 5 presents the results and discussion of the proposed algorithms. Finally, Sect. 6 presents the conclusion and future directions.

2 Literature Review

This section presents a brief review of CNN principles and gaps that may need to be addressed using BA to improve the performance of CNN models.

2.1 Definition and Method of Working

CNNs start with an input layer and end with an output layer with many hidden layers in between and are responsible for learning the image features using the layers shown below:

- Convolution: This step is responsible for image feature activation using convolutional filters that consist of weights that are multiplied using the dot product by the pixel brightness input layer to create a feature map [4].
- Rectified linear unit (ReLU): the role of the layer is maximising the training speed as negative values are mapped to zeros and positive values are maintained, which is helpful as an activation after each convolutional layer.

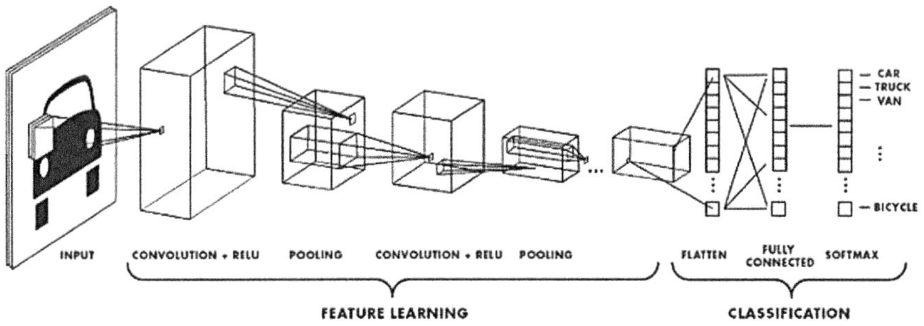

Fig. 1 CNN structure for Classification problems [9]

- Batch normalisation: The overfitting risk is mitigated by normalising the values resulting from the previous layer using this layer that is added after each convolutional layer [5].
- Pooling: responsible for output dimensionality reduction without losing the features that need to be extracted [6], which minimises the computational cost. There are two types of this layer: max pooling, which selects the maximum value (most activated feature if the dark background is used), and average pooling, which calculates the average presence of the feature, so it is better with a lighter background [7].

Furthermore, CNNs use two classification layers:

- Fully connected: responsible for image classification probability.
- SoftMax: This layer provides the output of the classification. It is used as a multi-class activation function, while the sigmoid logistic function is added with binary classification cases [6],it may not have parameters as mentioned in [8].

Figure 1 demonstrates the method of working for CNN showing feature learning and classification layers:

2.2 *Gaps and Open Issues*

One of the most important challenges in training CNNs is the generalisation to unused data. Overfitting is a common issue in training CNNs when the model fits well enough for training data used to build the model, so it has high training accuracy but is not able to generalise to other datasets, which means that the validation and testing accuracies are low [10]. In addition, the authors reported a learning process instability issue resulting from the exploding gradient problem. Redesigning the network topology might overcome this issue, but it is the most challenging step in training CNNs, as mentioned in [11–14], and [15]. The traditional approach of applying trial and error depends highly on user experience, so it is less accurate. Using nature-inspired optimisation algorithms such as BA to optimise the parameters of the CNN might result in better topology. However, no approach has yet been found that is used to design the best CNN topology. The following section will discuss the impact and recent applications of developing a hybrid DL model with nature-inspired algorithms.

3 Impact and Recent Applications of Parameter Optimisation Using Nature-Inspired Algorithms

Chiroma [16] discussed the use of nature-inspired algorithms in designing DL topology. The author mentioned that it contributes to solving local minima problems and increases the model accuracy as human intervention is eliminated. However, it is still an ongoing problem in the DL field.

Furthermore, other publications discussed hybrid CNNs with nature-inspired sward-based optimisation techniques, such as CNNs with evolutionary algorithms (EA-CNNs) and CNNs with genetic algorithms (GA-CNNs), in addition to integrating CNNs with DL networks called long short-term memory (LSTM-CNNs) and finally the hybridisation of CNNs and fuzzy logic (FL-CNNs). Table 1 below summarises the accuracy of the original CNN and improved accuracy after hybridisation.

Bernard and Leprévost [21] stated the process of evolution in EA-CNN by crossing members and inducing random mutation that reproduces the population through generation, and the feature activation is maximised by input image evolution. The genetic algorithm has been used to design better CNN topology by optimising its parameters, and the model was used to predict the stock market. The natural selection of the population's fittest individual improved the network accuracy from 71.69% to 75.95% as offspring are produced to inherit the parents' characteristics, which increases the probability of survival if their parents have better fitness [22]. The algorithm includes five stages: initialising the population, evaluating the fitness function, performing the selection, crossover, and mutation.

In addition to nature-inspired algorithms, CNNs can be hybridised with other DL algorithms called long short-term memory (LSTM) that recognise worker unsafe

Table 1 CNN accuracy for different datasets

Hybrid Algorithm	Accuracy		Data	References
	CNN (%)	Hybrid CNN (%)		
(EA-CNN)	–	98.88	Handwritten digits	[17]
	–	62.37	Cifar10DataDir	
(GA-CNN)	71.69	75.95	Stock market	[18]
(LSTM-CNN)	82.1	97	Motion data and site video	[19]
(FL-CNN)	97.35	99.10	Handwritten digits	[20]

actions in motion data automatically. The learning feature sequence is enabled using LSTM, which is the advantage of using this algorithm, as stated in [23]. It avoids the vanishing gradient problem, as it is an algorithm effective at capturing dependencies in the long term. Furthermore, CNN can be integrated with FL by adding one more layer called a fuzzy self-organisation layer. Korshunova [24] stated that it is used to distribute inputs into several clusters different from the number of output classes. This layer improved the accuracy of classifying handwritten digit recognition from 97.35% for the original CNN to 99.10% for the hybrid FL-CNN.

4 Proposed Hybrid BA-BO-CNN & BA-CNN Algorithms

A new approach to designing better CNN topology by optimising its parameters using BO and BA will be explained. In the BA-BO-CNN algorithm, the BO technique is used to find the optimal CNN hyperparameter values for the section depth, initial learning rate, momentum, and regularisation that minimise the classification error on the validation set. The objective function is a complex function of the mentioned four variables, and it is complicated to formulate. BA is applied to optimise the weight learning rate factor that adjusts the global learning rate obtained by the BO algorithm in each of the three convolutional layers and the fully connected layer. The optimal learning rate means the optimum amount of weight update, as stated in [25]. The learning rate will be adjusted in the convolutional filters and fully connected layer that perform the classification task, so the validation error is minimised.

The other hybrid algorithm (BA-CNN) uses BA to optimise the four CNN parameter values section depth, initial learning rate, momentum, and regularisation that yielded the minimum classification error on the validation set; they are the same four parameters that were optimised using the BO technique in the BA-BO-CNN algorithm. The four variables are selected based on the topology of the CNN and its learning capability. Ref [26] stated the function of each CNN parameter that will be optimised using BO in the BA-BO-CNN algorithm and using BA in the BA-CNN algorithm are shown in the following Table 2:

The following Fig. 2 illustrates where BO and BA are inserted in CNN:

The BO algorithm builds a probabilistic model of the objective function to select the network hyperparameters based on it and then evaluates the selection in the true objective function. It maintains a Gaussian process model as a surrogate model in

Table 2 CNN optimisation Variables

CNN parameter	Function	Range	Type
Section depth	Network depth control	1–3	Integer
Initial learning rate	Features learning	1e-2–1	Logarithm
Momentum	Hyperparameters update	0.8–0.98	Logarithm
Regularisation	Overfitting prevention	1e-10–1e-2	Logarithm

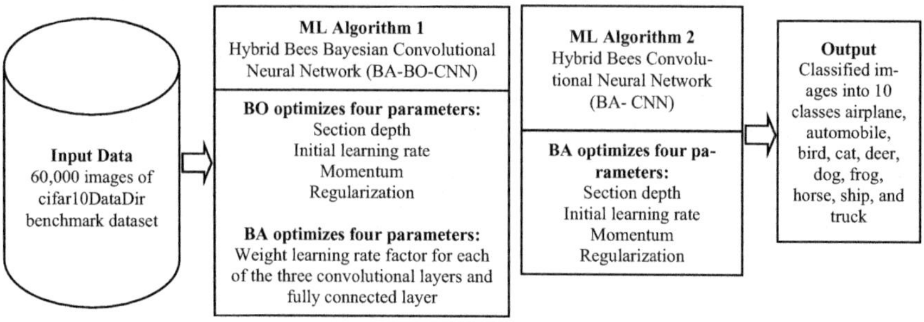

Fig. 2 Illustrative diagram for BA-BO-CNN and BA-CNN Algorithms

the objective function that uses network variables as inputs to specify the network architecture. Nag [27] mentioned that initial runs of the function are used as starting points "prior" and then enriched with "posterior" data points in each iteration. The acquisition function is used as a metric function to decide the optimal parameters, and the formula of the expected improvement (EI) function is as follows [27]:

$$\mathrm{EI(x)} = [\mu(\mathrm{f(x)}) - \max\{\mathrm{f(x)}\}]\,\Phi(\mathrm{z}) - \sigma(\mathrm{f(x)})\,\phi(\mathrm{z}) \text{ if } \sigma(\mathrm{f(x)}) > 0 \tag{1}$$

$$\mathrm{EI(x)} = 0 \text{ if } \sigma(\mathrm{f(x)}) <= 0 \tag{2}$$

where:

$$\mathrm{Z} = \mu(\mathrm{f(x)}) - \max\{\mathrm{f(x)}\}/\sigma(\mathrm{f(x)}) \tag{3}$$

$\mu(f(x))$ is the prediction from the Gaussian process for new data point x.

$\max\{f(x)\}$ is the maximum prediction from the entire prior list at the current stage

$\phi(z)$ is the standard normal cumulative probability density.
$\sigma(f(x))$ is the standard deviation of the prediction for new data point x.
$\emptyset(z)$ is the standard normal probability density.

The improvement is checked by the difference between the prediction at the current stage and the maximum prediction of the entire list of prior. If the difference is high, it means that the new prediction is significantly higher than the maximum obtained thus far. Multiplying the difference by the standard normal cumulative probability density gives the overall or expected mean improvement, which is the exploitation part. The standard deviation gives the uncertainty and is the secret of the exploration part. Applying BO in conjunction with stochastic gradient descent momentum (SGDM) is one of the training options in CNN architecture.

BA is a most popular swarm-based optimisation algorithm that has the advantage of global, local, and intense search [28], so adding BA to the BO approach will contribute to designing better CNN topology by optimising its parameters to minimise the validation error. The algorithm mimics honeybees' foraging behaviour. The parameters of the algorithm are as follows:

- Scout bees (n)
- Selected bees (m)
- Elite sites (e)
- Recruited bees for elite sites (nep)
- Recruited bees for other best sites (nsp)
- Neighbourhood size (ngh)

The first step of the algorithm is the global search when the scout bees evaluate different sites based on classification errors on the validation set. Then, the best sites are selected, and the remaining sites are abandoned in the local search. Then, the best of the best sites are selected in the intense search. The next step is neighbourhood search space selection to assign more bees for the elite sites and fewer bees for non-elite sites to perform local search simultaneously with the global search. The stopping rules for this iterative process are as follows:

- Reaching the optimal values
- Exceeding the maximum iteration number
- Not producing improvement in specific sequential iterations

Yuce [29] presented a flow chart showing the steps of BA as follows (Fig. 3):

In addition to the basic BA, Baronti [30] presented shrinking and standard algorithm types. The shrinking technique conducts the local search within a small neighbourhood size, while the standard algorithm has a stage for abandoning the sites during the local search. Lindfield and Penny [31] showed further modification for BA, which counts the number of failed times in exploring an improved site, which can be used as a metric to improve the efficiency and effectiveness of the exploration search process. Imanguliyev [32] enhanced BA with an early neighbourhood search strategy to start the search from promising patches. Additionally, the author presented another enhancement with an efficient recruitment strategy that dynamically changes the number of recruited bees based on the site's efficiency. In addition, the authors showed the hybrid Tabu and autonomous BA. Packianather [33] proposed using rules' automatic discovery when applying BA. The author added two new parameters, namely, quality weight and coverage, to avoid ambiguous situations during the prediction and to carry out meta-pruning, making the algorithm more suitable for performing classification tasks. The newly proposed model achieved higher classification accuracy and reduced the number of rules, making it a more efficient algorithm than other classification techniques, such as evolutionary algorithms and the Jrip.

The hybrid BA-BO-CNN uses the basic BA and BO to design a better CNN topology by optimising the section depth, initial learning rate, momentum and regularisation parameters using the BO technique in addition to optimising the weight learning rate factor to adjust the global learning rate obtained by the BO algorithm in each convolutional layer and fully connected layer.

The values of the hyperparameters for BA are assigned based on the computer capability since the maximum number of iterations that can be performed for the

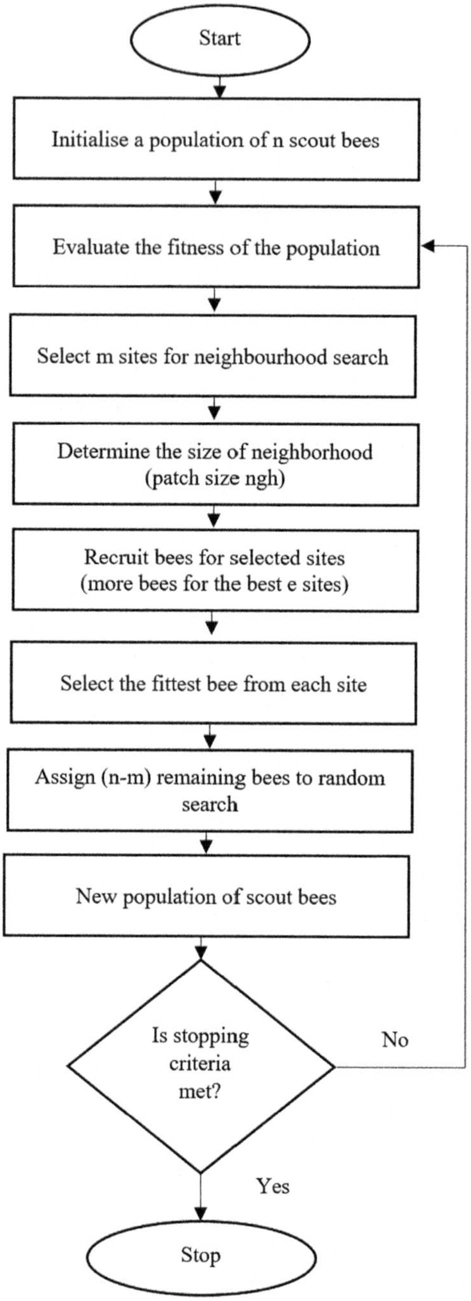

Fig. 3 Flowchart for basic BA (adapted from Yuce [29])

complicated CNN function is one and the number of scout bees that can be assigned is six with one elite site to be selected. The remaining parameters are selected based on the equations in MathWorks-3 [34], which are shown below:

- Maximum number of iterations = 1
- Scout bees (n) = 6
- Selected bees (m) = 0.5 * n = 3
- Elite bees (e) = 1
- Recruited bees for elite sites (nep) = 2 * m = 6
- Recruited bees for other best sites (nsp) = 0.5 * n = 3
- Neighbourhood size (ngh) = 0.1 * (Variable Maximum – Variable Minimum) = 0.02

The CNN architecture for cifar10DataDir image classification (10 classes) is composed of 15 layers, of which one input layer and 11 feature learning layers (three convolutional layers, rectified linear unit layers, batch normalisation layers, and two max-pooling layers). The topology ends with one fully connected layer, one SoftMax layer, and one classification layer.

The input layers are represented by a matrix of size height by width (32 × 32) that consists of pixel brightness numbers between 0 and 255 (0 for black and 255 for white). The convolutional layers contain filters represented by a matrix of weights that use a special dot product to slide along the pixel brightness input matrix, which creates a feature map matrix [4]. Rectified linear unit layers are used after each convolutional layer to increase the training speed by changing negative values to zero and maintaining positive values [9]. In addition, batch normalisation layers are used after each convolutional layer to mitigate the risk of overfitting by normalising the resulting values of the previous layers [5]. Pooling layers are used between the convolutional layers to reduce the output dimensions [6] while keeping the important features, which contributes to minimising the computational cost. The classification layer performs the classification with a fully connected layer that shows the classification probability, and the SoftMax layer is used as an activation to provide the classification output [9].

The starting number of convolutional layers is between two and three layers with a filter size of 3 × 3 or 5 × 5, as advised in [4]. The CNN was designed with three convolutional layers and two max pooling layers in between, as mentioned previously. The number of filters ranges between 11 for the first layer and 44 for the last layer, and each layer has twice as many filters as the previous layer [25]. The filter size is 3 × 3, the section depth is 3, and the padding that helps to detect the edges of the images is set as 'same' so that the software calculates the size of the padding at the training time automatically and produces an output size equal to the input size if the number of pixel shifts (stride) is one. The pooling type is 'max', which selects the maximum value of the matrix, so the most activated feature is selected when using a dark background, while the average pooling layer calculates the average presence of the feature, so the average pooling is better with a lighter background [7]. The default size of the pooling layer is 2 × 2 [4], but it yielded high computational cost since the image input size is large (32 × 32), and the size was increased to 3 × 3 with a stride value of 3 to minimise the training time.

The classification layer that performs the classification contains a fully connected layer with a parameter value of 10 as the benchmark image data consists of 10 classes and a SoftMax activation function that works better with multiclass classification problems rather than a binary classification problem that requires a sigmoid logistic function [6].

The CNN was trained using SGDM, which is the most common training algorithm [5]. The default values for this algorithm are an initial learning rate of 0.01, a momentum of 0.9 and a regularisation of 1e-04 with a maximum number of epochs of 20 [35]. After applying some experiments and monitoring the validation accuracy, the initial learning rate was changed to 0.1, and the regularisation became 1e-10 with 30 epochs.

In the first hybrid BA-BO-CNN algorithm, BA is applied to optimise the weight learning rate factor to adjust the global learning rate obtained by the BO algorithm in each of the three convolutional layers and the fully connected layer. Optimising the learning rate in the convolutional layers and fully connected layers results in an optimum update for the weight [25] in these layers, so the classification accuracy on the validation set is improved. The four learnable parameters are weight learning rate factors for three convolutional layers and the fully connected layer. They are updated after a local search with a range between 0.9 and 1.1 to adjust the global learning rate obtained by the BO algorithm.

Figure 4 below shows a workflow diagram explaining the method of working for the BA-BO scheme to optimise the CNN parameters in the BA-BO-CNN algorithm:

In the BA-CNN algorithm, BA is used to optimise the section depth, initial learning rate, momentum, and regularisation, which are the same parameters that were optimised using BO in the previous BA-BO-CNN algorithm. BA starts with six scout bees arriving at random positions and evaluating them based on the validation error value. Then, a local search starts by selecting the three best sites and abandoning the remaining sites. The next step is selecting the size of the neighbourhood search space, which is 0.2 for section depth, 0.099 for initial learning rate, 0.018 for momentum, and 0.001 for regularisation to recruit six bees for the elite sites and three bees for non-elite sites to conduct a local search simultaneously with the global search.

5 Results and Discussion of the Proposed BA-BO-CNN & BA-CNN Algorithms

The MATLAB platform is used to apply the hybrid BA-BO-CNN and BA-CNN algorithms on three benchmark datasets, cifar10DataDir benchmark image data, which consists of 60,000 images classified uniformly into 10 classes: truck, automobile, aeroplane, ship, frog, bird, deer, dog, cat and horse. The second dataset is 10,000 handwritten digit images with ten uniform classes from 0 to 9. The last set of data

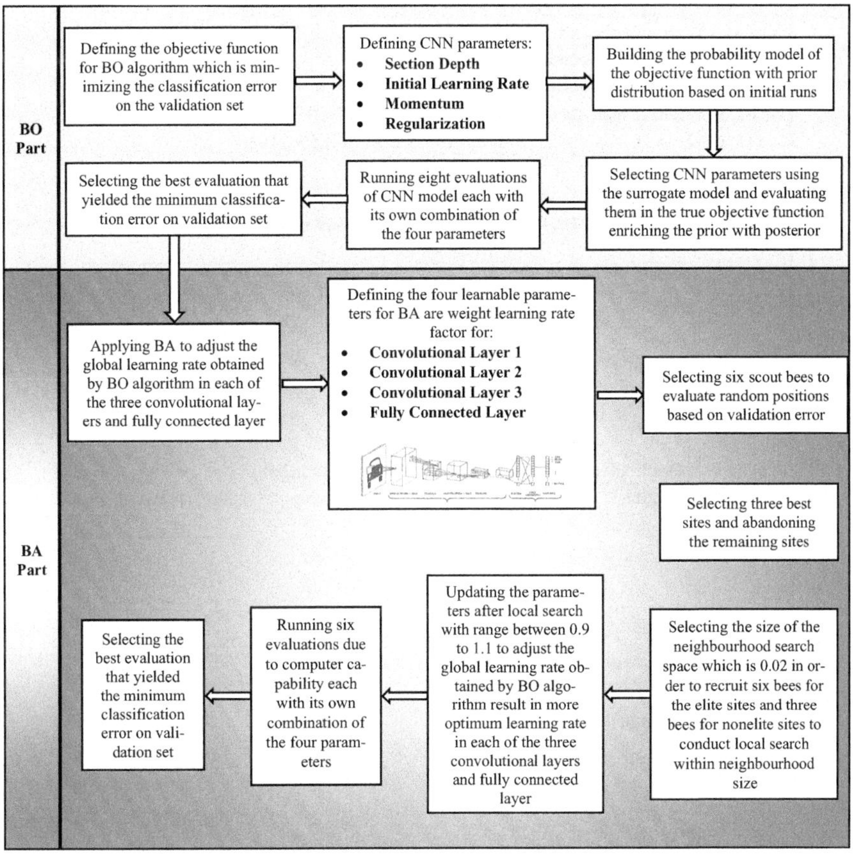

Fig. 4 Workflow diagram for BA-BO-CNN

is concrete crack images with two classes: 5000 positive images with cracks present in the road and 5000 negative images without cracks.

First, the BO technique is applied to the cifar10DataDir dataset to find the optimum CNN parameters. Figure 5 shows the optimum values for section depth, initial learning rate, momentum, and regularisation that yielded the minimum classification error:

The minimum classification error value is 0.1928, which is a result of a section depth value of 3, an initial learning rate of 0.68934, a momentum of 0.84074, and a regularisation of 4.5857e-05.

Then, BA is added to adjust the global learning rate in each of the three convolutional layers and fully connected layer. Table 3 shows the optimal weight learning rate factors optimised by BA along with classification error on the validation set:

Therefore, the global learning rate of 0.68934 obtained using the BO algorithm is adjusted in the first convolutional layer by multiplying it by 0.9308 to become 0.6416. In the second layer, the adjustment factor is 1.0924, which results in a learning rate

```
Optimization completed.
MaxTime of 14400 seconds reached.
Total function evaluations: 8
Total elapsed time: 14670.6496 seconds
Total objective function evaluation time: 14660.1188

Best observed feasible point:
    SectionDepth    InitialLearnRate    Momentum    L2Regularization
    ____________    ________________    ________    ________________

         3              0.68934         0.84074        4.5857e-05

Observed objective function value = 0.1928
Estimated objective function value = 0.19318
Function evaluation time = 2650.7429

Best estimated feasible point (according to models):
    SectionDepth    InitialLearnRate    Momentum    L2Regularization
    ____________    ________________    ________    ________________

         3              0.68934         0.84074        4.5857e-05

Estimated objective function value = 0.19318
Estimated function evaluation time = 3081.7537
```

Fig. 5 Optimal values for CNN parameters [36]

Table 3 Optimal weight learning rate factors for BA-BO-CNN [36]

Optimised variable using BA	Value
Optimal weight learning rate factor for convolutional layer 1	0.9308
Optimal weight learning rate factor for convolutional layer 2	1.0924
Optimal weight learning rate factor for convolutional layer 3	1.0753
Optimal weight learning rate factor for fully connected layer	0.9977
Validation error	0.1778

value of 0.7530. The weight factor for the third convolutional layer is 1.0753, so the adjusted learning rate is 0.7412. Finally, the new value for the fully connected layer is 0.6877 after multiplying the global learning rate by a factor of 0.9977. Figure 6 shows the training progress after adding BA in the blue line and validation in the black line. The training progress curves for CNN, BO-CNN and BA-BO-CNN are similar:

The following three figures show the confusion matrix for the BA-BO-CNN algorithm for all three sets. Recall and miss out in the blue and red columns, respectively, while precision and false alarm in the blue and red rows, respectively (Figs. 7, 8, 9):

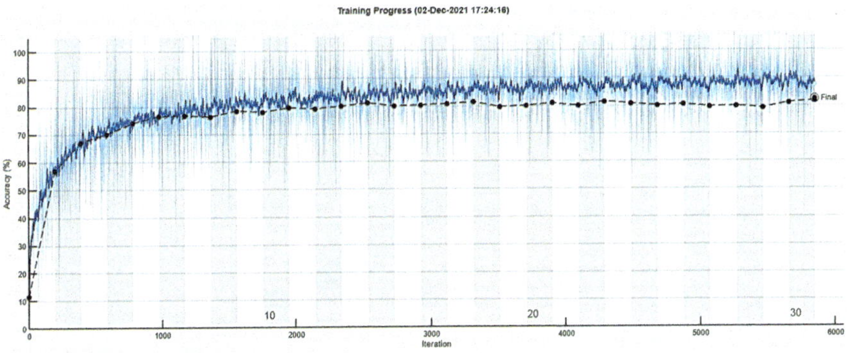

Fig. 6 Training progress for BA-BO-CNN [36]

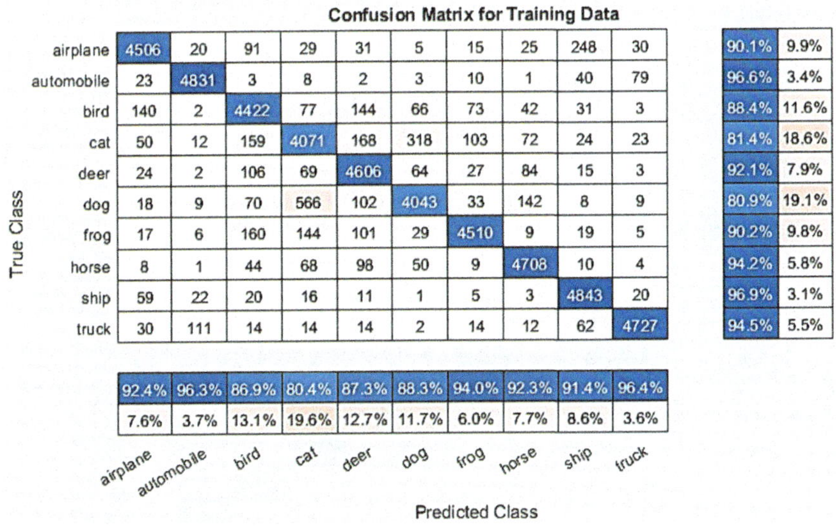

Fig. 7 Confusion matrix for training data for BA-BO-CNN [36]

Additionally, BA-CNN is applied to the same cifar10DataDir benchmark datasets. BA is used to find the optimal parameter values for section depth, initial learning rate, momentum, and regularisation that yield the minimum validation error. They are the same four parameters that were optimised using the BO in the BA-BO-CNN algorithm. The following Table 4 shows the optimal parameters optimised by BA along with classification error on the validation set:

Figure 10 shows the training progress after adding BA to the blue line and validation to the black line.

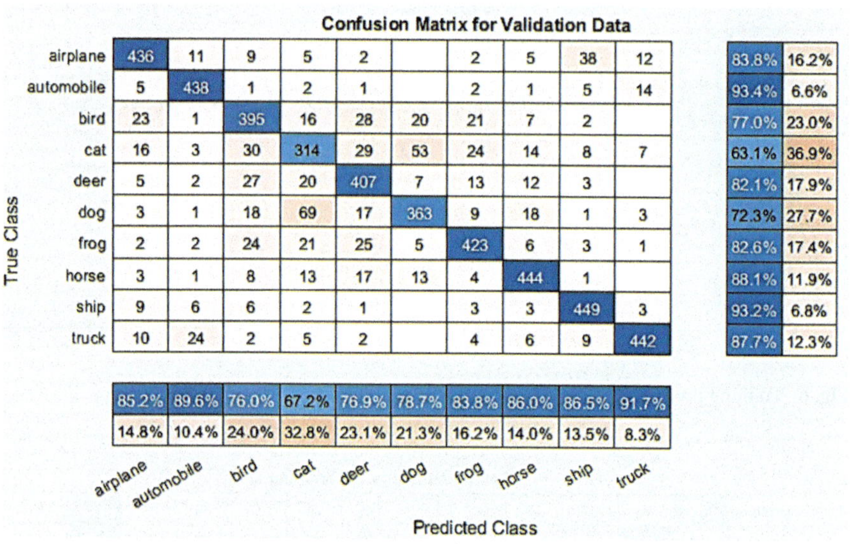

Fig. 8 Confusion matrix for validation data for BA-BO-CNN [36]

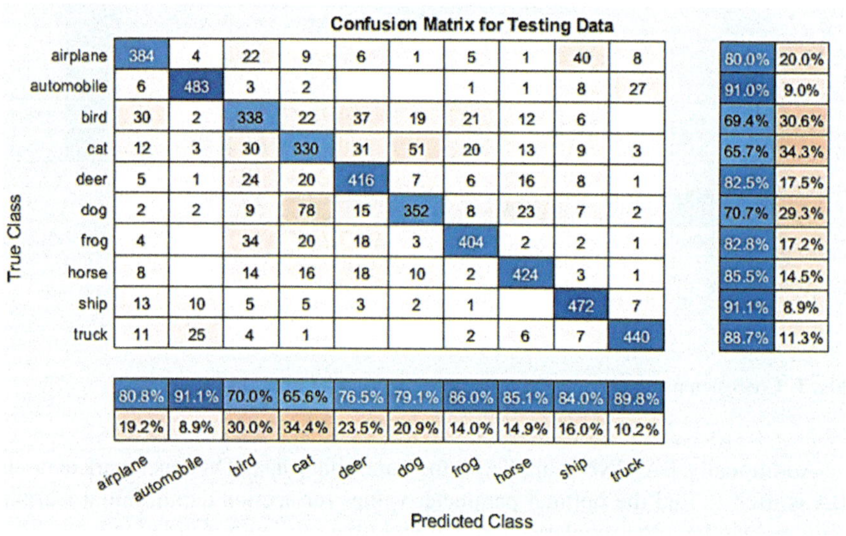

Fig. 9 Confusion matrix for testing data for BA-BO-CNN [36]

The minimum classification error value is 0.1928 achieved by a section depth value of 3, an initial learning rate of 0.0289, a momentum of 0.8734 and a regularisation of 0.0049.

Table 4 Optimal CNN parameter values for BA-CNN

Optimised variable using BA	Value
Optimal section depth	3
Optimal initial learning rate	0.0289
Optimal momentum	0.8734
Optimal regularisation	0.0049
Validation error	0.1928

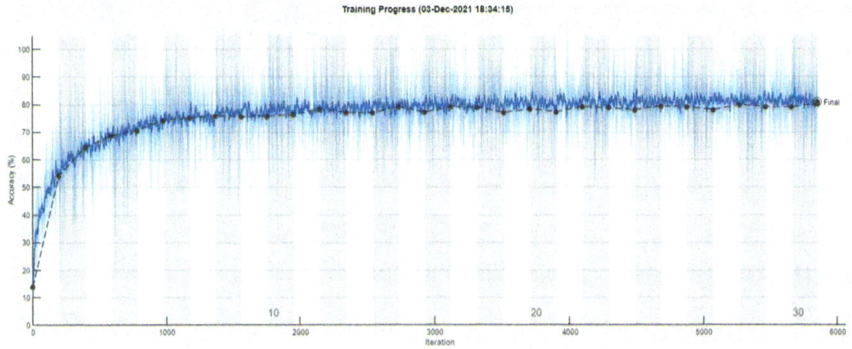

Fig. 10 Training progress for BA-CNN

The following three figures show the confusion matrix for the BA-CNN algorithm for all three sets. Recall and miss out in the blue and red columns, respectively, while precision and false alarm in the blue and red rows, respectively (Figs. 11, 12, 13):

The following Table 5 shows the accuracy for training, validation, and testing sets along with computational time for existing CNN and BO-CNN in addition to hybrid BA-BO-CNN and BA-CNN:

It is seen that BA-CNN performs slightly better than the existing CNN in terms of validation accuracy, but it is the same as the existing BO-CNN. In addition, it is more accurate than the EA-CNN model developed in [17], which achieved an accuracy of 62.37% on the same cifar10DataDir benchmark dataset, as presented in Table 1. The hybrid BA-BO-CNN has the best validation and testing accuracy, so it has the best classification performance and generalisation capability to unseen data. In addition, it is the best algorithm in terms of cost-effectiveness since it achieved lower computational time than BO-CNN by 2 min and 11 s.

The same procedure is followed to apply the existing CNN and BO-CNN in addition to the hybrid BA-BO-CNN and BA-CNN on the other two benchmark datasets. The results are shown in the following two Tables 6 and 7:

The table shows that the classification accuracy of the hybrid BA-BO-CNN is the same as that of the existing BO-CNN because of the simple features in the digit datasets, so the existing algorithms have high classification performance in terms of accuracy, but the computational time of the hybrid BA-BO-CNN algorithm is less

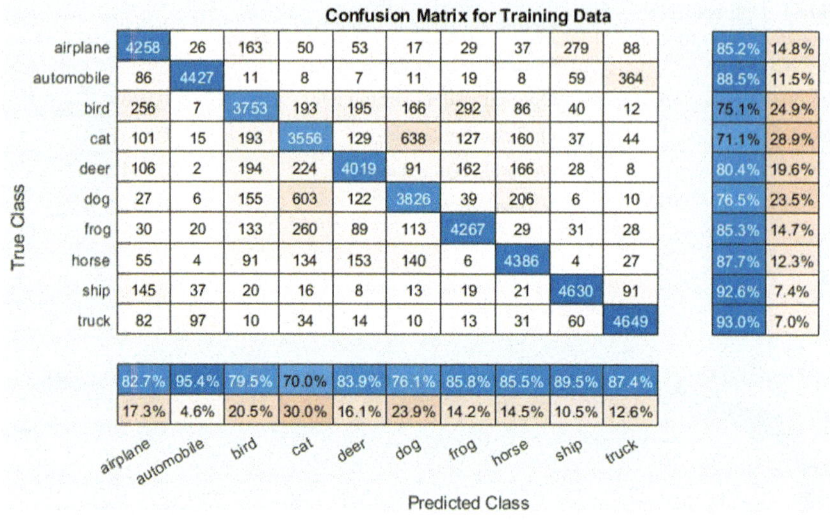

Fig. 11 Confusion matrix for training data for BA-CNN

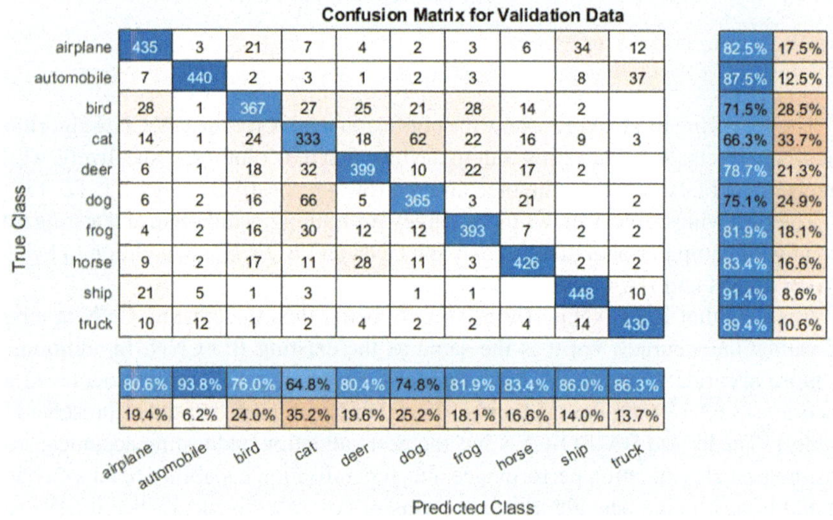

Fig. 12 Confusion matrix for validation data for BA-CNN

than that of the existing BO-CNN by 3 min and 12 s, so it is concluded that BA-BO-CNN is a cost-effective algorithm. The other hybrid BA-CNN algorithm produces almost similar results to existing BO-CNN in terms of accuracy, but it has the lowest computational time with 4 min and 14 s reductions compared to BO-CNN, so it is the best algorithm in terms of cost-effectiveness.

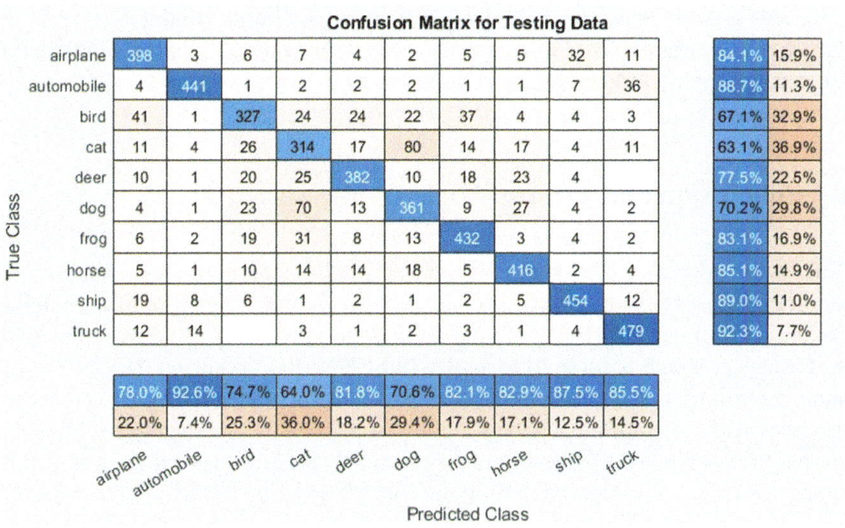

Fig. 13 Confusion matrix for testing data for BA-CNN

Table 5 The algorithms' accuracy and time (Cifar10DataDir Data)

	CNN	BO-CNN	BA-BO-CNN	BA-CNN
Training (%)	92.68	90.27	90.53	83.54
Validation (%)	80.34	80.72	82.22	80.72
Testing (%)	80.54	80.69	80.74	80.02
Time	42 Minutes and 56 Seconds	43 Minutes and 24 Seconds	41 Minutes and 13 Seconds	41 Minutes and 46 Seconds

Table 6 Algorithm accuracy and time (Digit Data)

	CNN	BO-CNN	BA-BO-CNN	BA-CNN
Training (%)	99.94	100	100	100
Validation (%)	98.80	100	100	99.90
Testing (%)	99.20	99.99	99.99	100
Time	2 Minutes and 35 Seconds	6 Minutes and 42 Seconds	3 Minutes and 30 Seconds	2 Minutes and 28 Seconds

Table 7 The algorithms' accuracy and time (Concrete Crack Data)

	CNN	BO-CNN	BA-BO-CNN	BA-CNN
Training (%)	99.96	99.90	99.95	98.90
Validation (%)	99.50	99.67	99.67	98.50
Testing (%)	99.83	99.30	99.35	98.85
Time	3 Minutes and 20 Seconds	2 Minutes and 46 Seconds	2 Minutes and 50 Seconds	2 Minutes and 22 Seconds

In this dataset, the new hybrid algorithms have almost similar results to the existing CNN and BO-CNN since they have only two classes, so applying the existing algorithms classifies the images with high accuracy.

6 Conclusion and Future Work

This work addressed designing a better convolutional neural network (CNN) topology issue by proposing a nature-inspired hybrid algorithm that combines Bayesian Optimisation (BO) with the Bees Algorithm (BA) to increase the CNN performance, which is referred to as BA-BO-CNN. BO optimised the CNN's four parameters of section depth, initial learning rate, momentum, and regularisation. BA was responsible for optimising the weight learning rate factor that adjusts the global learning rate obtained by the BO algorithm in each convolutional layer and fully connected layer. Applying the hybrid algorithm (BA-BO-CNN) to Cifar10DataDir data increased the validation accuracy from 80.72% for BO-CNN to 82.22% for BA-BO-CNN, while applying it on digit datasets showed the same accuracy as the existing BO-CNN because the features in the digit images are simple, but there was a reduction in the computational time by 3 min and 12 s. Finally, applying it to concrete crack images produced almost similar results to existing algorithms, as this dataset has only two classes.

Similarly, a new hybrid Bees Convolutional Neural Network (BA-CNN) algorithm was proposed that uses BA to optimise the CNN's four parameters (section depth, initial learning rate, momentum, and regularisation) to increase the classification accuracy of the network. Applying the hybrid algorithm on Cifar10DataDir data performed slightly better than the original CNN in terms of validation accuracy, but it is the same as the existing BO-CNN. Applying it to the digit dataset produced the lowest computational time with 4 min and 14 s reductions compared to BO-CNN, so it is the best algorithm in terms of cost-effectiveness. Applying it to concrete crack images yielded almost similar results to existing algorithms.

The developed hybrid algorithms can be improved further in the future by optimising the weight regularisation factor to adjust the regularisation in the convolutional layers and fully connected layer using BA, which contributes to improving the CNN performance, particularly in preventing overfitting.

Acknowledgements I would like to thank the Saudi Arabian Cultural Bureau and Cardiff University for their support.

References

1. Li BH, Hou BC, Yu WT, Lu XB, Yang CW (2017) Applications of artificial intelligence in intelligent manufacturing: a review. Front Inform Technol Electron Eng 18(1):86–96

2. Wuest T, Weimer D, Irgens C, Thoben KD (2016) Machine learning in manufacturing: advantages, challenges, and applications. Product Manufact Res 4(1):23–45
3. Singh AK, Ganapathysubramanian B, Sarkar S, Singh A (2018) Deep learning for plant stress phenotyping: trends and future perspectives. Trend plant Sci. https://doi.org/10.1016/j.tplants.2018.07.004
4. Hui J (2017) Convolutional neural networks (CNN) tutorial [Online] Available at https://jhui.github.io/2017/03/16/CNN-Convolutional-neural-network Accessed 15 Nov 2021
5. Yamashita R, Nishio M, Do RKG, Togashi K (2018) Convolutional neural networks: an overview and application in radiology. Insights Imaging 9(4):611–629
6. McDermott, J (2021). Convolutional neural networks—image classification w. Keras [Online] Available at https://www.learndatasci.com/tutorials/convolutional-neural-networks-image-classification Accessed 15 Nov 2021
7. Ouf H (2017) Maxpooling versus minpooling vs average pooling [Online] Available at https://hany-ouf.blogspot.com/2020/08/maxpooling-vs-minpooling-vs-average.html Accessed 15 Nov 2021
8. Wu J (2017) Introduction to convolutional neural networks. National Key Lab for Novel Software Technology. Nanjing University. China 5:23
9. (MathWorks-1), Convolutional Neural Network. [Online] Available at https://www.mathworks.com/discovery/convolutional-neural-network.html Accessed 18 Sept 2024
10. Joshi S, Verma DK, Saxena G, Paraye A (2019) Issues in training a convolutional neural network model for image classification. In: International conference on advances in computing and data sciences (pp 282–293). Springer, Singapore
11. Zhang H, Kiranyaz S, Gabbouj M (2018) Finding better topologies for deep convolutional neural networks by evolution. arXiv preprint arXiv:1809.03242.
12. Baldominos A, Saez Y, Isasi P (2018) Evolutionary convolutional neural networks: an application to handwriting recognition. Neurocomputing 283:38–52
13. Sinha T, Verma B, Haidar A (2017) Optimization of convolutional neural network parameters for image classification. In: 2017 IEEE symposium series on computational intelligence (SSCI) (pp 1–7). IEEE
14. Ma N, Zhang X, Zheng HT, Sun J (2018) Shufflenet v2: practical guidelines for efficient cnn architecture design. In: Proceedings of the european conference on computer vision (ECCV) (pp 116–131)
15. Panwar M, Padmini J, Acharyya A, Biswas D (2017) Modified distributed arithmetic based low complexity CNN architecture design methodology. In; 2017 European conference on circuit theory and design (ECCTD) (pp 1–4). IEEE
16. Chiroma H, Gital AYU, Rana N, Shafi'I MA, Muhammad AN, Umar AY, Abubakar AI (2019) Nature inspired meta-heuristic algorithms for deep learning: recent progress and novel perspective. In: Science and information conference (pp 59–70). Springer, Cham
17. Badan F (2019) Evolutionary algorithms in convolutional neural network design. [Online] Available at http://excel.fit.vutbr.cz/submissions/2019/033/33.pdf Accessed 16 Mar 2020
18. Chung H, Shin KS (2019) Genetic algorithm-optimized multi-channel convolutional neural network for stock market prediction. Neural Comput Appl 1–18
19. Ding L, Fang W, Luo H, Love PE, Zhong B, Ouyang X (2018) A deep hybrid learning model to detect unsafe behavior: integrating convolution neural networks and long short-term memory. Autom Constr 86:118–124
20. Popko EA, Weinstein IA (2016) Fuzzy logic module of convolutional neural network for handwritten digits recognition. In: Journal of physics: conference series (Vol 738, No 1, p 012123). IOP Publishing
21. Bernard N, Leprévost F (2018) Evolutionary algorithms for convolutional neural network visualisation. In: Latin american high performance computing conference (pp 18–32). Springer, Cham
22. Mallawaarachchi V (2017) Introduction to genetic algorithms. [Online] Available at https://towardsdatascience.com/introduction-to-genetic-algorithms-including-example-code-e396e98d8bf3 Accessed 2 Apr 2020

23. Motepe S, Hasan AN, Stopforth R (2019) Improving load forecasting process for a power distribution network using hybrid AI and deep learning algorithms. IEEE Access 7:82584–82598
24. Korshunova KP (2018) A convolutional fuzzy neural network for image classification. In: 2018 3rd russian-pacific conference on computer technology and applications (RPC) (pp 1–4). IEEE
25. Brownlee J (2020) Understand the impact of learning rate on neural network performance [Online] Available at https://machinelearningmastery.com/understand-the-dynamics-of-learning-rate-on-deep-learning-neural-networks/ Accessed 29 Nov 2021
26. (MathWorks-2), Deep Learning Using Bayesian Optimization. [Online] Available at https://www.mathworks.com/help/deeplearning/ug/deep-learning-using-bayesian-optimization.html Accessed 4 Apr 2020
27. Nag A (2021) Bayesian optimization: a step by step approach [Online] Available at https://towardsdatascience.com/bayesian-optimization-a-step-by-step-approach-a1cb678dd2ec Accessed 9 Aug 2022
28. Al-Musawi A (2019) The development of new artificial intelligence based hybrid techniques combining bees algorithm, data mining and genetic algorithm for detection, classification and prediction of faults in induction motors (Doctoral dissertation, Cardiff University)
29. Yuce B, Packianather MS, Mastrocinque E, Pham DT, Lambiase A (2013) Honey bees inspired optimization method: the bees algorithm. Insects 4(4):646–662
30. Baronti L (2020) Analysis and development of the bees algorithm for primitive fitting in point cloud models (Doctoral dissertation, University of Birmingham)
31. Lindfield G, Penny J (2017) Introduction to Nature-Inspired Optimization. Academic Press
32. Imanguliyev A (2013) Enhancements for the bees algorithm (Doctoral dissertation, Cardiff University)
33. Packianather MS, Al-Musawi AK, Anayi F (2019) Bee for mining (B4M)–a novel rule discovery method using the Bees algorithm with quality-weight and coverage-weight. Proc Inst Mech Eng C J Mech Eng Sci 233(14):5101–5112
34. (MathWorks-3), Bees Algorithm (BeA) in MATLAB. [Online] Available at https://www.mathworks.com/matlabcentral/fileexchange/52967-bees-algorithm-bea-in-matlab Accessed 18 Sept 2024
35. (MathWorks-4), TrainingOptionsSGDM. [Online] Available at https://uk.mathworks.com/help/deeplearning/ref/nnet.cnn.trainingoptionssgdm.html Accessed 15 Nov 2021
36. Alamri NMH, Packianather M, Bigot S (2022) Deep learning: parameter optimization using proposed novel hybrid bees Bayesian convolutional neural network. Appl Artif Intell 1–25

Ergonomic Risk Assessment Combining the Bees Algorithm and Simulation Tools

Abhijeet Singh, Mario Caterino, Marta Rinaldi, Marcello Fera, Roberto Macchiaroli, and D. T. Pham

Abstract This study proposes a methodology for integrating the use of the Bees Algorithm (BA) and simulation to reduce the workload for workers on production and assembly lines. Simulation is first employed for retrieving ergonomics data for different workers. Then, the BA is used as a tool for optimising the job rotations of the workers among the workstations, considering as objective the minimisation of the 2 indices, the Ovako Working Posture Analysing System (OWAS) and Energy Expenditure (EE). A case study related to a car assembly line is then proposed to test the methodology. Six workstations (WSs) are considered in this study, and the BA is validated by comparing its results with the exact solution evaluated by analysing all the possible combinations. The results show the good accuracy of the BA in finding solutions for this type of problem and the capacity to reduce the computational times compared with those of the exact solution.

A. Singh (✉)
Department of Computer Science and Engineering, Indian Institute of Technology Kharagpur, Kharagpur, India
e-mail: Abhijeet.singh@kgpian.iitkgp.ac.in

M. Caterino
Department of Industrial Engineering, University of Salerno, Salerno, Italy
e-mail: macaterino@unisa.it

M. Rinaldi · M. Fera · R. Macchiaroli
Department of Engineering, University of Campania Luigi Vanvitelli, Aversa, Italy
e-mail: marta.rinaldi@unicampania.it

M. Fera
e-mail: marcello.fera@unicampania.it

R. Macchiaroli
e-mail: roberto.macchiaroli@unicampania.it

D. T. Pham
Department of Mechanical Engineering, College of Engineering and Physical Sciences, The University of Birmingham, Birmingham B15 2TT, UK
e-mail: D.T.Pham@bham.ac.uk

D. T. Pham and N. Hartono (eds.), *Intelligent Engineering Optimisation with the Bees Algorithm*, Springer Series in Advanced Manufacturing,
https://doi.org/10.1007/978-3-031-64936-3_14

Keywords Ergonomic assessment · Job rotation · Ergonomic optimisation · Bees Algorithm

1 Introduction

In recent years, the study of work-related musculoskeletal disorders (WMSDs) has gained much attention. WMSDs are injuries that might affect several parts of the human body (nerves, muscles, joints, etc.) and often may arise because of the type of job performed [1]. Awkward postures, job repetition, vibrations and material handling are among the most common causes generating WMSDs; thus, the need to reduce such conditions is evident and can lead to numerous benefits, such as a reduction in absenteeism or an increase in operational performance [2] and productivity [3]. Within the manufacturing and assembling context, the problem of limiting WMSDs is addressed by a particular subject, i.e., ergonomics. In recent years, the attention of scientists and practitioners to ergonomics has increased, and the application of risk assessment methods has become a common practice starting from the process design phase [4]. According to the type of company and the type of tasks to perform, it is necessary to identify the most suitable methods to assess ergonomics. Each method provides an ergonomic indicator as output, which provides information on the severity of the task for the workers. One of the objectives for a correct workplace design is to minimise or maximise (depending on the chosen method) the ergonomics indicators, which means reducing the efforts of the workers and the probability of developing WMSDs.

Many researchers have proposed the use of algorithms for real-life applications related to picking [5], manual assembly [6], human–robot cooperation [7] and many other aspects [8, 9] to optimise ergonomics. The problem related to this type of approach is twofold: first, ergonomics indices are often evaluated using observational risk assessment methods, which are based on continuous and systematic observation of the worker while he or she is performing the task and the compilation of checklists to assign scores by expert ergonomists. This activity is very time-consuming and is also affected by the subjectivity of the judgements of the ergonomists [10, 11] due to the low sensitivity to the variety of workers' motions. The second problem is related to the fact that many researchers do not consider a wide population of workers, limiting the analysis to a few workers with similar anthropometric characteristics. Ergonomics indicators depend on the anthropometric characteristics of the workers; hence, considering a limited number of workers' anthropometric characteristics may represent a great limitation for ergonomics analysis. Indeed, the majority of companies (especially large companies) have a very large population of workers with different anthropometric characteristics, i.e., heterogeneous workers.

Thus, a methodology that allows us to overcome such problems and assess ergonomics for heterogeneous workers by eliminating the subjectivity of the judgments and speeding up the ergonomics evaluation process needs to be introduced.

Caterino et al. [12] developed a model in which digital twins (DTs) and simulations were used as tools for evaluating the ergonomics of heterogeneous workers, considering different sexes and heights. DT and simulation have been used by many researchers in the literature for ergonomics assessments [13, 14], and their effectiveness as reliable tools for performing very complex analyses in a limited time was demonstrated. Much commercial software allows reproducing the real working environment in which digital human models (DHM) can perform the working tasks. Such software offers a wide range of ergonomic risk assessment methods that can be used by researchers and practitioners for their analysis.

In this chapter, simulation is used to evaluate the ergonomic indices of workers performing their working tasks.

The objective of the chapter is to introduce the Bees Algorithm to further develop the methodology proposed by Caterino et al. [12], in which they evaluated the ergonomics for heterogeneous workers in 2 WSs of a car manual assembly line considering job rotation (JR). JR is an organisational strategy that creates variability in activities where workers apply forces and can increase muscular activity variability [15]. The diversification of tasks can be beneficial for workers because it contributes to varying muscle movements and reducing the physical load during the working day.

The idea of this chapter is to increase the number of WSs considered in [12] and validate the use of the Bees Algorithm (BA) for optimising ergonomics. The BA was proposed by Pham et al. [16] in 2005 and has been demonstrated to be suitable for many different applications, both continuous [17] and combinatorial [18, 19]. To the best of the authors' knowledge, no studies use the BA as an optimisation algorithm for ergonomics purposes, although the ergonomics optimisation topic has been addressed by the scientific literature [6, 20]. Thus, a model considering the combination of simulation and BA is first proposed in this chapter and validated using a case study.

Six WSs will be considered, and comparisons between the results of the BA and the exact solutions obtained by enumeration will be carried out. This will represent the basis for further future developments in which the higher number of WSs will not allow us to evaluate the exact solutions because of the size of the problem. In fact, by increasing the number of WSs, the rotations of workers increase as a factorial, and when the number of WSs becomes too high, the exact solutions cannot be calculated.

The outline of the chapter is as follows: in Sect. 2, a literature review on the use of optimisation algorithms in the industrial sector will be presented, with a particular focus on ergonomics risk assessment; in Sect. 3, the framework representing the model for ergonomics assessment will be illustrated; Sect. 4 will include the application of the model for the case study object of this chapter; in Sect. 5, the results will be reported and discussed; finally, conclusions will be depicted.

2 Related Work

Optimisation problems aim to find the best feasible solution that maximises or minimises an objective function with a process that consists of finding the optimal values of the variables affecting the objective function [21].

When the computational time for solving an optimisation problem increases exponentially with the size of the problem, the problems belong to the NP-complete or NP-hard classes [22] and require the use of iterative procedures to find solutions in a finite time. The most common procedures used for such a scope are represented by optimisation techniques which can explore different regions of the search space and provide good or near-optimal solutions to the specific problem considered.

In manufacturing and production contexts, the use of optimisation techniques covers many topics.

The assembly line balancing problem (ALBP) represents one of the most important topics in which the use of optimisation algorithms is fundamental. ALBP has various ramifications in various industry sectors, including automotive, aerospace and electronics. In its basic form, the ALBP consists of assigning tasks to an ordered sequence of workstations by respecting the precedence constraints to optimise the performance of the assembling systems in terms of times, costs or other variables [23]. According to the review by Li et al. [24], the most common objective functions concern the minimisation of cycle times and the number of workstations.

Despite the use of exact algorithms being possible in some cases [25], the ALBP becomes intractable when the size of the problem increases. For this reason, several heuristics and metaheuristic approaches have been used by researchers for solving ALBPs. The Genetic Algorithm, Ant Colony Optimisation, Tabu Search, Simulated Annealing, Bees Algorithm and others have been proposed for ALBP, demonstrating a good capability to solve both single and multiobjective problems [24].

A fundamental activity for each manufacturing and production company is represented by maintenance. Good management of maintenance activities can allow minimising production losses by improving equipment availability [26] while also guaranteeing safety and quality. Maintenance represents a strategic function for firms and can determine the level of competitiveness of companies because of its costs, which can represent up to 70% of the production costs [27]. Thus, there is a necessity to optimise maintenance activities according to the type of equipment. The use of heuristics and metaheuristic approaches is particularly suitable for the problem of minimising the time of a series of activities. Cassidy and Kutanoglu [28] proposed a heuristic to schedule maintenance activities to minimise the total completion time. Other authors used heuristics to study the optimal inspection intervals to minimise the costs of replacing components [29]. On the other hand, the optimal choice of components to replace may represent great savings for firms. De Smidt-Destombes et al. [30] developed a heuristic algorithm to jointly optimise the fixed number of spare parts and the maintenance frequency for multi-component systems, while Diallo et al. [31] formulated a two-stage heuristic algorithm to select components for maintenance considering a limited amount of time available. Workforce scheduling is another important topic in maintenance optimisation, such as in the case of López-Santana et al. [32], who defined a two-stage heuristic method to determine the assignation of maintenance technicians when multiple technicians are available.

Maintenance also represents a crucial activity of the remanufacturing process, which has emerged in recent years as a very promising strategy for returning used products in a state similar to the new state while achieving the goals of sustainability and circular economy principles [33]. The remanufacturing process is composed of different macro-phases (first level) and sub-phases (second level), and the level of uncertainty in the operations to perform can sometimes be extremely high because of the different quality of the returned products [34]; thus, the optimisation of the remanufacturing process is often needed. Caterino et al. proposed the use of the BA [35] and Tabu Search [36] for minimising the total completion time of the whole remanufacturing process considering that the various phases and sub-phases could be accomplished by different companies, as in cloud environments. Jiang et al. [37] used a genetic algorithm to optimise remanufacturing process planning to minimise costs considering the reliability of the returned products. Other authors focused on the optimisation of specific phases of the remanufacturing process. The optimisation of the disassembly process is a widely debated topic in the literature since a correct disassembly planning sequence is fundamental to reducing the costs and time of the remanufacturing process [38]. An improved genetic algorithm was used by Meng and Zhang [39] to schedule the disassembly tasks with the aim of reducing the costs of the process, while Xu et al. [40] and Liu et al. [41] implemented a BA to optimise the disassembly sequence planning of a human–robot disassembly cell and a robot cell, respectively. Fu et al. [42] developed a fruit fly optimisation algorithm to minimise the total completion time and the total tardiness of the disassemble-reprocessing (maintenance) reassembly sequence.

Optimisation may be needed for many other manufacturing processes, such as additive manufacturing, where process planning may allow for minimising the used materials and wastes [43, 44], or the design of new products by optimising shapes according to the type of production technology [45].

Most of the activities cited above require human labour to be accomplished. Human labour is fundamental in assembling or disassembling activities, maintenance operations or preprocess and postprocess activities in additive manufacturing. When human labour is considered, the ergonomics problem cannot be neglected, especially for production lines. Several methods are used for ergonomics assessment by professionals according to the type of job to accomplish. Among them, observational methods are the most used, but as highlighted in the previous section, they present 2 main problems related to (i) the subjectivity in the judgments of the ergonomics scores, which depends on the ergonomists' evaluation, and (ii) the time spent for the analysis, which can be very high, especially for companies having a large population of workers (i.e., heterogeneous workers). These 2 factors are the main ones responsible for errors in ergonomics final scores. A way to reduce errors is the use of modern technologies, such as simulation. Many authors have demonstrated that simulation is a very effective tool for assessing ergonomics, starting from the design phase of a production cell or line. Caputo et al. [46] used simulation and digital twins to demonstrate that preliminary ergonomics studies can be useful for companies to avoid errors during the design phase of a production line. Simulation can also be used to evaluate structural modifications to existing workstations to improve ergonomics [4].

Moreover, simulation allows for reducing the computational times for ergonomics assessments because of its capability to provide multi-scenario analysis by changing a few parameters. As demonstrated by Caterino et al. [12], this could help practitioners find solutions that minimise ergonomics risks when a large population of heterogeneous workers is considered. Although simulation is a very powerful tool for improving ergonomics in manufacturing and production lines, it does not allow finding optimised solutions in which the ergonomic risks of all heterogeneous workers are minimised. On the other hand, many authors optimised ergonomics without leveraging the powerful simulation technologies of the modern ages [47, 48]. Thus, based on the chapter proposed by Caterino et al. [12], this research aims to combine the use of simulation and optimisation tools by introducing and validating the BA as an optimisation algorithm to minimise the ergonomics risks of heterogeneous workers employed in manufacturing and production lines when JR is considered.

3 The Proposed Model

In this article, a new approach that combines the use of BA with simulation tools to minimise ergonomic risks in manufacturing and production lines is proposed. The model is represented in Fig. 1 and originated from the model proposed by Caterino et al. [12] in 2022.

The first step of the model consists of analysing the physical system and the working context. This step is needed to identify all the characteristics of the WSs and the ergonomics risk factors that may be critical.

The second step consists of building a digital system reproducing the real working environment by using simulation software. The ergonomics simulations must be realised within the digital environment by reproducing the real actions that workers perform during the working tasks. Thus, during this step, it is important to be accurate in simulating the working tasks by using digital human models (DHM) with the same anthropometric characteristics as the real workers. The advantage of using simulation for this type of analysis consists of the possibility of rapidly changing the anthropometric characteristics of the DHM without the necessity of building a new environment. This aspect allows for speeding up the ergonomics risk evaluation process, reducing the times needed in comparison to observational methods. Moreover, simulation allows for obtaining objective scores of the ergonomics risk factors, not influenced by the subjectivity of the observational methods.

Once the digital environment has been set, the third step of the model consists of determining the most suitable ergonomics risk assessment methods to assess according to the type of jobs performed. Such methods must be chosen among those

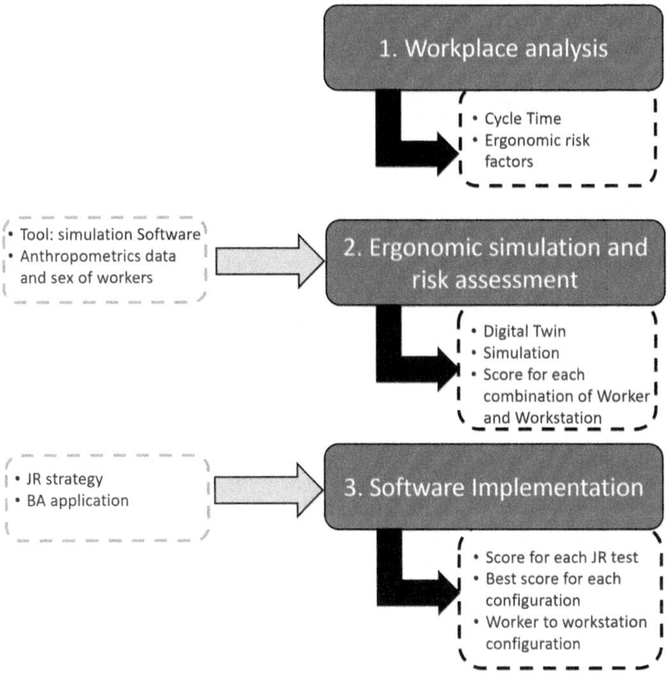

Fig. 1 Framework of the proposed model

implemented by the simulation software. At this point, simulations can be carried out.

The output of the third step of the model is a database that includes, for each combination of worker and WS, a score for each ergonomics analysis needed.

The last step of the model consists of setting several scenarios considering different JR strategies and fixing the objective functions to optimise the system. Some common objective functions used in the literature consist of the minimisation of ergonomics scores for each worker [49] or balancing ergonomics scores for workers [50]. In the model proposed here, the optimisation algorithm to use for optimising the objective function is the BA, which is implemented in the last step of the procedure. Although the application of BA is new for this type of problem, it has been chosen by the authors because of its capability to adapt to different kinds of problems, both continuous and discrete.

The outputs of the last step can be multiple. For each JR strategy, the scores will be calculated according to the objective functions, allowing us to evaluate the JR strategy that better improves the workers' ergonomics.

4 Application of the Model

In this section, the model described in Sect. 3 will be tested for a real case study. Starting from the results of Caterino et al. [12], who developed a case study for the application of the model previously described for 2 WSs, this chapter aims to implement and validate the BA for ergonomics optimisation. Thus, the results obtained by applying the BA will be compared with the exact solutions for the proposed case study obtained by enumeration.

4.1 System Description

The application of the proposed model begins with the analysis of the working environment. The design and organisation of each WS and the analysis of workers' actions must be assessed to (i) develop the digital environment that reproduces the real system (second step of the model in Fig. 1) and (ii) identify the physical ergonomics risk factors that affect the working environment to properly set the ergonomics assessment methods.

The working environment presented in this chapter concerns an automotive assembly line in which six WSs are considered. The tasks to accomplish in each WS are reported in Table 1.

The cycle time is the same for all the WSs and is 60 s. After this time, if workers have not finished their tasks, the line must be stopped. The working shift has a duration of 9 h, including a lunch break of 1 h and three official breaks of 10 min distributed during the day. Thus, the networking time is 7.5 h (450 min).

According to the methodology, the ergonomic risk factors must be identified by observing the type of job carried out in the single WSs. For the jobs considered in this case study, the main risk factors are related to the postures assumed by the workers and the energy needed to accomplish the tasks. The repetitiveness of the task can be neglected because of the introduction of the JR strategies. The risk factors identified must then be associated with ergonomics risk assessment methods. In this chapter, the OWAS and energy expenditure assessment methods have been considered the most

Table 1 Tasks performed in the WSs

	Description of the task	Cycle time (s)
Workstation #1	Grasp, carry and assemble the left side of the armrest	60
Workstation #2	Assemble the right side of the armrest	60
Workstation #3	Grasp, carry and assemble the lower steering panel	60
Workstation #4	Screw 2 parts of the boot panel	60
Workstation #5	Assemble the left seat belt rod	60
Workstation #6	Assemble the right seat belt rod	60

appropriate. The OWAS method is used to assess the awkward postures of the whole body [51] using an index that provides information on the severity of the postures assumed during the working tasks. In particular, this method provides as output a final score composed of the scores of 4 parts, i.e., back, arms, head and legs and considering the loads given by the weights of objects. For an in-depth understanding of the application of the method, the authors can refer to Caterino et al. [12]. The Energy Expenditure method allows for measuring the amount of energy spent by a person doing their job for a specific length of time. It was developed by Garg et al. [52] and uses oxygen or metabolic consumption to estimate the quantity of energy (in kilocalories) for each movement of the worker.

4.2 Ergonomics Simulation

The analysis of the system is fundamental for the second step of the proposed model, which consists of realising a digital mirror of the real working environment for running the ergonomics simulations. The software used in this study for reproducing the working environment is Tecnomatix Process Simulate® by Siemens. Such a tool has been chosen for 2 main reasons: (i) it allows changing the anthropometric characteristics of workers very easily, providing several datasets of people whose anthropometric characteristics are typical of different populations; (ii) it has various ergonomic assessment methods already implemented, including OWAS and energy expenditure methods. Moreover, Tecnomatix Process Simulate® allows importing CAD models of real objects in the simulated environment. This is a very important characteristic because, to correctly model the digital environment, it is necessary that the design of the objects present in the digital environment perfectly reproduces the real objects.

Once the CAD of the objects needed in the working scene is created and the DHMs for simulations are set, the third step of the model can start.

Figure 2 shows the simulated working environment and the workers performing some operations reported in Table 1.

4.3 Ergonomic Risk Assessment

After the definition of the digital mirror, ergonomics risk assessments can be carried out through simulations. The OWAS and EE methods have been used in this article as ergonomics risk assessment methods. The ergonomics scores related to such methods vary according to the anthropometric characteristics of the workers. The advantage of using simulation is the possibility of assessing the ergonomics scores for workers with different anthropometric characteristics by simply changing the DHM parameters in the simulation. In fact, the simulation software can adapt the postures of the DHMs to the assembly points (i.e., the points that workers have to reach to complete the tasks)

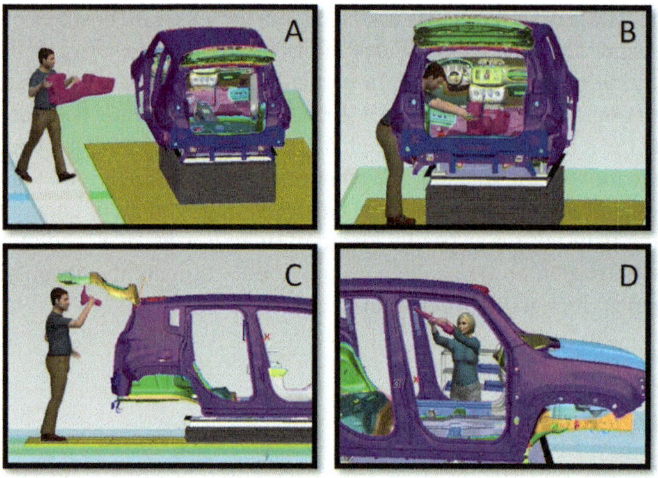

Fig. 2 Digital environment

according to their anthropometric characteristics. This allows setting the simulation only once, without the necessity of building a different simulation for each DHM.

In this chapter, 10 DHMs were used, changing the sexes and heights. The anthropometric characteristics of the DHMs were defined according to the height percentile, which is a parameter provided by the software to evaluate and compare DHMs with different body sizes. Five percentiles were considered for males and females, and the anthropometric characteristics of the workers are reported in Table 2. The ANSUR II (Anthropometric Survey of US Army Personnel) database has been used for the analysis in this chapter. It includes data from approximately 6000 adult US military personnel.

Once the DHMs were set, the ergonomics simulations were run. The outputs of the simulation are the OWAS and EE scores for each combination of worker and workstation. A total of 60 (2 sexes $\times$ 5 percentiles $\times$ 6 WSs) scores were computed by using simulation. For such simulations, JR strategies have not yet been considered. Figure 3 reports an excerpt of the results of the simulations, which are fully reported in the Appendix section.

Table 2 Anthropometric characteristics of the workers considered

Percentile (P)	Male height (cm)	Female height (cm)
5	164	152
25	171	159
50	175	163
75	180	167
95	187	173

	Worker		Workplace	Results of the simulation	
id_simulation	Sex	Percentile	Workstation	$OWAS_{w,i}$	$EE_{w,i}$ [Kcal/min]
1	M	5P	1	217.9	2.3
2	M	25P	1	206.2	2.5
3	M	50P	1	153.0	2.8
4	M	75P	1	153.0	2.8
5	M	95P	1	152.8	2.9
26	F	5P	3	248.8	1.4
27	F	25P	3	237.9	1.5
28	F	50P	3	221.9	1.6
29	F	75P	3	173.8	1.8
30	F	95P	3	100.0	1.9

Fig. 3 Results of the ergonomics simulations for different workers

As seen from Fig. 3, the ergonomics scores obtained by simulation may be very different according to the anthropometric characteristics of the workers; thus, the minimisation of such scores using a correct assignation of workers to WSs is fundamental to reducing the risks of developing WMSDs.

4.4 Software Implementation

In the final step of the methodology proposed and reported in Fig. 1, the results of the ergonomics simulations obtained in the previous phase are used as inputs to evaluate the possible JR strategies that can minimise the efforts of workers. The JR strategies consist of the number of JR the workers have to do during the day, and they allow defining different JR scenarios, which consists of determining the WS to which each worker is assigned at each rotation during the work shift. The objective is to determine the JR scenarios that minimise the ergonomics scores. To achieve this purpose, the BA is introduced in this chapter as the optimisation algorithm. For each JR scenario, it is possible to calculate the final $OWAS_w$ and EE_w scores for each worker w by means of the following formulas:

$$OWAS_w = \sum_{i=1}^{n} t_{i\%} \cdot OWAS_{w,i} \quad \forall w \tag{1}$$

$$EE_w = \sum_{i=1}^{n} t_{i\%} \cdot EE_{w,i} \quad \forall w \tag{2}$$

where i represents the WS, n is the total number of WSs considered, and $t_{i\%}$ is the percentage of time spent on the single WS by the worker and depends on the JR strategies.

Two objective functions are considered in this chapter. They concern the minimisation of the sum of all the $OWAS_w$ and EE_w scores, namely, $OWAS_t$ and EE_t, as reported in (3) and (4).

$$\min!OWAS_t = \sum_{w=1}^{n} OWAS_w \tag{3}$$

$$\min!EE_t = \sum_{w=1}^{n} EE_w \tag{4}$$

Subject to:

$$OWAS_w < 250 \tag{5}$$

$$EE_w < 2.5 \tag{6}$$

A further constraint is that workers cannot visit the same workplace twice during the work shift.

As previously stated, the objective of this chapter is to validate the use of the BA for ergonomics applications. Thus, both the BA and an exact algorithm that achieves the exact solution by enumerations have been implemented in Python. Regarding the BA, three parameter settings (PS) have been used in this application, as reported in Table 3.

The results will be shown and discussed in the next section.

Table 3 Parameter settings for BA parameters

Parameters	PS#1	PS#2	PS#3
Number of scout bees (n)	20	30	50
Number of best or selected sites (m)	8	12	20
Number of elite sites (e)	2	3	5
Number of best bees (nsp)	10	15	25
Number of elite site bees (nep)	40	60	100

5 Results and Discussion

JR strategies affect the final ergonomics scores of workers. When a JR strategy is introduced, the final scores for OWAS and EE are obtained using (1) and (2), which take into account the time spent in each WS. In this chapter, 2 job rotation strategies have been considered: in JR = 1, workers rotate only once among the WSs; in JR = 2, workers rotate twice. Considering these strategies, the size of the problem for 6 workers and 6 WSs allows for evaluating the exact solutions that minimise the objective functions (3) and (4). In fact, the maximum number of combinations in each JR is 6!; thus, the size of the problem makes it possible to evaluate all the possible combinations.

Only one set of workers has been tested in this chapter. The workers are F50, F75, F95, M5, M25, and M50. They have been chosen randomly to guarantee the generality of this application.

Starting from the ergonomics scores obtained by simulations, the two JR strategies were implemented, and the final ergonomics scores were computed by using (1–4). The solutions that minimise $OWAS_t$ and EE_t are reported in Table 3 together with the computational times (expressed in seconds) for the two JR strategies.

The BA has been implemented and tested for the same cases with 3 different PSs, as reported in Table 3. For each PS, 20 experiments were run to calculate the average solutions and errors with respect to the exact solution. The average error is defined as follows:

$$Error\% = \frac{Avg - Exact}{Exact} \cdot 100 \quad (7)$$

where *Avg* represents the average of the 20 BA experiments and *Exact* is the exact solution reported in Table 4.

The stopping criterion was set to be the maximum number of iterations.

Tables 5 and 6 report the results for the OWAS score considering the 3 PSs for JR = 1 and JR = 2, respectively, and a stopping criterion fixed to 200 iterations.

Table 4 Exact solutions considering 2 JR strategies

JR strategy	$min!OWAS_t$	$min!EE_t$	Computational time (s)
JR = 1	882.22	12.8	1.41
JR = 2	897.56	12.8	137

Table 5 Results of BA application considering $OWAS_t$ for JR = 1 and number of maximum iterations 200

Parameters setting	Avg for $OWAS_t$	BA best solution	Computational time (s)	Error (%)
PS#1	882.22	882.22	0.36	0
PS#2	882.22	882.22	0.42	0
PS#3	882.22	882.22	1.45	0

Table 6 Results of BA application considering $OWAS_t$ for JR = 2 and 200 maximum iterations

Parameters setting	Avg for $OWAS_t$	BA best solution	Computational time (s)	Error (%)
PS#1	899.27	897.56	0.48	0.19
PS#2	898.71	897.56	1.17	0.13
PS#3	898.02	897.56	2.73	0.05

From Tables 5 and 6, it is clear that the performance of the BA is very good. The average solution always coincides with the exact solution in the case of JR = 1 (the average error is always zero). For JR = 2, the average solutions are very close to the exact ones, and overall, for each PS, the BA always found the best solution in at least one out of 20 experiments. The average error is up to 0.19%. In this last case, the average error changes according to the BA parameters, decreasing when intensifying the search activity of the algorithm. The computational times for the BA remain very similar in the cases of JR = 1 and JR = 2 (they mainly depend on the iterations, which do not change) for the same PS, and they are greatly reduced in comparison to the computational times of the exact solution.

Tables 7 and 8 report the results for the EE_t score considering the 3 PSs for JR = 1 and JR = 2, respectively, and a stopping criterion fixed to 200 iterations.

The results for EE confirm the effectiveness of the algorithm in being able to find good solutions for the minimisation of the objective function. In fact, for the case of EE, the average error is 0% for both JR strategies, and the computational time is still reduced in comparison with the exact solution.

The results for EE confirm the effectiveness of the algorithm in being able to find good solutions for the minimisation of the objective function. In fact, for the case of EE, the average error is 0% for both JR strategies, and the computational time is still reduced in comparison with the exact solution. It is worth noting that, in this case, the average error is 0% because the measure of the EE (one decimal place) is less precise than the measure of OWAS (two decimal places). It is not possible to increase

Table 7 Results of BA application considering EE_t for JR = 1 and 200 maximum iterations

Parameters setting	Avg for EE_t	BA best solution	Computational time (s)	Error (%)
PS#1	12.8	12.8	0.54	0
PS#2	12.8	12.8	0.91	0
PS#3	12.8	12.8	1.98	0

Table 8 Results of BA application considering EE_t for JR = 2 and 200 maximum iterations

Parameters setting	Avg for EE_t	BA best solution	Computational time (s)	Error (%)
PS#1	12.8	12.8	0.73	0
PS#2	12.8	12.8	1.38	0
PS#3	12.8	12.8	2.63	0

Table 9 Results of BA application considering different numbers of iterations for OWAS, JR = 2 and PS#1

Number of iterations	Avg for $OWAS_t$	BA best solution	Computational time (s)	Error (%)
50	901.36	897.56	0.36	0.42
200	898.71	897.56	0.48	0.13
1000	898.09	897.56	2.92	0.06

the precision of the EE index in the Process Simulate software, but this aspect does not affect the final results obtained, which validated the BA.

The last analysis concerns an evaluation of the behaviour of the BA by changing the number of maximum iterations. The JR = 2 strategy for $OWAS_t$ minimisation has been assessed because, from the previous analysis, it resulted in the only one with an average error greater than 0.

The same population of workers and PS#1 were considered, while three different numbers of iterations were set as the stopping criterion: 50, 200, and 1000. The experiments were run 20 times, and the results are reported in Table 9.

Once again, the results demonstrate the accuracy of the BA in achieving very good results, near the exact solution. As expected, by increasing the number of iterations, the average error decreases, while the computational time increases.

In light of these results, it can be concluded that the BA can achieve very good performance for the 2 objective functions proposed in this chapter. In fact, the BA can often find the exact solution, and the average error is always under 1%, also in the worst-case scenario (PS#1 and 50 iterations for OWAS and JR = 2). Additionally, the computational times are very low, allowing for fast analyses. For these reasons, the BA can be considered validated for this application, in which ergonomics efforts were to be minimised.

6 Conclusions

This chapter presented a new application for BA related to the optimisation of ergonomic factors for heterogeneous workers that can rotate among workstations. The chapter extends the work done by Caterino et al. [12], in which a framework for ergonomics improvements has been proposed by introducing the BA as an optimisation algorithm for minimising and balancing the ergonomics efforts of workers. The objective was to validate the BA by demonstrating that the results achieved by the BA are consistent with the results achieved by the exact solution obtained by enumeration. The OWAS and EE methods were chosen as ergonomics assessment methods for this purpose, according to the type of working activity performed by the workers. The results of the experiments validated the BA, highlighting its accuracy in finding solutions very close to the exact one in very few computational times.

Since the objective of this chapter was not to discuss ergonomics, the results reported in the previous section about ergonomics due to the JR strategies have not been commented upon. Future work will extend this chapter by increasing the size of the problem and by introducing factors for evaluating other aspects of ergonomics, such as the correct balancing of the ergonomics scores of the workers and the JR strategies that should be used for improving the working conditions of different populations of heterogeneous workers.

Appendix

Id_simulation	Sex	Percentile (P)	Workstation	$OWAS_{w,i}$	$EE_{w,i}$
1	M	5	1	217.9	2.3
2	M	25	1	206.2	2.5
3	M	50	1	153.0	2.8
4	M	75	1	153.0	2.8
5	M	95	1	152.8	2.9
6	F	5	1	235.1	2.0
7	F	25	1	223.6	2.1
8	F	50	1	223.6	2.3
9	F	75	1	206.2	2.5
10	F	95	1	153.4	2.5
11	M	5	2	212.8	2.1
12	M	25	2	203.7	2.3
13	M	50	2	153.3	2.6
14	M	75	2	151.9	2.6
15	M	95	2	151.9	2.7
16	F	5	2	234.0	1.8
17	F	25	2	218.8	1.9
18	F	50	2	218.8	2.1
19	F	75	2	198.4	2.3
20	F	95	2	151.9	2.3
21	M	5	3	220.0	1.6
22	M	25	3	171.7	1.8
23	M	50	3	223.3	2.0
24	M	75	3	289.2	2.1
25	M	95	3	289.2	2.2
26	F	5	3	248.8	1.4
27	F	25	3	237.9	1.5

(continued)

(continued)

Id_simulation	Sex	Percentile (P)	Workstation	$OWAS_{w,i}$	$EE_{w,i}$
28	F	50	3	221.9	1.6
29	F	75	3	173.8	1.8
30	F	95	3	169.5	1.9
31	M	5	4	121.3	1.7
32	M	25	4	121.3	1.8
33	M	50	4	120.7	2.0
34	M	75	4	120.0	2.1
35	M	95	4	120.0	2.2
36	F	5	4	128.7	1.5
37	F	25	4	122.0	1.6
38	F	50	4	122.0	1.7
39	F	75	4	121.3	1.8
40	F	95	4	120.0	1.9
41	M	5	5	163.0	2.1
42	M	25	5	159.1	2.3
43	M	50	5	159.1	2.5
44	M	75	5	135.7	2.6
45	M	95	5	120.7	2.6
46	F	5	5	180.0	1.9
47	F	25	5	170.0	2.0
48	F	50	5	171.5	2.1
49	F	75	5	135.7	2.3
50	F	95	5	135.7	2.3
51	M	5	6	163.0	2.1
52	M	25	6	159.1	2.3
53	M	50	6	159.1	2.5
54	M	75	6	135.7	2.6
55	M	95	6	120.7	2.6
56	F	5	6	180.0	1.9
57	F	25	6	170.0	2.0
58	F	50	6	171.5	2.1
59	F	75	6	135.7	2.3
60	F	95	6	135.7	2.3

References

1. Otto A, Scholl A (2013) Reducing ergonomic risks by job rotation scheduling. OR Spect 35:711–733
2. Neumann WP, Winkel J, Palmerud G, Forsman M (2018) Innovation and employee injury risk in automotive disassembly operations. Int J Prod Res 56:3188–3203
3. Glock CH, Grosse EH, Neumann WP, Sgarbossa F (2017) Editorial: human factors in industrial and logistic system design. Comput Ind Eng 111:463–466
4. Greco A, Caterino M, Fera M, Gerbino S (2020) Digital twin for monitoring ergonomics during manufacturing production. Appl Sci 10:7758
5. Diefenbach H, Glock CH (2019) Ergonomic and economic optimization of layout and item assignment of a U-shaped order picking zone. Comput Ind Eng 138:106094
6. Dalle Mura M, Dini G (2019) Optimizing ergonomics in assembly lines: a multi objective genetic algorithm. CIRP J Manuf Sci Technol 27:31–45
7. Van der Spaa L, Gienger M, Bates T, Kober J (2020) Predicting and optimizing ergonomics in physical human-robot cooperation tasks. In: 2020 IEEE International conference on robotics and automation (ICRA). IEEE, pp 1799–1805
8. Wu SQ, Shen B, Tang YZ, Wang JH, Zheng DT (2018) Ergonomic layout optimization of a smart assembly workbench. Assem Autom 38:314–322
9. Gomes W, Maurice P, Dalin E, Mouret JB, Ivaldi S (2021) Multi-objective trajectory optimization to improve ergonomics in human motion. IEEE Robot Autom Letters 7:342–349
10. Takala EP, Pehkonen I, Forsman M, Hansson GÅ, Mathiassen SE, Neumann WP, Sjogard G, Veiersted KB, Westgaard RH, Winkel J (2010) Systematic evaluation of observational methods assessing biomechanical exposures at work. Scand J Work Environ Health 3–24
11. Massiris Fernández M., Fernández JÁ, Bajo JM, Delrieux CA (2020) Ergonomic risk assessment based on computer vision and machine learning. Comput Industr Eng 149:106816
12. Caterino M, Rinaldi M, Fera M (2022) Digital ergonomics: an evaluation framework for the ergonomic risk assessment of heterogeneous workers. Int Journal Comput Integr Manuf 1–21
13. Laudante E, Greco A, Caterino M, Fera M (2020) Human-robot interaction for improving fuselage assembly tasks: a case study. Appl Sci 10:5757
14. Kadir BA, Broberg O (2021) Human-centered design of work systems in the transition to industry 4.0. Appl Ergon 92:103334
15. Moussavi SE, Zare M, Mahdjoub M, Grunder O (2019) Balancing high operator's workload through a new job rotation approach: application to an automotive assembly line. Int J Ind Ergon 71:136–144
16. Pham DT, Ghanbarzadeh A, Koc E, Otri S, Rahim S, Zaidi M (2005) The Bees algorithm. Manufacturing Engineering Centre, Cardiff University, UK, Technical Note
17. Pham DT, Castellani M (2014) Benchmarking and comparison of nature-inspired population-based continuous optimisation algorithms. Soft Comput 18:871–903
18. Zeybek S, Ismail AH, Hartono N, Caterino M, Jiang K (2021) An Improved Vantage Point Bees Algorithm to Solve Combinatorial Optimization Problems from TSPLIB. Macromol Symp 396:2000299
19. Ismail AH, Hartono N, Zeybek S, Caterino M, Jiang K (2021) Combinatorial Bees algorithm for vehicle routing problem. In: Macromolecular symposia, vol 396, p 2000284
20. Freitas AA, Lima TM, Gaspar PD (2022) Ergonomic risk minimization in the Portuguese wine industry: a task scheduling optimization method based on the Ant Colony optimization algorithm. Processes 10:1364
21. Pham DT, Ghanbarzadeh A (2007) Multiobjective optimisation using the Bees algorithm. In: 3rd International virtual conference on intelligent production machines and systems, p 6
22. Hidalgo-Herrero M, Rabanal P, Rodriguez I, Rubio F (2013) Comparing problem solving strategies for NP-hard optimisation problems. Fund Inform 124:1–25
23. Salveson ME (1955) The assembly-line balancing problem. Trans Am Soc Mech Eng 77:939–947

24. Li Z, Kucukkoc I, Nilakantan JM (2017) Comprehensive review and evaluation of heuristics and meta-heuristics for two-sided assembly line balancing problem. Comput Oper Res 84:146–161
25. Baybars I (1986) A survey of exact algorithms for the simple assembly line balancing problem. Manage Sci 32:909–932
26. Abbate R, Caterino M, Fera M, Caputo F (2022) Maintenance digital twin using vibration data. Procedia Comput Sci 200:546–555
27. Bevilacqua M, Braglia M (2020) The analytic hierarchy process applied to maintenance strategy selection. Reliab Eng Syst Saf 70:71–83
28. Cassady CR, Kutanoglu E (2005) Integrating preventive maintenance planning and production scheduling for a single machine. IEEE Trans Reliab 54:304–309
29. De Jonge B, Scarf PA (2020) A review on maintenance optimization. Eur J Oper Res 285(3):805–824
30. De Smidt-Destombes KS, van der Heijden MC, van Harten A (2009) Joint optimisation of spare part inventory, maintenance frequency and repair capacity for k-out-of-N systems. Int J Prod Econ 118:260–268
31. Diallo C, Venkatadri U, Khatab A, Liu Z (2018) Optimal selective maintenance decisions for large serial k-out-of-n: G systems under imperfect maintenance. Reliab Eng Syst Saf 175:234–245
32. López-Santana E, Akhavan-Tabatabaei R, Dieulle L, Labadie N, Medaglia AL (2016) On the combined maintenance and routing optimization problem. Reliab Eng Syst Saf 145:199–214
33. Sakao T, Sundin E (2019) How to improve remanufacturing? A systematic analysis of practices and theories. J Manuf Sci Eng 141
34. Guide VDR Jr (2000) Production planning and control for remanufacturing: industry practice and research needs. J Oper Manag 18:467–483
35. Caterino M, Fera M, Macchiaroli R, Pham DT (2022) Task optimisation for a modern cloud remanufacturing system using the Bees algorithm. In: Intelligent production and manufacturing optimization—the Bees algorithm approach. Springer series in advanced manufacturing (in press)
36. Caterino M, Fera M, Macchiaroli R, Pham DT (2022) Cloud remanufacturing: remanufacturing enhanced through cloud technologies. J Manuf Syst 64:133–148
37. Jiang Z, Zhou T, Zhang H, Wang Y, Cao H, Tian G (2016) Reliability and cost optimization for remanufacturing process planning. J Clean Prod 135:1602–1610
38. Liu J, Zhou Z, Pham DT, Xu W, Ji C, Liu Q (2020) Collaborative optimization of robotic disassembly sequence planning and robotic disassembly line balancing problem using improved discrete Bees algorithm in remanufacturing☆. Robot Comput-Integr Manuf 61:101829
39. Meng W, Zhang X (2020) Optimization of remanufacturing disassembly line balance considering multiple failures and material hazards. Sustainability 12:7318
40. Xu W, Tang Q, Liu J, Liu Z, Zhou Z, Pham DT (2020) Disassembly sequence planning using discrete Bees algorithm for human-robot collaboration in remanufacturing. Robot Comput-Integr Manuf 62:101860
41. Liu J, Zhou Z, Pham DT, Xu W, Yan J, Liu A, Ji C, Liu Q (2018) An improved multi-objective discrete bees algorithm for robotic disassembly line balancing problem in remanufacturing. Int J Adv Manuf Technol 97:3937–3962
42. Fu Y, Zhou M, Guo X, Qi L (2021) Stochastic multi-objective integrated disassembly-reprocessing-reassembly scheduling via fruit fly optimization algorithm. J Clean Prod 278:123364
43. Druzgalski CL, Ashby A, Guss G, King WE, Roehling TT, Matthews MJ (2020) Process optimization of complex geometries using feed forward control for laser powder bed fusion additive manufacturing. Addit Manuf 34:101169
44. Jiang J, Xu X, Stringer J (2019) Optimization of process planning for reducing material waste in extrusion based additive manufacturing. Robot Comput-Integr Manuf 59:317–325
45. Porziani S, Groth C, Biancolini ME (2018) Automatic shape optimization of structural components with manufacturing constraints. Procedia Struct Integrity 12:416–428

46. Caputo F, Greco A, Fera M, Macchiaroli R (2019) Digital twins to enhance the integration of ergonomics in the workplace design. Int J Ind Ergon 71:20–31
47. Rinaldi M, Fera M, Bottani E, Grosse EH (2022) Workforce scheduling incorporating worker skills and ergonomic constraints. Comput Ind Eng 168:108107
48. Raolji VG, Siddiqui N, Nandan A, Pandya K (2018) A case study on optimization of manual activities through ergonomics interventions. Int J Appl Eng Res 13:5075–5080
49. Gebennini E, Zeppetella L, Grassi A, Rimini B (2018) Optimal job assignment considering operators' walking costs and ergonomic aspects. Int J Prod Res 56:1249–1268
50. Mossa G, Boenzi F, Digiesi S, Mummolo G, Romano VA (2016) Productivity and ergonomic risk in human based production systems: a job-rotation scheduling model. Int J Prod Econ 171:471–477
51. Karhu O, Kansi P, Kuorinka I (1977) Correcting working postures in industry: a practical method for analysis. Appl Ergon 8:199–201
52. Garg A, Chaffin DB, Herrin GD (1978) Prediction of metabolic rates for manual materials handling jobs. Am Ind Hyg Assoc J 39:661–674

A Knowledge Transfer-Based Bees Algorithm for Expert Team Formation Problem in Internet Company

Yanjie Song, Yangyang Guo, Jiting Li, Jian Wu, Qinwen Yang, and Yingwu Chen

Abstract How to build a team of experts plays an important role in the smooth development of a new project. The Internet company was investigated to determine how to form a team that can communicate effectively. This problem is known as the problem of expert team formation in Internet companies (TFPICs). We construct a mathematical optimisation model based on an undirected graph and propose a bees algorithm using knowledge transfer, named KT-BA. KT-BA uses the best information obtained through cooperative search of multiple subpopulations and transfers the dominant subpopulations to other subpopulations. A reproduction mechanism is also used in the algorithm to generate new individuals to enhance population diversity. The experimental results verify the solution performance of KT-BA, and the obtained expert team formation scheme outperforms the traditional bees algorithm and neighbourhood search algorithm. Thus, KT-BA is suitable for application in large Internet companies to provide scientific decisions for project development.

Keywords Knowledge transfer · Bees algorithm · Team formation · Internet companies · Evolutionary algorithm

---

Y. Song
National Engineering Research Center of Maritime Navigation System, Dalian Maritime University, Dalian, China

Y. Guo (✉)
School of Systems Science, Beijing Jiaotong University, Beijing, China
e-mail: g2002dmu@163.com

J. Li
Chinese Academy of Military Science, Beijing, China

J. Wu
Key Laboratory of Collaborative Intelligence Systems, Ministry of Education, Xidian University, Xi'an, China

Q. Yang
Department of Computer Science, The University of Auckland, Auckland, New Zealand

Y. Chen
College of Systems Engineering, National University of Defense Technology, Changsha, China

D. T. Pham and N. Hartono (eds.), *Intelligent Engineering Optimisation with the Bees Algorithm*, Springer Series in Advanced Manufacturing,
https://doi.org/10.1007/978-3-031-64936-3_15

1 Introduction

The team formation problem (TFP) is a combinatorial optimisation problem often faced in enterprise recruitment, social networking, and large-scale projects [1]. Large Internet companies develop operating systems, software and websites through cooperation with multiple subsidiaries and across countries, which can make the best use of human resources within the companies to achieve established goals by communicating and collaborating efficiently. The selection of experts is the most critical step in forming a team of experts and has a significant impact on the subsequent development of the project.

We focus on the team formation problem in an Internet company (TFPIC), which considers which experts to select to form an efficient team of experts in a situation with multiple branches worldwide. We focus on the team formation problem in Internet companies (TFPIC), which considers which experts to select to form an efficient team of experts in a situation where there are multiple branches around the world. This problem has similarities to the set covering problem and the TFP in social networks. However, since the selected experts come from one specific company, there is also the need to ensure that the company's normal business is not affected by the new project. The optimisation model and solution algorithm we use can effectively solve the long-term perplexity of enterprise managers by transforming the leadership decision to select a team of experts into a combinatorial optimisation problem, allowing for a more scientific team formation scheme.

In recent years, a series of studies have attempted to solve the team formation problem [1–4]. Bachrach et al. used a deep reinforcement learning method to solve the negotiation team formation problem and trained the intelligent body through a data-driven approach to negotiate and form a team [5]. The experimental results showed that the proposed method performed better than the cooperative game theory method. Rahmanniyay et al. constructed a multiobjective stochastic optimisation model and verified the algorithm's effectiveness in solving TFP using many real-world cases [6]. In related studies, only a few of them have used metaheuristic algorithms to solve TFP. El-Ashmawi et al. proposed an improved particle swarm algorithm to find team formation solutions [7]. This algorithm used a new swap operator to explore the solution space efficiently. Gutiérrez et al. used a variable neighbourhood search algorithm to solve the multiteam formation problem [8]. The results proved that the metaheuristic algorithm had better performance in solving the problem. Selvarajah et al. designed the Trust Score index for TFP in social networks to measure the relationship between the experts [9]. A multiobjective cultural algorithm was proposed, in which belief space was adopted to achieve effective information retention.

Based on the above related studies, it is clear that the design of the solution algorithm is particularly important for the scheme of TFP. The team formation problem is known to be an NP-hard problem; hence, using heuristic or evolutionary algorithms to solve this problem is more suitable for the application's needs. We use the classical bees algorithm in evolutionary algorithms to solve TFPIC. The bees algorithm is well designed with global and local search strategies and efficiently solves multiclass combinatorial optimisation problems. We add the knowledge transfer mechanism into the BA framework to effectively use the information obtained from the search and help the algorithm find the approximate optimal solution through knowledge guidance. The main contributions of this chapter are as follows.

1. We construct a 0–1 programming model for TFPIC. The model is constructed with the optimisation objective of minimising the total distance, and the constraint part limits the needed skills and the composition of experts.
2. We propose a bees algorithm based on knowledge transfer. We transfer the optimal location information to the subpopulation search and transfer the optimal population information to other subpopulations after the subpopulation search is completed.

The rest of the chapter is structured as follows. The second part introduces the mathematical model of TFPIC. The third part introduces the bees algorithm based on knowledge transfer, with an introduction to the algorithm coding, knowledge transfer mechanism, population search and individual reproduction mechanism. The fourth part designs simulation experiments to verify the results of the proposed algorithm. The fifth part summarises the main work of this chapter and future research directions.

2 Mathematical Model

A multinational company C with several branches has a new software development task M, that requires a range of skills $NS = \{ns_1, ns_2, \ldots, ns_{|NS|}\}$ to complete the development work successfully. The company needs to select several experts from the pool of experts E to form a development team *Team*. The set of skills possessed by expert e_i is $S_i = \{s_1, s_2, \ldots, s_{|S_i|}\}$, the set of experts with the needed skill ns_k is $C(ns_k)$, and the set RS_i is the needed skill of expert i. Software development can move forward when the selected experts meet all the skill requirements. In the software development process, different experts need to communicate with each other to advance the progress of software development and conduct the corresponding test to ensure that the software can go online smoothly. The time difference between different regions of the world can make some communication difficult because the multinational company has branches in many parts of the world. We use the distance d_{ij} to measure the cooperation difficulty between expert i and expert j, where d_{ij} belongs to [0, 1].

The graph model clearly describes the relationship between experts in the problem. For an undirected graph $G(N, E)$, the set of nodes represents experts, and the set of edges represents distances. The distance between adjacent experts in the model depends on how similar their skills are and whether they have communicated or worked before. At the same time, the shortest path calculation determines the actual distance between nonadjacent experts. The shorter the distance, the easier it is for the two experts to work on a project together.

The problem is to decide which experts are involved in the project development team. We use the decision variable x_i to indicate whether expert i participates in the development team. The value of x_i is shown as follows.

$$x_i = \begin{cases} 1 \; if \; the \; expert \; i \; is \; chosen \\ 0 \; else \end{cases}$$

TFPIC needs to meet the following assumptions:

1. The team comprises experts who work until the end of the project and will not be transferred to other projects or leave midway.
2. No additional skills will be needed to develop the project.
3. The distance between experts is defined and will not change during the process.
4. All experts can be selected to work on developing the new project.

In the TFPIC, our optimisation objective is to find a group of experts who can meet the requirements of the software development team and minimise the total cost of communication during development.

Objective function:

$$\min \sum_{i \in E} d_{ij} \cdot x_i \tag{1}$$

where d_{ij} denotes the distance between expert i and expert j.

Constraints:

1. The selected experts have all the skills to meet the needs of the project development.

$$\sum_{i \in E} \sum_{o \in RS_i} s_o \cdot x_i \geq |NS| \tag{2}$$

2. At least one expert meets each needed skill in the project.

$$\sum_{i \in C(ns_k)} x_i \geq 1, \forall k \in NS \tag{3}$$

3 Bees Algorithm Based on Knowledge Transfer

3.1 Framework

To solve the TFPIC, we propose a knowledge transfer-based bees algorithm (KT-BA). The bees algorithm is a metaheuristic algorithm that mimics the behaviour of the bees colony searching flowers for high-quality nectar sources [10, 11]. The traditional bees algorithm (BA) has strong global and local search ability but is not efficient for searching complex combinatorial optimisation problems [12]. We add the idea of knowledge transfer into the BA framework to further explore other regions while using the optimal search information. The individual reproduction mechanism is also adopted in the proposed Bees Algorithm to increase the diversity of the population search. When the mechanism trigger condition is met, a completely new location will randomly replace an existing one. In the following, the main steps and operations of the algorithm are described.

3.2 Coding

We used a method of coding individual sequences of bees based on the skills needed by the team. Each position in the individual sequence is represented by $\langle ns_k, C(ns_k)\rangle$, and ns_k represents the k-th skill needed by the team. Based on the skill corresponding to the individual sequence position, one of the experts with that skill is randomly selected as a member of the development team. Figure 1 shows a coding diagram of three skills and five experts.

Figure 1 shows that the first skill has two optional experts, the second skill has three optional experts, and the third skill has one optional expert.

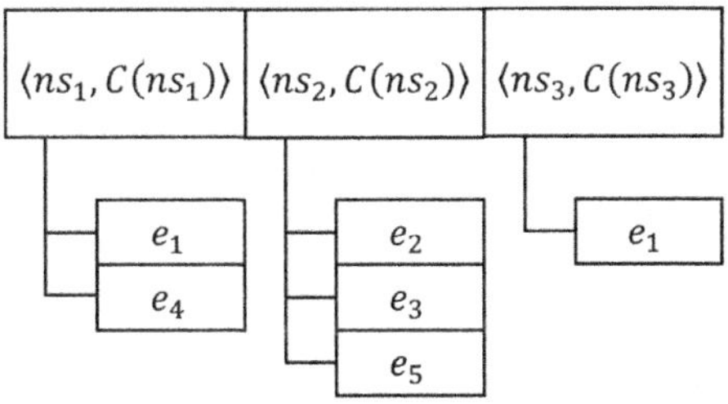

Fig. 1 Diagram of coding

3.3 Knowledge Transfer for Scout Bee Search

Scout bee search is the main search operation of KT-BA, where scout bees search for new nectar sources through populations based on existing locations. We have introduced a knowledge transfer mechanism into the scout bee search process to retain useful information through transference [13]. The high-quality sites found by the bee population search can be of great help in subsequent searches. However, there is a possibility of inadequate solution space search if overreliance is placed on existing information. We use a multisubpopulation search, make use of the obtained search information, and transfer the optimal subpopulation to other subpopulations after the search is completed according to the specific search performance. KT-BA transfers once at the beginning and once at the end of each search, allowing useful information to be used in subsequent searches.

For the scout bee search, we simultaneously use the number of *subpopsize* subpopulations, each of which has a size of N_{sub}. The search operation is carried out separately in each subpopulation. A number of $(1/N_{sub}) * sites$ individuals in the subpopulation are composed of the best nectar sources found after the previous population search. A corresponding number of individuals in the subpopulation are randomly replaced, ensuring that the individuals replaced in multiple subpopulations are different. The individual sequence structure is updated by swapping the positions of the corresponding sequence fragment between two individuals. After the searches of multiple subpopulations are completed, the search performance of each subpopulation is evaluated, and the subpopulation with the best search performance is selected to replace the remaining subpopulations. Finally, a number of N_s best sites are selected according to the individual performance to complete the scout bee search process.

3.4 Recruited Bees Search

Recruited bees also search for high-quality nectar sources in the bees algorithm. A recruited bee search takes place after the scout bee has completed its search and is based on the best current site. We randomly select a location of the best sequence of site individuals and then replace the original expert with other experts with that skill. When the search is completed, it is necessary to determine whether the quality of the newly found nectar source has improved. If there is an improvement, the previous best site will be replaced.

3.5 Individual Reproduction Strategy

With the continuous development of population search, diversity plays an important role in finding better solutions. We add a reproduction mechanism in KT-BA to generate a completely new individual with a certain probability [14, 15]. After the bee population search is completed, a random number belonging to the interval [0, 1] is randomly generated and compared with the set control parameter ε. If the random number is less than the threshold, a site will be randomly selected for replacement. For example, if the control parameter ε is set to 0.6 in advance and the generated random number is 0.2, a site will be randomly selected for replacement.

3.6 The KT-BA

This section introduces the KT-BA procedure. First, an initial population is randomly obtained for population initialisation. Afterward, the algorithm searches for nectar sources using scout bees and recruited bees. Each individual bee performance is evaluated using the objective function proposed in the second part. To increase the diversity of the population, a reproduction mechanism is also used in the KT-BA. The pseudocode for the proposed algorithm is shown in Algorithm 1.

Algorithm 1: Knowledge Transfer-based Bees Algorithm (KT-BA)

Input: number of subpopulations $subpopsize$, subpopulation size N_{sub}, number of sites N_s, control parameterε
Output: Solution S

Initialise the algorithm parameters;
Generate the initial population randomly;
While the termination condition is not met **do**
  //Scout bees search phase;
  $subpop$ ←Transfer sites information to subpopulations;
  Subpopulations search separately;
  Transfer the optimal subpopulation to other subpopulations;
  $sites, local_{best}$ ←Select sites from the optimal subpopulation according to the value of the fitness function;
  If $local_{best} < gobal_{best}$ **then**
    $gobal_{best} \leftarrow local_{best}$;
  End If
  //Recruited bees search phase;
  $site^*$ ←Select the optimal site;
  fit ←Generate new individuals and evaluate the fitness function values;
  If $fit < local_{best}$ **then**
    Update the optimal site;
  End If
  //Reproduction mechanism;
  If $rand() \leq \varepsilon$ **then**
    Regenerate new individuals;
  End If
End While

4 Experiments

To verify the solution performance of the proposed KT-BA, we conduct experiments on a Core (TM) i5-8265U CPU 1.60 GHz, Windows 10 OS computer with MATLAB R2020b as its coding environment. The running environment for all algorithms is kept consistent.

Due to the lack of benchmarks for the TFPIC, we use random generation to obtain a series of test scenarios. In the randomly generated scenarios, the number of experts is larger than the number of skills needed to ensure that the team can be formed successfully. In the experiments, N_{NS} denotes the number of skills and N_E denotes the number of experts.

We use the traditional bees algorithm (TBA) and the local search algorithm (LS) as comparison algorithms to verify the performance of the proposed algorithm in solving the problem. The local search algorithm randomly selects one skill at a time and replaces the selected expert from the optional experts [16]. KT-BA, TBA and LS run 30 times in each scenario, and the results of each algorithm are recorded. We evaluate the optimisation performance of each algorithm using the best value (denoted as Best) and the average value (denoted as Ave).

First, we test the optimisation performance of the algorithm using various scenarios with different skill requirements and experts, and the experimental results are shown in Table 1. As seen from the results, KT-BA outperforms both TBA and LS for comparison in terms of both the best and average values. TBA performs better than LS, which reflects the advantage of the EC population search.

Then, we test the effectiveness of the strategies used in KT-BA, using the bees algorithm without the knowledge transfer mechanism (denoted KT-BA/W1) and the bees algorithm without the reproduction mechanism (denoted KT-BA/W2). The experimental results are shown in Fig. 2, with the horizontal coordinate No. corresponding to each of the above scenarios. Figure 2 shows that the knowledge transfer mechanism and reproduction mechanism used in KT-BA can help the algorithm perform a better solution space search. The knowledge transfer mechanism plays a more significant role in the algorithm than the reproduction mechanism, and useful search information can guide the algorithm toward obtaining better-quality teams.

In addition, we analyse the effect of different subpopulation number settings on the KT-BA optimisation performance. The number of subpopulations in the experiment is set to 2, 3, and 4, and the experimental results are shown in Fig. 3. The figure shows that an increase in the number of subpopulations can improve the algorithm search

Table 1 Performance of different algorithms in solving the problem

N_{NS}	N_E	KT-BA		BA		LS	
		Best	Ave	Best	Ave	Best	Ave
10	50	**55**	**63.37**	57	64.80	81	89.43
14	70	**123**	**151.73**	129	146.90	177	189.93
18	90	**245**	**266.97**	241	272.60	306	321.83
22	110	**423**	**438.53**	435	446.47	477	496.83
26	130	**575**	**594.90**	580	608.07	643	670.67
30	150	**812**	**838.33**	828	856.13	898	919.47
34	170	**1069**	**1087.77**	1089	1112.67	1153	1181.57
38	190	**1323**	**1350.87**	1354	1383.63	1420	1466.80
42	210	**1616**	**1653.53**	1666	1691.33	1726	1770.20
46	230	**1952**	**1988.57**	1999	2032.03	2079	2118.77

Bolded content in the table indicates the best result in the comparison

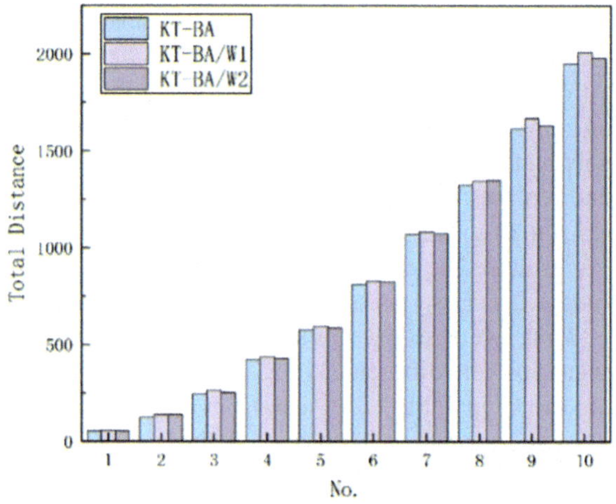

Fig. 2 Results of different strategies

performance to a certain extent. However, it also faces a corresponding increase in calculation cost, and from this perspective, the number of subpopulations selected should be considered in light of factors such as the skills needed for the problem and the scale of the experts.

The above experiments show that KT-BA can scientifically form teams of experts in various types of complex problem scenarios. The knowledge transfer and reproduction mechanisms used in the algorithm are effective for improving the BA search performance. The excellent performance of our proposed algorithm in large-scale problem scenarios can well meet the needs of large Internet companies in forming expert teams.

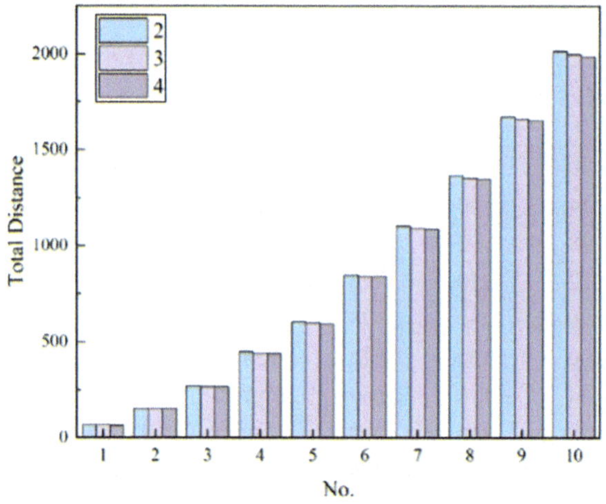

Fig. 3 Results for different subpopulation sizes

5 Conclusion

Expert team formation is a problem that many Internet companies often encounter when conducting new projects. We construct an optimisation model for the TFPIC and propose a KT-BA that introduces a knowledge transfer mechanism. The knowledge transfer mechanism allows the found high-quality solutions to further search the neighbourhood in the subsequent search process and interact with the information obtained in the subpopulation searches, thus improving the global search efficiency of the algorithm. The algorithm also adopts a reproduction mechanism to increase the diversity of the population by generating a new individual with a small probability. The experiments have verified the algorithm's performance, and the search strategies' design in KT-BA can effectively improve the quality of the expert teams.

In the future, we will consider situations that are more closely related to the actual expert team formation process, such as when some experts are unable to leave the current project team to participate in a new project or when the participation of experts in a new project may affect the original work to some extent. In addition, the algorithm is also worthy of further design, such as a data-driven approach to mine useful information in the problem [17], and guide the algorithm to search for expert team formation solutions that fit the application scenarios.

References

1. Anagnostopoulos A, Becchetti L, Castillo C, Gionis A, Leonardi S (2012) Online team formation in social networks. In: Proceedings of the 21st international conference on World Wide Web, pp 839–848
2. Farasat A, Nikolaev AG (2016) Social structure optimisation in team formation. Comput Oper Res 74:127–142
3. Flores-Parra JM, Castañón-Puga M, Evans RD, Rosales-Cisneros R, Gaxiola-Pacheco C (2018) Towards team formation using Belbin role types and a social networks analysis approach. In: 2018 IEEE technology and engineering management conference (TEMSCON). IEEE, pp 1–6
4. Keane P, Ghaffar F, Malone D (2020) Using machine learning to predict links and improve Steiner tree solutions to team formation problems-a cross company study. Appl Net Sci 5(1):1–19
5. Bachrach Y, Everett R, Hughes E, Lazaridou A, Leibo JZ, Lanctot M, Graepel T (2020) Negotiating team formation using deep reinforcement learning. Artif Intell 288:103356
6. Rahmanniyay F, Yu AJ, Seif J (2019) A multi-objective multi-stage stochastic model for project team formation under uncertainty in time requirements. Comput Ind Eng 132:153–165
7. El-Ashmawi WH, Ali AF, Tawhid MA (2019) An improved particle swarm optimization with a new swap operator for team formation problem. J Ind Eng Int 15(1):53–71
8. Gutiérrez JH, Astudillo CA, Ballesteros-Pérez P, Mora-Melià D, Candia-Véjar A (2016) The multiple team formation problem using sociometry. Comput Oper Res 75:150–162
9. Selvarajah K, Zadeh PM, Kobti Z, Palanichamy Y, Kargar M (2021) A unified framework for effective team formation in social networks. Expert Syst Appl 177:114886
10. Gao H, Shi Y, Pun CM, Kwong S (2018) An improved artificial bee colony algorithm with its application. IEEE Trans Industr Inf 15(4):1853–1865

11. Zhou J, Yao X, Lin Y, Chan FT, Li Y (2018) An adaptive multi-population differential artificial bee colony algorithm for many-objective service composition in cloud manufacturing. Inf Sci 456:50–82
12. Pham DT, Ghanbarzadeh A, Koç E, Otri S, Rahim S, Zaidi M (2006) The bees algorithm—a novel tool for complex optimisation problems. In: Intelligent production machines and systems. Elsevier Science Ltd, pp 454–459
13. Vapnik V, Izmailov R (2015) Learning using privileged information: similarity control and knowledge transfer. J Mach Learn Res 16(1):2023–2049
14. Caraveo C, Valdez F, Castillo O (2018) A new optimisation meta-heuristic algorithm based on self-defense mechanism of the plants with three reproduction operators. Soft Comput 22(15):4907–4920
15. Wang Z, Fang X, Li H, Jin H (2020) An improved partheno-genetic algorithm with reproduction mechanism for the multiple traveling salesperson problem. IEEE Access 8:102607–102615
16. Cao Y, Zhang H, Li W, Zhou M, Zhang Y, Chaovalitwongse WA (2018) Comprehensive learning particle swarm optimization algorithm with local search for multimodal functions. IEEE Trans Evol Comput 23(4):718–731
17. Sutharssan T, Stoyanov S, Bailey C, Yin C (2015) Prognostic and health management for engineering systems: a review of the data-driven approach and algorithms. J Eng 2015(7):215–222

Green Vehicle Routing Optimisation Using the Bees Algorithm

Aryan Satpathy, Millon Madhur Das, Natalia Hartono, and D. T. Pham

Abstract The green vehicle routing problem (GVRP) aims to find a set of vehicle tours that minimise the total distance travelled to service a subset of customers while incorporating stops at alternative fuel stations because of the vehicle's limited fuel capacity. This is the first study to investigate using the Bees Algorithm to find a solution to the GVRP. The aim is to apply the Bees Algorithm to the GVRP and compare the results of six optimisation algorithms. A grid search was used to find the best parameters of the algorithms. For a fair comparison, the same number function of evaluation was used as the stopping criterion for all algorithms in this study. Statistical analysis is used to evaluate the performance of the algorithms. The results demonstrate that the Bees Algorithm outperforms other algorithms and can find near-optimal solutions.

Keywords Bees algorithm · Green vehicle routing problem · Optimisation

---

A. Satpathy (✉)
Department of Electronics and Electrical Communication Engineering, Indian Institute of Technology Kharagpur, Kharagpur, West Bengal 721302, India
e-mail: satpathy.gvrp@gmail.com

M. M. Das
Department of Mechanical Engineering, Indian Institute of Technology Kharagpur, Kharagpur, West Bengal 721302, India
e-mail: millonmadhurdas@kgpian.iitkgp.ac.in

N. Hartono
Department of Engineering, University of Exeter, Exeter EX4 4QJ, UK
e-mail: nxh886@alumni.bham.ac.uk

D. T. Pham
Department of Mechanical Engineering, The University of Birmingham, Birmingham B15 2TT, UK
e-mail: d.t.pham@bham.ac.uk

D. T. Pham and N. Hartono (eds.), *Intelligent Engineering Optimisation with the Bees Algorithm*, Springer Series in Advanced Manufacturing,
https://doi.org/10.1007/978-3-031-64936-3_16

1 Introduction

To reach net zero in 2050, emissions must be reduced by 45% by 2030 [1]. The transportation sector is the largest contributor to greenhouse gas emissions (GHGs) in the United States, at 27%, and the fourth largest globally, at 14% in 2020 [2]. Similarly, in the EU, road transportation will contribute 77% of GHGs in 2020 [3]. During a pandemic, transportation decreases [2, 3]; however, this decline was temporary, and since 2021, transportation activity has been on the rise [3]. Despite the switch to cleaner energy sources, petroleum-based fuels still make up a significant portion of the world's transportation energy at 95% [2]. As a result of planned decarbonisation efforts, GHGs from road transport are predicted to decline more rapidly than those from other modes of transport [3]. Medium- and heavy-duty trucks are responsible for 26% of greenhouse gas (GHG) emissions in the US transportation sector, and switching to cleaner fuels is one way to reduce these emissions [2].

Efficient transportation routes reduce emissions, which helps reduce the environmentasl burden. Vehicle routing is a combinatorial optimisation problem that, given a set of delivery locations and vehicle capacities, tries to find the optimal route for a single vehicle or group of vehicles. The objective of the vehicle routing problem (VRP) is to determine the most efficient route from a depot to customers with each vehicle having capacity and the customer having demand [4–7]. Since 2010, attention to greener fuel sources has led many US companies to convert their truck fleets to alternative fuel vehicles (AFVs) voluntarily or to meet new government regulations to reduce GHGs [4]. AFVs consist of biodiesel, CNG, electricity, ethanol, hydrogen, LNG, propane fuel, and renewable diesel [8]. However, the limited refueling infrastructure places barriers on AFVs. Erdogan and Miller-Hook introduced the Green Vehicle Routing Problem (GVRP) in 2012 as a variant of VRP with the aim of eliminating the risk of running out of fuel while maintaining low-cost routes with a plan for refueling and incorporating stops at alternative fueling stations (AFSs) [4]. After the initial introduction of this problem, GVRP variants have evolved to include additional constraints and enhancements with the same objective of incorporating environmental considerations into their goals [9, 10].

VRP are NP-hard problems [5, 7, 9, 11], and as a special case of VRP, GVRP is also NP-hard [4, 12, 13]. Thus, metaheuristic algorithms were the preferred choice to find near optimal solutions to these problems within a strict time bound [4, 11, 12]. Methods used in previous research to solve the GVRP were exact, heuristic, metaheuristic, and hybrid methods [9, 10]. Since 2014, the popularity of the metaheuristic algorithms used to solve the GVRP has surged, and they have become the most widely used methods. Additionally, previous GVRP research indicates that metaheuristics are the preferred approach for resolving larger, more complex problems [4, 9–13]. The GVRP has been solved using metaheuristics such as the genetic algorithm (GA), ant colony algorithm (ACO), artificial bee colony (ABC), and particle swarm optimisation (PSO) [9, 10]; however, the Bees Algorithm (BA) has never been used. Numerous real-world applications and complex problems utilising BA demonstrate its reliability in producing optimal results [14]. Previous research using BA in vehicle routing problems [5–7] shows its ability to solve combinatorial problems by yielding a good solution faster. This encourages us to expand upon the work that has been done previously on the VRP using BA. The Bees Algorithm is a population-based metaheuristic that was first introduced in 2005 by Pham et al. [15]. It was inspired by foraging bees' behaviour when finding the best route between the hive and food source [15, 16]. To the best of our knowledge, this is the first study to optimise GVRP utilising BA. The aim is to investigate the BA's efficacy and enhanced BA on an environmentally conscious vehicle routing problem. This research proposes two local search operators, permuteAFS and removeAFS, for the Bees Algorithm.

The rest of the chapter is organised as follows: Sect. 2 briefly explains the relevant literature on the GVRP and the Bees Algorithm in VRP. The method and research steps are explained in Sect. 3. Section 4 presents and discusses the results. Finally, the conclusion and further research recommendations are presented in the last section.

2 Literature Review

2.1 Green Vehicle Routing Problem

The GVRP illustration is as follows: several fleets with tank capacity must determine the optimal location of AFS stops along the route from depot to customers and back to the depot [4]. This study defines and formulates the problem using the Erdogan formulation. The GVRP is a complete graph with an undirected route $G = (V, A)$. The depot v_0, customer set $C = \{v_1, v_2..., v_n\}$, AFSs set $F = \{v_{n+1}, v_{n+2}, ..., v_{n+s}\}$. The vertex set $V = \{v_0\} \cup C \cup F = \{v_0, v_1, v_2..., v_{n+s}\}$, $|V| = n + s + 1$. The edges connecting vertices of V are $A = \{(v_i, v_j): v_i, v_j \in V, i < j\}$. Each edge (v_i, v_j) has a nonnegative travel time t_{ij}, cost c_{ij} and distance d_{ij}. The following assumptions were made: a constant transport speed, unlimited number of refuelling stops, tank is filled to maximum capacity each refuelling, all refuelling stations have infinite capacity, all customers can be supplied by a fleet whose route starts and ends from depot within T_{max}, the depot can serve as a refuelling station, and the customer must be visited exactly once. The graph G is augmented $G' = (V', A')$ with a set of s' dummy vertices $\Phi = \{v_{n+s+1}, v_{n+s+2}, ..., v_{n+s+s'}\}$ to allow numerous (or zero) visits to a subset of the vertices while requiring exactly one visit to other vertices. $V' = V \cup \Phi$. For each refuelling station, $v_f \in F$ consists of the n_f dummy vertices for $f = 0, ..., n + s$. To minimise network size, the number of dummy vertices, n_f, should be as minimal as possible while still being large enough to allow for multiple visits.

The objective is to minimise the fleet's total distance travelled, presented by Eq. (1). Equations (2, 3) guarantees that each customer and AFS has exactly one successor, which may be a customer, AFS, or depot vertex. The number of arrivals equal to departures for all vertices guaranteed by Eq. (4). Equations (5, 6) ensure that no more than m vehicles leave the depot and return to depot each day. Each node's arrival time is determined by Eqs. (7)–(9) along with the requirement that each vehicle return to the depot no later than T_{max}. Fuel levels are measured by Eq. (10) based on the vertex sequence and type. For example, if vertex *i* is a customer vertex and vertex *j* is the next visited vertex ($x_{ij} = 1$), then Eq. (10) reduces the fuel level at vertex *j* based on the distance travelled and fuel consumption rate from vertex *i*. Additionally, Eqs. (7) and (10) eliminate the possibility of subtour formation without depot nodes. When the vehicle arrives at any AFS, Eq. (11) restores the vehicle's fuel tank level to *Q*. Equation (12) ensures that there will be enough fuel left over to travel back to the depot using an AFS from any customer location. Equation (13) to ensure the binary travel decision. Table 1 lists the variable descriptions.

$$OS = \min \sum_{\substack{i,j \in V' \\ i \neq j}} d_{ij} x_{ij} \tag{1}$$

Subject to

$$\sum_{\substack{j \in V' \\ j \neq i}} x_{ij} = 1, \forall i \in C \tag{2}$$

$$\sum_{\substack{j \in V' \\ j \neq i}} x_{ij} \leq 1, \forall i \in F_0 \tag{3}$$

Table 1 Variable description

Variable	Description
c_{ij}	Cost
C_0	Set of customer vertices and depot, $C_0 = \{v_0\} \cup C$
d_{ij}	Distance *i* to *j*
F_0	Set of AFS vertices and depot, $F_0 = \{v_0\} \cup F'$, where $F' = R \cup \Phi$
p_i	Service time at vertex *i* (if $i \in C$, then p_i is the service time at the customer vertex, if $i \in F$, p_i is the refuelling time at the AFS vertex
Q	Fuel tank capacity
r	Fuel consumption rate (gallons per mile)
t_{ij}	Travel time (nonnegative)
v_0	Depot
x_{ij}	Indicator taking the value 1 if a fleet travels from vertex *i* to *j* and 0 otherwise
y_j	Fuel level upon arrival to vertex *j*
τ_j	Time of arrival of a fleet at vertex *j*

$$\sum_{\substack{i \in V' \\ j \neq i}} x_{ji} - \sum_{\substack{i \in V' \\ j \neq i}} x_{ij} = 0, \forall j \in V' \quad (4)$$

$$\sum_{j \in V'\{0\}} x_{0j} \leq m \quad (5)$$

$$\sum_{j \in V'\{0\}} x_{j0} \leq m \quad (6)$$

$$\tau_j \geq \tau_i + (t_{ij} - p_j)x_{ij} - T_{max}(1 - x_{ij}), i \in V', \forall j \in V'\{0\} and i \neq j \quad (7)$$

$$0 \leq \tau_0 \leq T_{max} \quad (8)$$

$$t_{0j} \leq \tau_j \leq T_{max} - (t_{j0} + p_j), \forall j \in V' \backslash \{0\} \quad (9)$$

$$y_j \leq y_i - r.d_{ij}x_{ij} + Q(1 - x_{ij}), \forall j \in C and i \in V', i \neq j \quad (10)$$

$$y_j = Q, \forall j \in F_0 \quad (11)$$

$$y_j \geq \min\{r.d_{j0}, r.(d_{jl} + d_{l0})\}, \forall j \in C, \forall l \in F' \quad (12)$$

$$x_{i,j} \in \{0, 1\}, \forall i, j \quad (13)$$

2.2 Bees Algorithm in VRP

Since it was first introduced in 2005 by Pham et al. [15] and subsequently published in 2006 [16], the Bees Algorithm has been widely known for its ability to solve complex problems in a short amount of time and give better results compared to other algorithms [7, 14].

The Bees Algorithm was developed to solve continuous domains and was first used to solve combinatorial domains in 2007 for job scheduling problems [17].

Previous research using the Bees Algorithm (BA) [5, 6] and two-parameter Combinatorial Bees Algorithm (BA_{C2}) [7] to solve VRP shows that the BA and its variant are able to give optimum results faster. BA_{C2} is the parameter reduction version of BA, and it statistically yields the same performance [7]. In addition, because of the reduced number of parameters that were used, BA_{C2} has a simpler structure.

3 Methodology

The research steps for this study are depicted in Fig. 1 as five steps. The initial phase involved determining the optimal settings for each algorithm. There were six algorithms used in this step: the Basic Bees Algorithm (BBA), the 2-parameter BBA (BBA2), the Genetic Algorithm (GA), the Standard Bees Algorithm (SBA), the 2-parameter SBA (SBA2), and the hybrid SBA with Simulated Annealing (SBA_SA). The second step was deciding on the stopping criteria. The experiments were then carried out on a small dataset [18] in the third step. The fourth step involved conducting statistical tests and analysing the experimental results to determine the best performing algorithms. The last steps use the best-performing algorithm on a larger dataset using the "goeke 2018 G-VRP" dataset [19].

3.1 Parameter Tuning

As the initial phase in this study, the algorithm's parameters were determined using a grid search. Grid search is an optimisation method that seeks to determine the best hyperparameter values [20, 21]. Exhaustive research is conducted on the exact parameter values of an optimiser. Since the search space is combinatorial, gradient-based approaches are inapplicable. Grid search is a widely used method for tuning machine learning model hyperparameters [20, 21]. The grid search consisted of the course and fine search. In the course search, thirty runs were used for each combination of a possible parameter. The best course search results are used in the fine search, which has a smaller step size and uses 100 runs for each possible parameter combination. Grid search allows three to four parameters that can be applied to fine-tune the algorithm. Having more parameters either exponentially increases the run time or degrades the solution quality by compromising the step size.

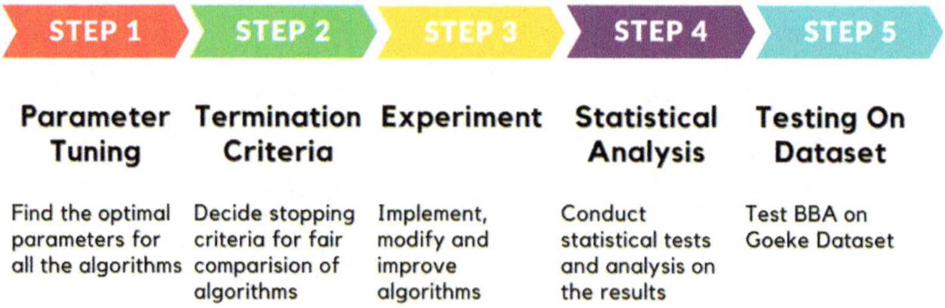

Fig. 1 Research roadmap

3.2 Stopping Criteria Decision

Comparing metaheuristics (especially population-based ones) is significantly more challenging than comparing other algorithms. Using the number of iterations as a stopping criterion yields inaccurate comparisons across population-based algorithms, as the outcome is directly reliant on the number of populations analysed per iteration. To compare methods, the number of function evaluations (NFE) is employed as the stopping criterion in this work. It enables a fair comparison and parameters that are independent and directly proportional to the algorithm's complexity.

3.3 The Bees Algorithm Variants in This Research

In the development of the basic Bees Algorithm (BBA), two procedures of neighbourhood shrinking and site abandonment were introduced by [22] and known as the standard Bees Algorithm (SBA) [14]. The pseudocode for the BBA and SBA is presented in Figs. 2 and 3. The two-parameter Bees Algorithm was proposed by [7], and this work adapted the BA variant as BBA2 and SBA2. The hybrid Bees Algorithm proposed in this research hybridises the Bees Algorithm with simulated annealing (SA). Recent publications in VRP that apply SA demonstrate that hybridisation with SA improves the algorithm's optimisation capability [23–25]. The concept of hybrid algorithms aims to achieve a balance between exploration and exploitation of the search space, thereby enhancing the quality of the solutions discovered. The simulated annealing (SA) used to refine the solutions found by the foraging bees and the pseudocode of this proposed hybrid algorithm are depicted in Fig. 4.

The local search operators used in the combinatorial basic Bees Algorithm were swap, insert and reserve [5, 26, 27]. The concept of swap is straightforward; two nodes were selected and then swapped. Insertion involves choosing two nodes, deleting the element at the first node, and inserting it at the second node. The reverse operator works by selecting two nodes and reversing their order. As depicted in Figs. 5, 6 and 7, this enables the discovery of a better solution in the vicinity.

In GVRP, the presence of optional AFS nodes makes the problem slightly more complex and necessitates the use of a distinct set of local search operators. As previously mentioned in the introduction, in this research, two additional local searches were proposed. The PermuteAFS and RemoveAFS developed for this GVRP enable better usage of AFS nodes in the solution. The AFS is permuted by PermuteAFS in an arbitrary order. As its name suggests, RemoveAFS removes an AFS at random. Figures 8 and 9 provide an illustration of how the two local searches work.

```
Algorithm 1: Basic Bees Algorithm
Input: n: Number of scout bees, m: Number of best sites, e: Number of
elite sites, nsp: Foraging bees for best sites, nep: Foraging bees for elite
sites, nfe_limit: Number function of evaluation
function BBA(n, m, e, nsp, nep, size_max, nfe_limit)
population ← GlobalSearch(n)
while (nfe < nfe_limit)
    population ← sorted(population)[ : n]
    for i = 1, …, e do
        bee ← population[i]
        population[i] ← argmin(LocalSearch(bee, nep, size_max))
    for i = e + 1, …, m do
        bee ← population[i]
        population[i] ← argmin(LocalSearch(bee, nsp, size_max))
    population ← population + GlobalSearch(n - m)
```

Fig. 2 Basic Bees Algorithm

3.4 *Statistical Analysis Decision*

A statistical test was conducted to find the best performing algorithm among the five Bees Algorithms and Genetic Algorithm (GA) using IBM SPSS version 27. The flow chart of the decisions on the statistical tests is shown in Fig. 10. After the statistical descriptive of the data, the assumption check lists are conducted to determine if the test will be a parametric or nonparametric test. To use a parametric test, the data need to be more than 30, the distribution is normal, and homogenous data. If one of the assumptions was violated, the nonparametric test was conducted. A post hoc test was conducted to find the best performing algorithm.

3.5 *Testing the Best Algorithm on a Larger Dataset*

As stated in the beginning of this section, the "Goeke 2018 G-VRP" dataset [19] was used to evaluate the best-performing algorithm as the final step of this study.

4 Results and Discussion

The first step was to identify the best parameters using the grid search for six algorithms. The GA parameters followed [18], and the same grid search procedure was used to determine the best parameter settings. The grid search results are displayed in Table 2.

Algorithm 2: Standard Bees Algorithm

Input: *n*: Number of scout bees, *m*: Number of best sites, *e*: Number of elite sites, *nsp*: Foraging bees for best sites, *nep*: Foraging bees for elite sites, *α*: Shrink factor, *stlim*: Number of iterations before site abandonment, *nfe_limit*: Number function of evaluation

```
function SBA(n, m, e, nsp, nep, α, stlim, size_max, nfe_limit)
population ← GlobalSearch(n)
while (nfe < nfe_limit)
    population ← sorted(population)[ : n]
    for i = 1, …, e do
        bee ← population[i]
        if stlim = 0 then
            delete population[i]
            continue
        end if
        best ← argmin(LocalSearch(bee, nep, size))
        if best < bee then
            bee ← best
        Else
            stlim ← stlim - 1
            size ← size * α
        end if
    for i = e + 1, …, m do
        bee ← population[i]
        if stlim = 0 then
            delete population[i]
            continue
        end if
        best ← argmin(LocalSearch(bee, nsp, size))
        if best < bee then
            bee ← best
        Else
            stlim ← stlim - 1
            size ← size * α
        end if
    population ← population + GlobalSearch(n - m)
```

Fig. 3 Standard Bees Algorithm

Algorithm 3: Hybrid Bees Algorithm

Input: *n*: Number of scout bees, *m*: Number of best sites, *e*: Number of elite sites, *nsp*: Foraging bees for best sites, *nep*: Foraging bees for elite sites, *α*: Shrink factor, *stlim*: Number of iterations before site abandonment, *nfe_limit*: Number of function evaluation

```
function SBA_SA(n, m, e, nsp, nep, α, stlim, size_max, T0, nfe_limit)
    population ← GlobalSearch(n)
    while (nfe < nfe_limit)
        population ← sorted(population)[ : n]
        for i = 1, …, e do
            bee ← population[i]
            if stlim = 0 then
                delete population[i]
                continue
            end if
            best ← argmin(LocalSearch(bee, nep, size))
            p ← random(0, 1)
            if exp((bee - best)/T) < p then
                bee ← best
            Else
                stlim ← stlim – 1
                size ← size * α
                T ← T
            end if
        for i = e + 1, …, m do
            bee ← population[i]
            if stlim = 0 then
                delete population[i]
                continue
            end if
            best ← argmin(LocalSearch(bee, nsp, size))
            p ← random(0, 1)
            if exp((bee - best)/T) < p then
                bee ← best
            Else
                stlim ← stlim – 1
                size ← size * α
            end if
        population ← population + GlobalSearch(n - m)
```

Fig. 4 Proposed Hybrid Bees Algorithm (SBA_SA)

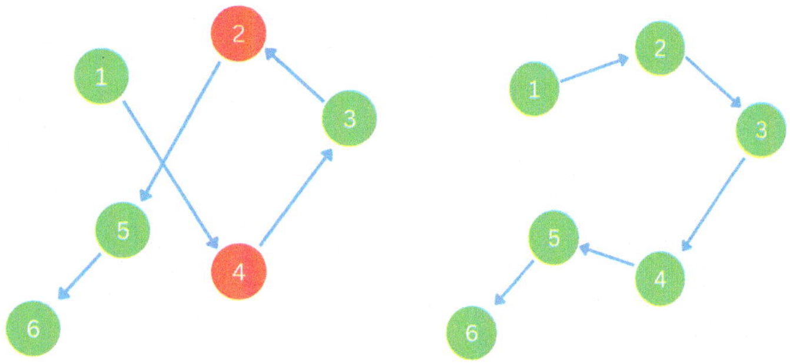

Fig. 5 Swap operator: a suboptimal solution gives a better result when nodes 2 and 4 are swapped

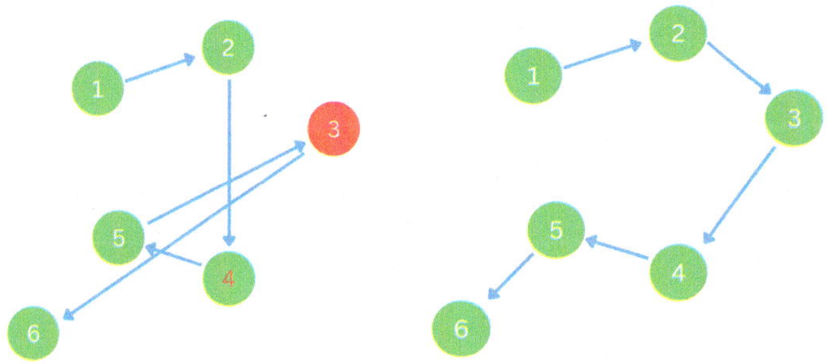

Fig. 6 Insert operator: a suboptimal solution gives a better result when node 3 is inserted at node 4's position

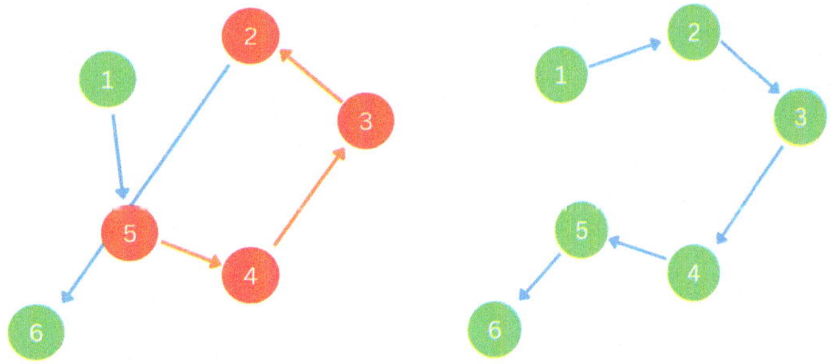

Fig. 7 Reverse operator: a suboptimal solution gives a better result when sequence nodes 5, 4, 3, and 2 are reversed

The termination criteria in the second steps decided to use the NFE for a fair comparison, as mentioned in Sect. 3. Figure 11 depicts how the BBA and convergence graph were analysed to determine the NFE. The graph shows that the method converges at approximately 40,000 NFE. Therefore, it is decided to select 60,000

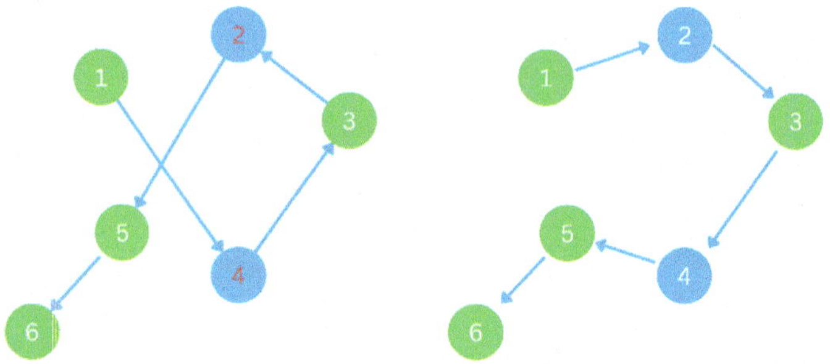

Fig. 8 PermuteAFS operator: changing the order of AFS 2 and 4 results in a smaller path

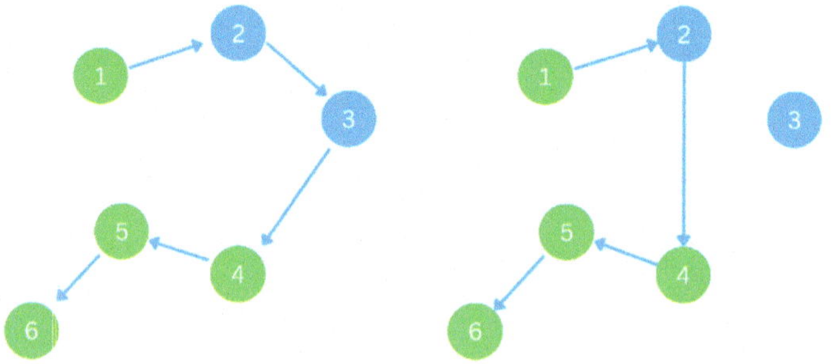

Fig. 9 RemoveAFS: unnecessary AFS can be removed to obtain a shorter path

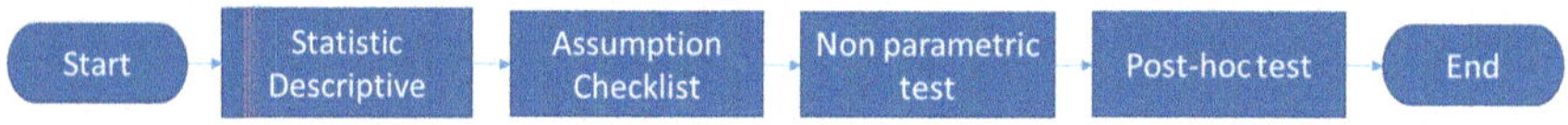

Fig. 10 Statistical decision in this research

Table 2 Parameter settings for the six algorithms in this study

Parameters	BBA	BBA-2**	GA	SBA	SBA_SA	SBA-2**
n	10	10		10	10	10
m	9			9	9	
e	5			5	5	
nsp	10			10	10	
nep	15	15		15	15	15
alpha				0.9025	0.925	0.9025
Stlim				27	27	27
Initial temperature					*	
Mutation rate			0.2			
Tournament size			3			
Elitism			True			

*The initial temperature is taken as any initial guess at the total cost/path length
**2 Parameter versions of BBA and SBA have two parameters: *n* and *nep*

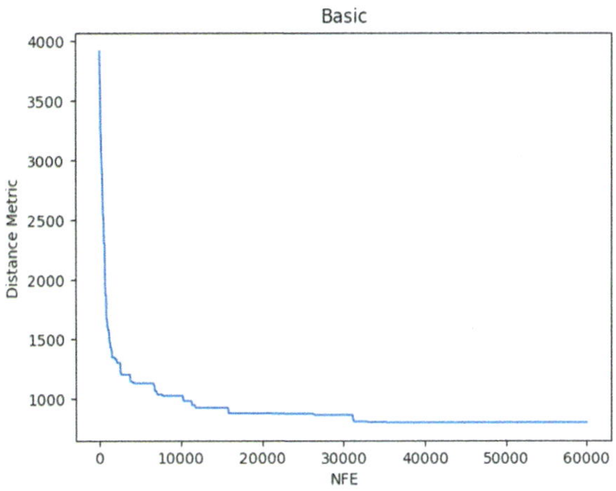

Fig. 11 BBA convergence graph to determine the NFE value for the stopping criteria

NFE as the stopping criterion for all algorithms by increasing the amount of NFE found.

Sample1, sample2, and sample3 from [18] were used to find the best parameters for each algorithm as the third step. For each algorithm, the experiment runs independently 100 times.

Figure 12 displays the outcomes of the fourth step. The statistical description is shown as a boxplot for the three samples calculated with the six optimisation algorithms.

The assumption check lists show that the data exceeded 30 and satisfied the first assumption. As shown in Figs. 13 and 14, the data are not normally distributed or homogeneous. Therefore, the nonparametric Kruskal–Wallis test was selected for the next statistical steps.

The hypothesis for the statistical test is as follows:

$\mathbf{H_0}$: The distribution of the sample is the same across BBA, BBA-2, GA, SBA, SBA-SA, SBA-2

$\mathbf{H_1}$: The distribution of the sample is not the same across BBA, BBA-2, GA, SBA, SBA-SA, SBA-2

As shown in Fig. 15, the results for all three samples rejected the null hypothesis ($\mathbf{H_0}$), which means that the alternative hypothesis ($\mathbf{H_1}$) was accepted. Based on the statistical test, there is a difference between the six algorithms.

However, this test only indicates that there is a difference and is not able to pinpoint which one is different (or the best). Therefore, the post hoc test needs to be conducted, and the results are shown in Figs. 16, 17 and 18. The result shows that the BBA and BBA-2 statistically yielded the best results, so the two algorithms were the best performing algorithms for this problem.

Algorithms that performed well at the preliminary stage were tested on a larger dataset [19] in this last research step. One hundred runs for each algorithm on each sample of the dataset are collected. BBA and BBA-2 statistically produce the best

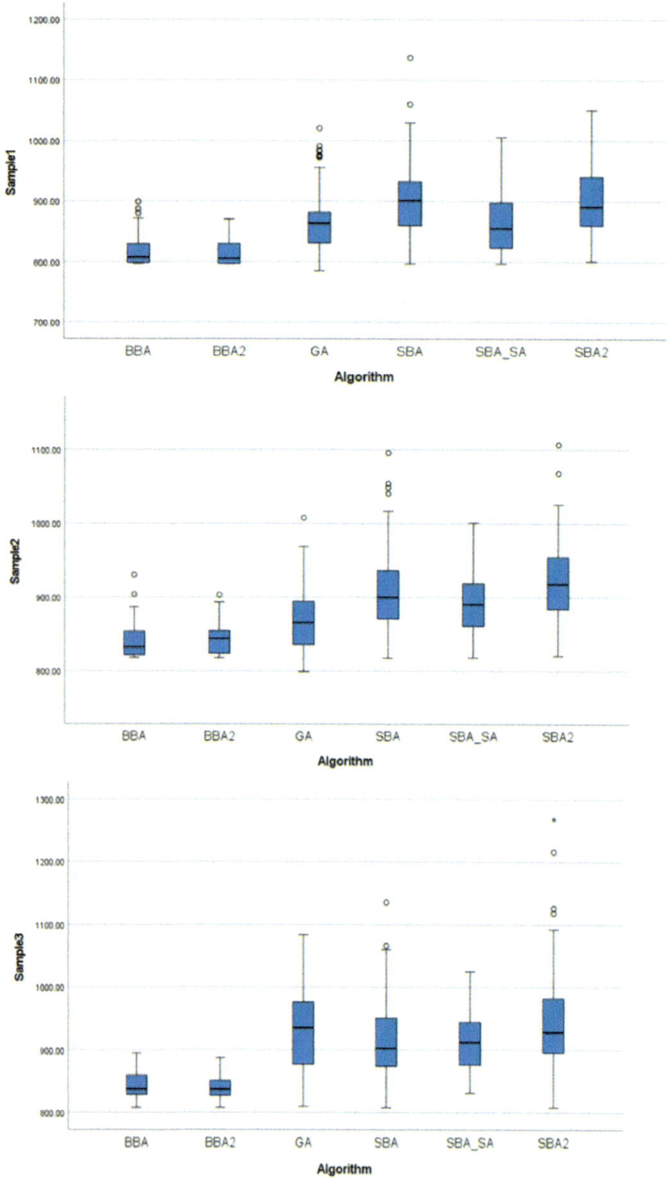

Fig. 12 Descriptive statistics

Tests of Normality

		Kolmogorov-Smirnov[a]			Shapiro-Wilk		
	Algorithm	Statistic	df	Sig.	Statistic	df	Sig.
Sample1	BBA	.205	100	<.001	.825	100	<.001
	BBA2	.233	100	<.001	.811	100	<.001
	GA	.152	100	<.001	.927	100	<.001
	SBA	.075	100	.178	.967	100	.013
	SBA_SA	.107	100	.006	.943	100	<.001
	SBA2	.081	100	.102	.973	100	.040
Sample2	BBA	.153	100	<.001	.863	100	<.001
	BBA2	.135	100	<.001	.923	100	<.001
	GA	.109	100	.005	.958	100	.003
	SBA	.079	100	.126	.951	100	<.001
	SBA_SA	.060	100	.200*	.979	100	.108
	SBA2	.072	100	.200*	.975	100	.053
Sample3	BBA	.111	100	.004	.958	100	.003
	BBA2	.116	100	.002	.953	100	.001
	GA	.072	100	.200*	.981	100	.164
	SBA	.092	100	.037	.945	100	<.001
	SBA_SA	.062	100	.200*	.977	100	.078
	SBA2	.103	100	.011	.928	100	<.001

*. This is a lower bound of the true significance.

a. Lilliefors Significance Correction

Fig. 13 Tests of normality result

Test of Homogeneity of Variance

		Levene Statistic	df1	df2	Sig.
Sample1	Based on Mean	21.106	5	594	<.001
	Based on Median	20.982	5	594	<.001
	Based on Median and with adjusted df	20.982	5	466.644	<.001
	Based on trimmed mean	21.203	5	594	<.001
Sample2	Based on Mean	21.158	5	594	<.001
	Based on Median	19.828	5	594	<.001
	Based on Median and with adjusted df	19.828	5	419.604	<.001
	Based on trimmed mean	20.721	5	594	<.001
Sample3	Based on Mean	31.444	5	594	<.001
	Based on Median	27.057	5	594	<.001
	Based on Median and with adjusted df	27.057	5	320.193	<.001
	Based on trimmed mean	29.864	5	594	<.001

Fig. 14 Test of homogeneity of variance results

Hypothesis Test Summary

	Null Hypothesis	Test	Sig.[a,b]	Decision
1	The distribution of Sample1 is the same across categories of Algorithm.	Independent-Samples Kruskal-Wallis Test	.000	Reject the null hypothesis.
2	The distribution of Sample2 is the same across categories of Algorithm.	Independent-Samples Kruskal-Wallis Test	.000	Reject the null hypothesis.
3	The distribution of Sample3 is the same across categories of Algorithm.	Independent-Samples Kruskal-Wallis Test	.000	Reject the null hypothesis.

a. The significance level is .050.

b. Asymptotic significance is displayed.

Fig. 15 Nonparametric test—Kruskal–Wallis

Pairwise Comparisons of Algorithm

Sample 1-Sample 2	Test Statistic	Std. Error	Std. Test Statistic	Sig.	Adj. Sig.[a]
BBA2-BBA	10.740	24.503	.438	.661	1.000
BBA2-SBA_SA	-177.750	24.503	-7.254	<.001	.000
BBA2-GA	-197.030	24.503	-8.041	<.001	.000
BBA2-SBA	-282.265	24.503	-11.520	.000	.000
BBA2-SBA2	-282.315	24.503	-11.522	.000	.000
BBA-SBA_SA	-167.010	24.503	-6.816	<.001	.000
BBA-GA	-186.290	24.503	-7.603	<.001	.000
BBA-SBA	-271.525	24.503	-11.081	.000	.000
BBA-SBA2	-271.575	24.503	-11.083	.000	.000
SBA_SA-GA	19.280	24.503	.787	.431	1.000
SBA_SA-SBA	104.515	24.503	4.265	<.001	.000
SBA_SA-SBA2	104.565	24.503	4.267	<.001	.000
GA-SBA	-85.235	24.503	-3.479	<.001	.008
GA-SBA2	-85.285	24.503	-3.481	<.001	.008
SBA-SBA2	-.050	24.503	-.002	.998	1.000

Each row tests the null hypothesis that the Sample 1 and Sample 2 distributions are the same.

Asymptotic significances (2-sided tests) are displayed. The significance level is .050.

a. Significance values have been adjusted by the Bonferroni correction for multiple tests.

Fig. 16 Post hoc test for sample 1

results on small datasets. To show the performance of the algorithm on the larger datasets, we tested the four Bees Algorithm variants used in this research. The results in Table 3 show the mean and the best value found (BVF).

Figure 19 displays the average outcomes of the four algorithms applied to each dataset. The results indicate that BBA and SBA yield comparable outcomes. The SBA was an improvement of the BBA in the continuous domain. The distinction between the BBA and the SBA is that the SBA utilises neighbourhood shrinking and site abandonment. It assumes objective functional continuity across the domain. However, in

Pairwise Comparisons of Algorithm

Sample 1-Sample 2	Test Statistic	Std. Error	Std. Test Statistic	Sig.	Adj. Sig.[a]
BBA-BBA2	-13.315	24.513	-.543	.587	1.000
BBA-GA	-103.210	24.513	-4.210	<.001	.000
BBA-SBA_SA	-199.720	24.513	-8.148	<.001	.000
BBA-SBA	-248.120	24.513	-10.122	.000	.000
BBA-SBA2	-284.395	24.513	-11.602	.000	.000
BBA2-GA	-89.895	24.513	-3.667	<.001	.004
BBA2-SBA_SA	-186.405	24.513	-7.604	<.001	.000
BBA2-SBA	-234.805	24.513	-9.579	.000	.000
BBA2-SBA2	-271.080	24.513	-11.059	.000	.000
GA-SBA_SA	-96.510	24.513	-3.937	<.001	.001
GA-SBA	-144.910	24.513	-5.912	<.001	.000
GA-SBA2	-181.185	24.513	-7.391	<.001	.000
SBA_SA-SBA	48.400	24.513	1.974	.048	.725
SBA_SA-SBA2	84.675	24.513	3.454	<.001	.008
SBA-SBA2	-36.275	24.513	-1.480	.139	1.000

Each row tests the null hypothesis that the Sample 1 and Sample 2 distributions are the same.
Asymptotic significances (2-sided tests) are displayed. The significance level is .050.

a. Significance values have been adjusted by the Bonferroni correction for multiple tests.

Fig. 17 Post hoc test for sample 2

Pairwise Comparisons of Algorithm

Sample 1-Sample 2	Test Statistic	Std. Error	Std. Test Statistic	Sig.	Adj. Sig.[a]
BBA2-BBA	13.870	24.514	.566	.572	1.000
BBA2-SBA	-236.580	24.514	-9.651	.000	.000
BBA2-SBA_SA	-244.680	24.514	-9.981	.000	.000
BBA2-GA	-273.320	24.514	-11.150	.000	.000
BBA2-SBA2	-287.910	24.514	-11.745	.000	.000
BBA-SBA	-222.710	24.514	-9.085	.000	.000
BBA-SBA_SA	-230.810	24.514	-9.415	.000	.000
BBA-GA	-259.450	24.514	-10.584	.000	.000
BBA-SBA2	-274.040	24.514	-11.179	.000	.000
SBA-SBA_SA	-8.100	24.514	-.330	.741	1.000
SBA-GA	36.740	24.514	1.499	.134	1.000
SBA-SBA2	-51.330	24.514	-2.094	.036	.544
SBA_SA-GA	28.640	24.514	1.168	.243	1.000
SBA_SA-SBA2	43.230	24.514	1.763	.078	1.000
GA-SBA2	-14.590	24.514	-.595	.552	1.000

Each row tests the null hypothesis that the Sample 1 and Sample 2 distributions are the same.
Asymptotic significances (2-sided tests) are displayed. The significance level is .050.

a. Significance values have been adjusted by the Bonferroni correction for multiple tests.

Fig. 18 Post hoc test for sample 3

Table 3 Results

Name	BBA		BBA-2		SBA–SA		SBA	
	Mean	BVF	Mean	BVF	Mean	BVF	Mean	BVF
lc101	1072.4	934.09	1001.71	838.04	1238.01	1098.83	1085.95	913.61
lc102	1062.76	908.28	992.83	891.32	1248.75	1113.17	1089.35	919.04
lc103	1060.82	911.76	992.24	847.94	1228.58	1089.00	1089.58	912.27
lc104	1071.83	908.32	994.6	865.43	1238.62	1067.54	1081.87	932.91
lc105	1077.93	921.14	996.18	888.23	1246.26	1080.78	1095.94	958.81
lc106	1062.89	931.46	993.78	857.02	1238.37	1074.31	1088.58	941.02
lc107	1073.79	942.63	994.14	856.39	1249.16	1104.20	1088.43	954.74
lc108	1077.42	900.24	998.45	855.99	1236.85	1051.54	1087.84	994.64
lc109	1075.65	921.96	996.35	865.55	1240.14	1103.23	1089.24	922.32
lc201	1124.1	1007.27	1005.31	857.96	1303.04	1159.31	1147.84	982.85
lc202	1121.61	1012.44	1005.1	855.80	1316.05	1114.95	1146.55	979.21
lc203	1124.8	968.51	1006.32	852.75	1306.21	1163.98	1135.04	974.27
lc204	1125.91	968.53	1018.96	844.09	1314.03	1125.46	1143.11	996.91
lc205	1130.2	1003.14	1013.49	906.65	1312.83	1155.62	1143.48	993.07
lc206	1122.01	1008.06	1008.69	878.81	1309.79	1101.07	1133.51	1025.88
lc207	1135.49	1035.67	1009.64	877.25	1297.23	1149.28	1141.32	1005.99
lc208	1120.65	985.21	1011.15	849.27	1318.03	1141.84	1134.6	995.95
lr101	1127.97	1026.25	1013.59	921.18	1293.77	1163.80	1130.52	1003.29
lr102	1129.07	1052.10	1016.5	933.48	1300.1	1180.50	1134.74	1036.80
lr103	1137.07	1059.20	1027.21	929.88	1301.05	1184.42	1140.55	1027.41
lr104	1156.34	1053.50	1033.95	901.32	1326.84	1214.58	1151.55	1079.89
lr105	1162.39	1060.81	1037.2	948.09	1333.96	1223.40	1158.05	1023.73
lr106	1163.98	1055.91	1043.36	951.96	1350.65	1221.92	1155.28	1057.84
lr107	1147.24	1035.41	1031.72	941.12	1325.48	1217.66	1152.91	1058.40
lr108	1156.91	1062.56	1036.49	961.09	1335.63	1185.78	1156.5	1035.45
lr109	1157.35	1070.19	1034.82	947.57	1333.2	1202.56	1149.01	1063.01
lr110	1156.48	1048.55	1034.34	951.15	1327.17	1212.77	1147.96	1012.70
lr111	1157.31	1072.28	1039.22	961.80	1334.5	1223.09	1154.1	1069.60
lr112	1155.75	1038.16	1038.5	950.97	1323.89	1202.75	1146.45	1001.22
lr201	981.41	917.33	879.47	806.51	1142.41	1041.77	937.43	846.85
lr202	965.61	915.12	864.69	798.75	1143.09	1042.82	923.97	819.00
lr203	978.92	890.72	884.56	819.74	1148.73	1069.16	940	840.29
lr204	961.27	889.60	862.43	781.52	1146.97	1048.81	922.4	828.86
lr205	973.4	911.91	874.84	809.63	1145.49	1063.81	929.57	845.22
lr206	980.23	913.36	878.51	801.11	1148.78	1046.37	947.64	833.74
lr207	954.75	882.27	863.33	764.73	1135.4	1049.50	909.39	797.64

(continued)

Table 3 (continued)

Name	BBA		BBA-2		SBA–SA		SBA	
	Mean	BVF	Mean	BVF	Mean	BVF	Mean	BVF
lr208	966.17	902.38	868.58	800.72	1144.88	1032.60	919.67	841.43
lr209	975.15	887.13	880.16	800.63	1151.97	1080.28	943.36	849.71
lr210	976.11	918.32	878.15	790.47	1151.09	1048.29	933.17	855.72
lr211	953.87	833.52	864.77	780.74	1135.64	1007.99	918.27	833.74
lrc101	1362.21	1210.77	1213.83	1042.20	1576.32	1416.86	1362.73	1235.35
lrc102	1370.44	1204.95	1213.06	1030.41	1587.66	1405.40	1364.49	1232.27
lrc103	1371.32	1208.76	1211.89	930.93	1570.71	1370.27	1349.23	1152.42
lrc104	1357.46	1216.59	1215.97	1063.56	1566.02	1416.52	1368.74	1227.18
lrc105	1357.07	1174.82	1197.96	1085.66	1556.51	1406.68	1352.81	1218.95
lrc106	1373.33	1188.17	1218.15	1069.40	1577.97	1436.24	1357.04	1222.17
lrc107	1357.51	1156.80	1209.67	1062.10	1570.55	1408.62	1357.53	1197.41
lrc108	1363.34	1210.16	1211.31	1052.82	1573.28	1416.99	1356.08	1185.03
lrc201	1124.82	1022.80	1001.19	864.31	1315.72	1132.96	1070.06	971.46
lrc202	1089.19	995.79	972.68	847.47	1284.66	1148.98	1027.69	902.75
lrc203	1129.15	1016.97	1002.43	874.18	1316.01	1163.26	1078.66	952.14
lrc204	1228.72	1067.99	1069.64	925.33	1421.46	1206.79	1207.51	1082.63
lrc205	1150.44	1031.28	1021.87	852.13	1335.97	1202.91	1098.18	946.60
lrc206	1112.44	1011.07	989.41	879.37	1305.53	1206.46	1055.49	979.28
lrc207	1132.5	1029.94	987.97	873.31	1308.13	1187.76	1074.47	925.71
lrc208	1205.85	1098.39	1054.26	882.68	1405.62	1276.14	1169.18	1029.75

regard to problems in the combinatorial domain, the idea of a neighbourhood becomes less relevant, as does the continuity of the objective function.

The results of lc101, lr201, and lrc208 are depicted in Figs. 20, 21, and 22, respectively, to illustrate the comparison in detail. The hybrid bees algorithm (SBA_SA) performs worst among BA variants. BBA2 yields the best results because it provides the best value and means. This finding indicates that BBA2 outperforms other BA variants.

5 Conclusion

Overall, this research strengthens the finding of previous research that the Bees Algorithm is the best performing algorithm for VRP and its variants. This work presents several variants of the Bees Algorithm as a solution to the Green Vehicle Routing Problem. Two local search parameters, PermuteAFS and RemoveAFS, were proposed in the Bees Algorithm. The Basic Bees Algorithm (BBA), Standard Bees

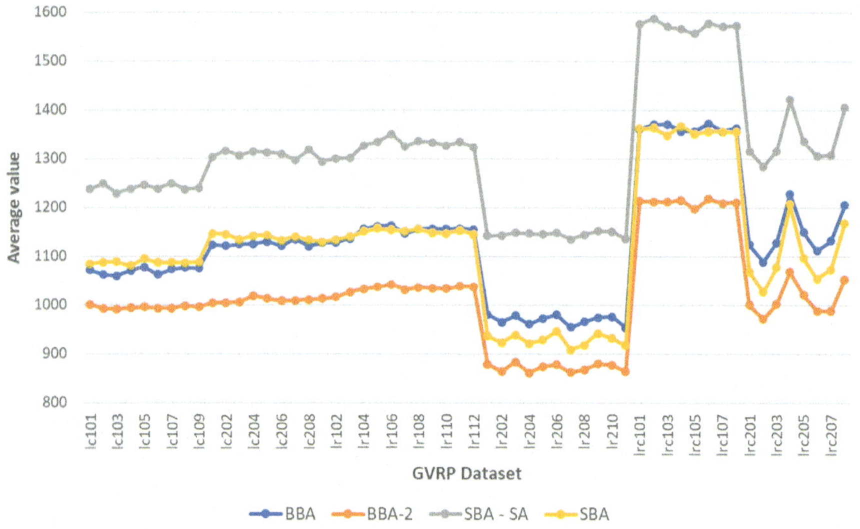

Fig. 19 Average results

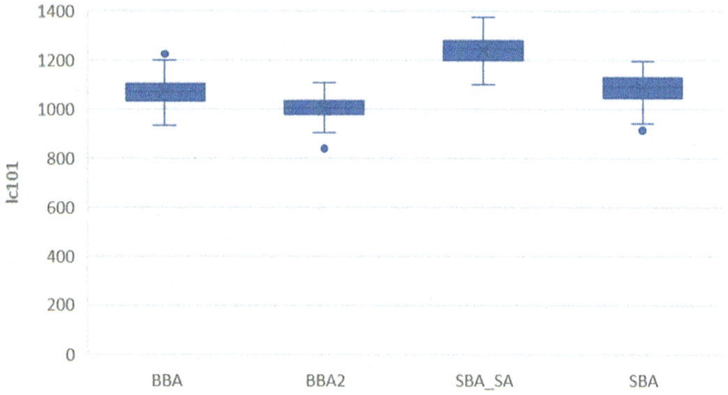

Fig. 20 lc101

Algorithm (SBA), Hybrid Standard Bees Algorithm with Simulated Annealing Algorithm (SBA_SA), 2-parameter version of Basic Bees Algorithm (BBA2) and a 2-parameter version of Standard Bees Algorithm (SBA2) were analysed, and it was found that BBA and BBA2 perform as the best performing algorithms on small datasets. Statistical tests are performed to confirm the significance of these findings. The findings show that the basic Bees Algorithm is the best choice for combinatorial optimisation problems. The four Bees Algorithms are tested on a larger dataset to confirm their performance. The results show that BBA2 was the best variant. In addition to producing the best results, the fact that BBA2 only requires two parameters as opposed to several parameters in other variants makes it more attractive for use

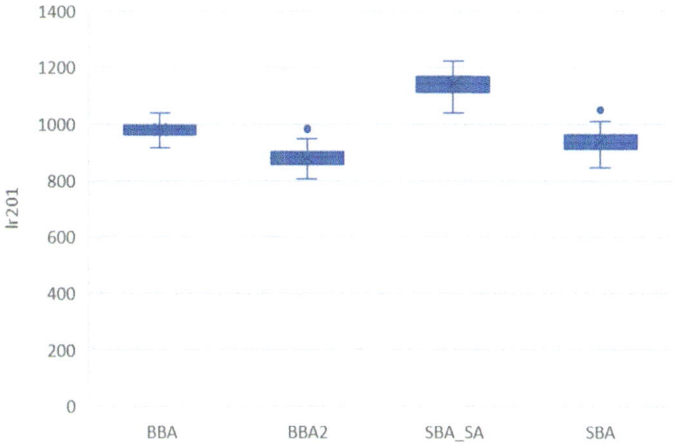

Fig. 21 lr201

Fig. 22 lrc208

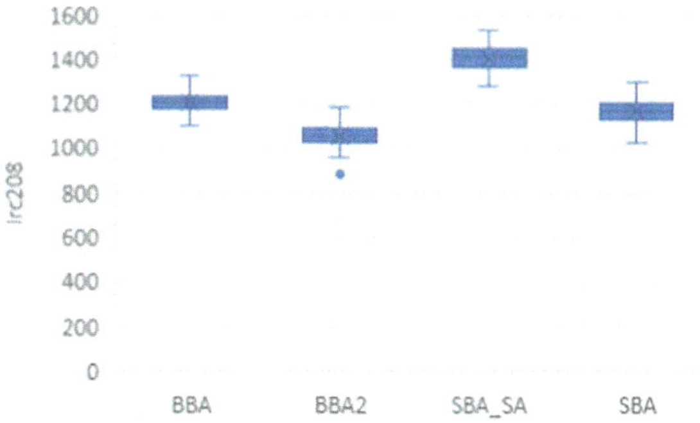

in other applications. The Bees Algorithm is flexible and can be applied to other combinatorial optimisation problems that can be framed in a similar way.

Acknowledgements The Indonesian Endowment Fund for Education (LPDP) supports Natalia's doctoral study under grant no. 201908222215155.

References

1. The United Nations Net Zero Coalition (2022) https://www.un.org/en/climatechange/net-zero-coalition. Last accessed 5 February 2023
2. US Environmental Protection Agency Global Greenhouse Gas Emissions Data (2022) https://www.epa.gov/ghgemissions/global-greenhouse-gas-emissions-data. Last accessed 5 February 2023
3. European Environment Agency (2022) https://www.eea.europa.eu/ims/greenhouse-gas-emissions-from-transport. Last accessed 5 February 2023

4. Erdoğan S, Miller-Hooks E (2012) A green vehicle routing problem. Transportation research part E: logistics and transportation review 48(1):100–114
5. Ismail AH, Hartono N, Zeybek S, Caterino M, Jiang K (2021) Combinatorial bees algorithm for vehicle routing problem. Macromol Symp 396(1):2000284
6. Natalia C, Triyanti V, Setiawan G, Haryanto M (2021) Completion of capacitated vehicle routing problem (CVRP) and capacitated vehicle routing problem with time windows (CVRPTW) using bee algorithm approach to optimise waste picking transportation problem. J Modern Manuf Syst Technol 5(2):69–77
7. Ismail AH, Pham DT (2023) Bees traplining metaphors for the vehicle routing problem using a decomposition approach. In: Pham DT, Hartono N (eds) Intelligent Production and Manufacturing Optimisation-The Bees Algorithm Approach. Springer Series in Advanced Manufacturing
8. US Department of Energy Alternative Fuel Data Center (n.d.). https://afdc.energy.gov/fuels/. Last accessed 5 February 2023
9. Sabet S, Farooq B (2022) Green vehicle routing problem: state of the art and future directions. IEEE Access
10. Moghdani R, Salimifard K, Demir E, Benyettou A (2021) The green vehicle routing problem: a systematic literature review. J Clean Prod 279:123691
11. Prakash R, Pushkar S (2022) Green vehicle routing problem: metaheuristic solution with time window. Expert Syst e13007.
12. Wang H, Li M, Wang Z, Li W, Hou T, Yang X, Zhao Z, Wang Z, Sun T (2022) Heterogeneous fleets for green vehicle routing problem with traffic restriction. IEEE Trans Intell Trans Syst
13. Erdoğdu K, Karabulut K (2022) Bi-objective green vehicle routing problem. Int Trans Oper Res 29(3):1602–1626
14. Castellani M, Pham DT (2023) The Bees Algorithm – A Gentle Introduction In: Pham DT, Hartono N (eds) Intelligent Production and Manufacturing Optimisation-The Bees Algorithm Approach. Springer Series in Advanced Manufacturing
15. Pham DT, Ghanbarzadeh A, Koc E, Otri S, Rahim S, Zaidi M (2005) The Bees Algorithm. Manufacturing Engineering Centre, Cardiff University, UK, Technical Note
16. Pham DT, Ghanbarzadeh A, Koç E, Otri S, Rahim S, Zaidi M (2006) The bees algorithm—a novel tool for complex optimisation problems. In: Intelligent production machines and systems, pp 454–459
17. Pham DT, Koc E, Lee J, Phrueksanant J (2007) Using the bees algorithm to schedule jobs for a machine. Proceedings Eighth International Conference on Laser Metrology, CMM and Machine Tool Performance, LAMDAMAP, Euspen, UK, pp 430–439
18. Anon. (2020). https://github.com/AdamMCG97/GVRP-GA/. Last accessed 5 February 2023
19. Goeke (2017) goeke 2018 G-VRP. http://www.vrp-rep.org/variants/item/g-vrp.html Last accessed 5 February 2023
20. Wilbur M, Kadir SU, Kim Y, Pettet G, Mukhopadhyay A, Pugliese P, Samaranayake S, Laszka A, Dubey A (2022) An online approach to solve the dynamic vehicle routing problem with stochastic trip requests for paratransit services. In: 2022 ACM/IEEE 13th International Conference on Cyber-Physical Systems (ICCPS) IEEE. pp 147–158
21. Santos AMP, Soares CG (2022) A hub-location periodic vehicle routing problem in offshore oil and gas logistics. Marit Econ Logistics pp 1–25
22. Pham DT, Castellani M (2015) A comparative study of the Bees Algorithm as a tool for function optimization, Cogent Engineering, Volume 2 (online). DOI. https://doi.org/10.1080/23311916.2015.1091540
23. Yu VF, Susanto H, Jodiawan P, Ho TW, Lin SW, Huang YT (2022) A Simulated Annealing Algorithm for the Vehicle Routing Problem with Parcel Lockers. IEEE Access 10:20764–20782
24. Yang T, Wang W, Wu Q (2022) Fuzzy Demand Vehicle Routing Problem with Soft Time Windows. Sustainability 14:5658
25. Ren T, Luo T, Jia B, Yang B, Wang L, Xing L (2023) Improved Ant Colony Optimization for the Vehicle Routing Problem with Split Pickup and Split Delivery. Swarm Evol Comput 77:101228

26. Ismail AH, Hartono N, Zeybek S, Pham DT (2020) Using the bees algorithm to solve combinatorial optimisation problems for TSPLIB. In: IOP conference series: materials science and engineering (Vol 847, No 1, p 012027). IOP Publishing
27. Zeybek S, Ismail AH, Hartono N, Caterino M, Jiang K (2021) An improved vantage point bees algorithm to solve combinatorial optimization problems from tsplib. In: Macromolecular symposia (Vol 396, No 1, p 2000299)

Utilising the Bees Algorithm for UAV Path Planning—A Simultaneous Collision Avoidance and Shortest Path Approach

Anubhab Dasgupta, Satyam Kumar, Aaditri Vaibhav, and Asrul Harun Ismail

Abstract Due to the significant interest in Unmanned Aerial Vehicles (UAVs), numerous metaheuristic or nature-inspired algorithms have been proposed for path-planning problems, allowing units to navigate through diverse threats effectively. By strategically avoiding threats identified by radar, these algorithms aim to determine the most optimal route. This chapter explores the application of two variants of the Bees Algorithm (BA) to address this issue and evaluates their performance against other metaheuristic algorithms. The comparison encompasses various configurations of waypoints and threats, with statistical analyses conducted to assess performance differences. The findings indicate that the Bees Algorithm with two parameters (BA2) surpasses the performance of the Basic BA and other metaheuristic algorithms. Statistical estimates derived from the data are utilised for comparisons, ensuring that optimisation parameters for all algorithms are selected unbiased and meaningfully.

Keywords Combinatorial optimisation · Bees algorithm · Metaheuristics · UAV · Collision avoidance · Path planning

A. Dasgupta (✉)
Mechanical Engineering Department, Indian Institute of Technology Kharagpur, Kharagpur, India
e-mail: anubhab.dasgupta@kgpian.iitkgp.ac.in

S. Kumar
Department of Civil Engineering, Indian Institute of Technology Kharagpur, Kharagpur, India
e-mail: satyamk1102@kgpian.iitkgp.ac.in

A. Vaibhav
Department of Metallurgical and Material Sciences, Indian Institute of Technology Kharagpur, Kharagpur, India
e-mail: aadi15@kgpian.iitkgp.ac.in

A. H. Ismail
Centre of Operational Research and Logistics, University of Portsmouth, Portsmouth, UK
e-mail: asrul.ismail@port.ac.uk

D. T. Pham and N. Hartono (eds.), *Intelligent Engineering Optimisation with the Bees Algorithm*, Springer Series in Advanced Manufacturing,
https://doi.org/10.1007/978-3-031-64936-3_17

1 Introduction

Autonomous unmanned aerial vehicles (UAVs) have attracted the curiosity of many people in recent years due to their ability to undertake a range of dangerous and demanding tasks over extended periods. In addition to being utilised for local surveillance, crowd monitoring, and other defence and military objectives, the transportation of specific products by UAVs has recently sparked the curiosity of many people.

Many combinatorial optimisation problems are NP-hard, and nature-inspired optimisation algorithms such as particle swarm optimisation [1] and ant colony optimisation [2] provide very good strategies for finding the optimal solution. In our work, we use the Bees Algorithm (BA). Pham et al [3] is the first version of the BA, developed by Pham et al. (see also references [4–7]). Following that, there were both application-based uses of the algorithm such as [8] in the training of neural networks and development-based work that focused on improving the algorithm such as in [9], which is the basic version of the BA we use in our work as a comparison benchmark, and the algorithm we use (referred to as BA_2) was developed and is described in [10].

Previously, BA has been used in the study of aircraft control, such as in the pitch control of aircraft [11] and for aircraft landing [12]. Several nature-inspired optimisation strategies have been used for UAV path planning and optimisation, such as [13] using angle-encoded swarm optimisation and [14] using pigeon-based optimisation and differential evolution. Aggarwal and Neeraj Kumar [15] provides a survey of UAV motion planning. We focus on [16], which uses the grey-wolf optimiser for UCAV path planning and compares it with other similar optimisation algorithms. We use a similar setup and integrate BA_2 with the setup for optimisation and give comparison results and analysis.

However, unlike other autonomous vehicles, UAVs cannot rely on physical markers on the road or infrastructures because they have a larger movement area than ground units, which are confined by the infrastructure itself. Even if the UAV course is less restricted in the sky, the unit must consider barriers in the field, such as topography, high-rise buildings, and other dangerous items. The metaheuristic will be the most appropriate method with this limitation on the information stored in advance.

In this chapter, we discuss our research on using the BAs for autonomous single UAV path planning in 2D in the presence of static circular obstacles of varying danger levels. A threat is a danger that the UAV should avoid when travelling. In addition to demonstrating the performance of the BA for UAV path planning, we elaborate on and compare the implementation of our proposed BA method. A series of experimental comparison results show that our proposed approach is applicable, effective, and long-lasting. Finally, we address the issue of computational complexity that arises when we incorporate numerous UAVs into our environment and provide a reasonable solution rooted in the principles of the Bees Algorithm.

2 Problem Formulation

In general, the main goal of this problem is to discover the shortest safe path for a UAV that allows a unit to navigate a wide range of threats. Certain parameters can represent the full model of the UAV(s) and the obstacles that surround them. The model considers the UAV(s) to be point objects and the obstacles to be circular, since in this study, we look at 2-dimensional UAVs.

We first define the dimensionality of the problem by the scalar, where $n_d = 2$. Thereafter, we define a bounding box within which the UAV(s) will be restricted to fly. This is denoted by B, which is a matrix with dimensions of 2×2, where the rows denote the x-coordinate and y-coordinate of the cartesian space and the 2 columnar scalars denote the lower limit and the upper limit of that coordinate in the order of left to right. The starting point and the ending point of the UAV(s) are denoted by S and E, respectively, which are vectors with dimensions of 1×2, where nv denotes the number of UAV(s) in the problem and S represents the starting coordinate of the UAV in the problem. Similarly, E denotes the ending coordinate of the UAV in the problem. Now, we define the obstacles. The number of obstacles in the problem is denoted by n_{obs}. The position of the obstacles is stored in a matrix P, with dimensionality $n_{\text{obs}} = 2$. In each row, i denotes the coordinate of the centre of the circle of the i-th obstacle. The radii of the obstacles are stored in the vector R, with dimensionality $n_{\text{obs}} \times 1$, where each row denotes the radius of the obstacle being represented by that particular row index. Another characteristic of importance related to the obstacles is the threat grade, which quantitatively tells a particular UAV which obstacle to make the most effort to avoid when there is more than one obstacle within the decision boundary. It is denoted by the vector T, having the dimensionality $n_{\text{obs}} \times 1$, with each row representing the threat grade of the UAV being indexed by that row number. Finally, we define n_{wp}, which denotes the number of waypoints that the optimisation algorithm must output excluding the starting and ending points.

The objective function, also referred to as the fitness function or the cost function in our problem setting and solution, takes as inputs at the instant at which it is applied to the system, the instantaneous position of the UAV, the matrices S and E and information about the obstacles, that is, P, R and T. It additionally takes a weighting parameter, the utility of which will be explained below. The most common way of modelling the cost function in a UAV-based setting, as in [16], is to divide the cost into two additive components. One of them is the economic cost, including the fuel cost due to the length of the path created by the optimisation algorithm. We want the algorithm to favour shorter paths for less fuel consumption. The other part is the collision avoidance term, which quantitatively indicates the joint threat produced by all the obstacles within the decision boundary of the problem. For the purpose of comparison with other algorithms, we set the parameter to be equal to 0.5, meaning that it will give equal importance to both threat avoidance and fuel consumption (length of the planned path).

$$J = \lambda J_t + (1 - \lambda) J_f \tag{1}$$

where J represents the overall cost of the flight of the UAV and J_f and J_t represent the respective fuel and threat costs for the total duration of travel. We expand the respective fuel and threat costs as an integral of time and instantaneous threats and rewrite the cost as:

$$J = \lambda \int_0^L w_t dl + (1 - \lambda) \int_0^L w_f dl \tag{2}$$

where J is the weighted sum of flight cost for this flight path, w_t and w_f are variables closely related to the current path point and changing along with the length, which, respectively present the threat cost and fuel cost of each line segment on the route, and L is the total length of the generated path. The optimal path is found only when J reaches its minimum value.

When the UAV is flying along path $L(i \rightarrow i + 1)$, the threat cost generated by all n_{obs} can be written as:

$$\int_0^{length_i} \sum_{k=1}^{n_{obs}} \frac{(T[k])}{\left((x - x_k)^2 + (y - y_k)^2\right)^2} \tag{3}$$

To simplify the integral operations in the detection cost w_t from L_i to L_{i+1} is calculated at five sample waypoints as:

$$w_{t,L_i \rightarrow L_{i+1}} = \frac{length_i}{5} \sum_{k=1}^{n_obs} T[k] \times \left(\frac{1}{d_{0.1,i,k}^4} + \frac{1}{d_{0.3,i,k}^4} + \frac{1}{d_{0.5,i,k}^4} + \frac{1}{d_{0.7,i,k}^4} + \frac{1}{d_{0.9,i,k}^4} \right) \tag{4}$$

where $d_{0.nv,i,k}$ is the distance from the $nv/10$ point on the path and the k-th threat centre. Figure 1 illustrates the above integral computation. The points where the expression in Eq. (4) is evaluated are marked along the path. The larger the number of points taken, the better the integral estimation. However, given the particular test cases, we considered only 5 such evaluation points since they approximate the actual integral to a very good extent.

3 Methodology

As previously stated, this UAV collision avoidance problem (UCAP) employs the BA_2. The case study utilised in this research is the 2D threat environment, in which the drone must avoid certain threats while navigating from origin to destination. After conducting a literature review, it was concluded that numerous metaheuristic algorithms have been applied to this problem but not to BA. There are also multiple

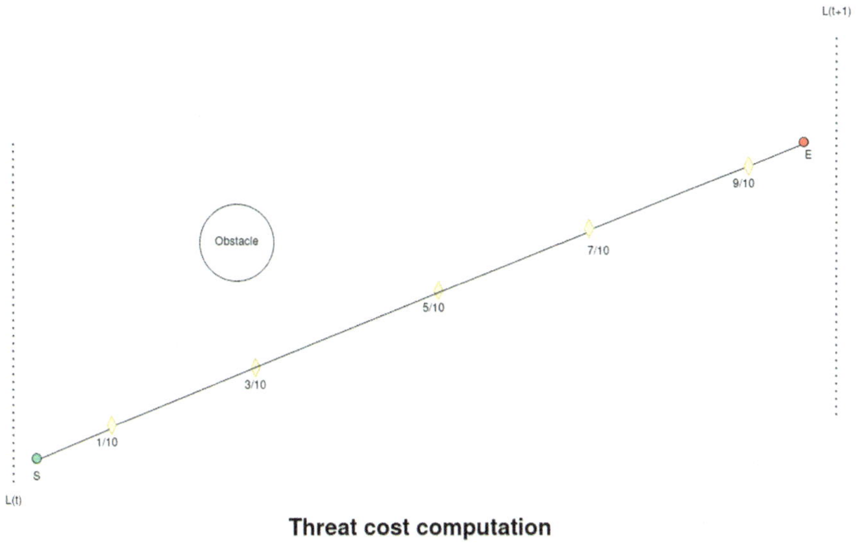

Fig. 1 An obstacle and the sample path that it follows with the evaluation points for the integral computation marked

datasets that can be utilised to solve this problem. However, we chose the 2D threat environment dataset because it is the most frequently used dataset for comparing the performance of various algorithms on this problem.

This study will compare the accuracy and convergence performance of the basic BA, grey wolf optimisation (GWO), artificial bee colony (ABC), flower pollination algorithm (FPA), cuckoo search (CS), Nelder–Mead algorithm (NBA), bird swarm algorithm (BSA) and guided gravitational search algorithm (GGSA). The experiment will be conducted with the same number of evaluations for each scenario, determined by the problem functions and the number of waypoints. This diagram illustrates the experimental procedure of a metaheuristic-based comparative study on UCAP.

After defining the research gap and selecting the appropriate datasets, the problem function was constructed using identical parameters as the UAV model. Using a computer with an AMD Ryzen 5000 series processor and 16 GB of memory, BA_2 was implemented in MATLAB version R2022a to solve UCAP. The development of BA_2 consisted of the modification of a worker bee recruitment regulator. This modification was made to investigate the nature of BA_2 in a space that is not linear, such as logarithmic.

To compare the proposed algorithm with previous versions and other metaheuristics, we run both the Basic BA and BA_2 algorithms 30 times with the same initial configuration and optimisation parameters as the results of other algorithms from the literature and compare them. The best obtained value and the average value from 30 separate experiments will be compared (Fig. 2).

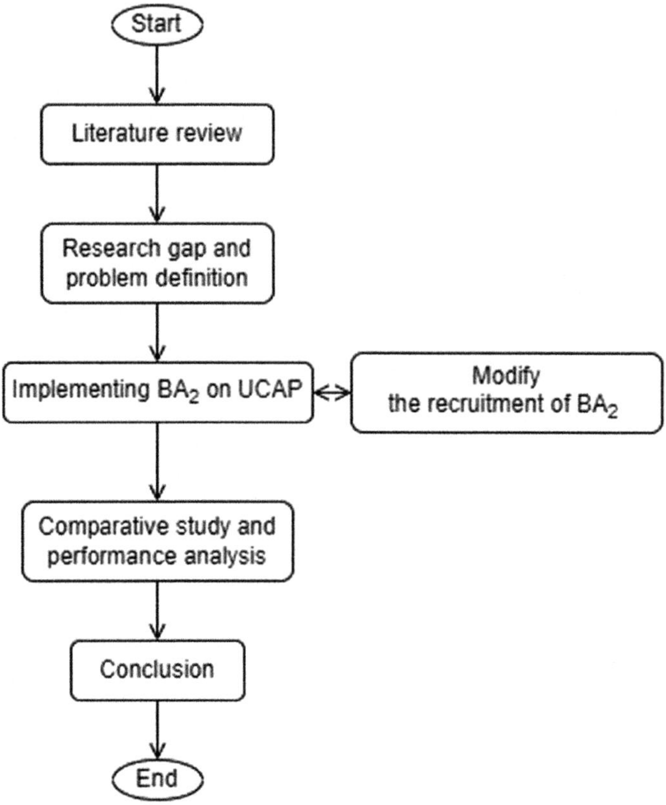

Fig. 2 An outline of the work

3.1 *Overview of* BA_2

This version of the BA poses only two initial parameters, *n* (scout bees) and *nep* (the maximum number of worker bees or the number of worker bees on the elite sites), by combining exploration and exploitation when foraging honey.

The quantity of worker bees will indicate the quality of the location. The best location will have the greatest number of employees, while the worst location will have the fewest.

In the search mechanism, there will be a ratio of exploitation to exploration. Therefore, the bees assigned to a particular location may be assigned to either exploit or explore the location. A triangular random number generator was employed to assign the worker bees. The greatest number of worker bees will exploit the best site, while on the worst site, most of the bees will explore it (Fig. 3).

This research modifies the procedure for worker bee recruitment and assignment. Basic BA and BA_2 both use a linear function, but in this study, we use nonlinear functions for recruitment and assignment.

Fig. 3 A flow chart describing the BA_2

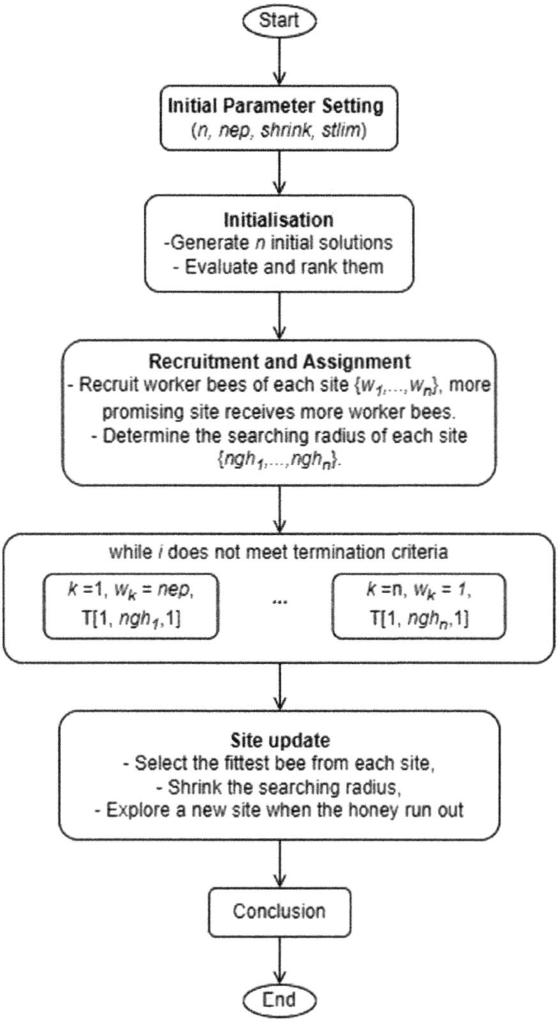

3.2 *Bees Algorithm and Its Integration with the Problem*

In our problem of UAV collision avoidance, we use a set of generated waypoints as a patch for search and initially generate 30 such patches. This is followed by the cost function evaluating the cost of the patches. Before doing so, the 10 waypoints of each patch are used as input to a MATLAB function to form a third-order spline, connecting all the waypoints, and then points are sampled from the spline uniformly to evaluate the cost. This prevents any constraint in the calculation from taking place due to the number of waypoints selected being low in number, since the higher the number of waypoints is, the better the cost estimation (the integral estimation done using summation). These steps are followed by the general BA_2 steps mentioned above in the flow algorithm, and the results are generated and analysed in the next section.

4 Results and Analysis

4.1 Results

First, we compare the results obtained from BA_2 and other nature-inspired optimisation algorithms: grey wolf optimisation (GWO), artificial bee colony (ABC), flower pollination algorithm (FPA), cuckoo search (CS), Nelder–Mead algorithm (NBA), bird swarm algorithm (BSA), guided gravitational search algorithm (GGSA) and basic BA. We performed the computation on a 16 GB AMD Ryzen 5 processor using MATLAB R2022a-academic.

Table 1 presents the initial conditions and parameters, which remained consistent across all algorithms during the comparison process. For the comparison to be meaningful and unbiased, we used the same optimisation parameters, dataset, cost function and initial configuration for all the different algorithms. The dataset and the initial configurations are given in Table 1. We use only case T1 here (Table 2). The dataset was obtained from [16] along with the different parameters listed below, which we kept the same when we executed Basic BA and BA_2. The comparison is performed based on statistical estimates over 30 independent runs for each of the algorithms listed in Table 3.

Given the above parameters, we averaged the results for 30 independent runs and recorded statistical metrics such as the best cost, worst cost, mean cost and standard deviation. These statistical metrics over 30 independent runs rule out any outliers and are important for the comparison since the optimisation is stochastic in nature. The results are given in Table 3:

From the table, we can observe that BA_2 performs the best in terms of the best cost, mean cost, standard deviation, and worst cost.

A sample run of the case "T1" and its results from BA_2 is as follows (Fig. 4):

Table 1 Initial parameters and conditions

Parameter	Value
Number of scout bees (n)	30
Maximum worker bees (nep)	10
Number of waypoints (n_{wp})	10
Shrink factor	0.8
Stagnation limit	5
Maximum iterations	200
Cost weight (λ)	0.5

Table 2 Obstacle configuration parameters

Case Number.	Start x	Start y	End x	End y	Threat x	Threat x	Threat radius	Thread grade
T1	10	10	55	100	45	52	10	2
					12	40	10	10
					32	68	8	1
					36	26	12	2
					55	80	9	3
T2	11	11	75	75	52	52	10	2
					32	40	10	10
					12	48	8	1
					36	26	12	2
					80	60	9	3
					63	56	7	5
					50	42	10	2
					30	70	10	4
					30	20	10	5
					50	15	21	5
T3	0	0	100	100	65	55	10	5
					0	24	17	5
					50	80	12	5
					75	90	10	5
					100	70	20	5
					50	36	11	5

Table 3 Performance comparison of the various algorithms applied to the same problem

Result	ABC	GWO	CS	FPA	NBA	BSA	GGSA	**BA$_2$**	BBA
Best cost	50.899	50.718	50.889	51.165	50.725	50.920	50.762	**50.347**	54.667
Worst cost	51.625	51.083	51.769	52.007	52.301	58.023	59.750	**50.363**	57.578
Mean cost	51.139	50.753	51.307	51.522	51.208	53.347	53.424	**50.352**	56.157
Standard deviation	0.162	0.088	0.246	0.224	0.502	2.014	2.440	**0.004**	0.765

The graph on the left shows the cost reduction with time, and the graph on the right shows the generated path in 2D in the presence of circular obstacles.

Note that all the data collected from [16] and that generated using Basic BA and BA_2 have the same parameters and settings as listed above. Thus, the comparison

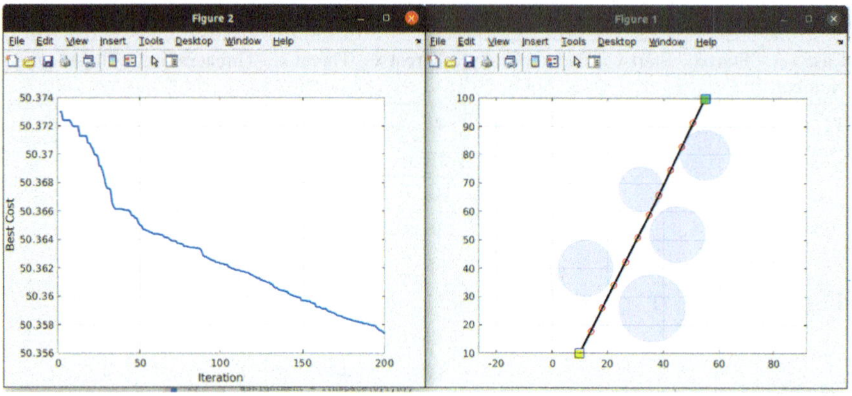

Fig. 4 Variation of cost with the increase in iterations (left) and the connected waypoint trajectory obtained from BA_2 (right)

is meaningful and provides no undue advantage to any of the results of a particular algorithm.

We went further to propose the logarithmic distribution of recruitment, and we observed that having the logarithmic distribution in the optimisation process reduces the execution time compared to BA_2. The results are provided below:

The above is the iteration result when BA_2 is run with a logarithmic distribution for recruitment. The time taken for 200 iterations is 11.8938 s. Earlier, we saw that BA_2 gives a mean cost of 50.3522 and in 200 iterations, but the time of execution varies between 14 and 15 s. Using the logarithmic distribution for recruitment, we can achieve lower execution times, which might be significant for higher dimensional settings (Fig. 5).

Command Window

```
Iteration 185: Best Cost = 50.3535 * --> Time = 11.0043
Iteration 186: Best Cost = 50.3532 * --> Time = 11.0635
Iteration 187: Best Cost = 50.3531 * --> Time = 11.1263
Iteration 188: Best Cost = 50.3531 * --> Time = 11.1873
Iteration 189: Best Cost = 50.353 * --> Time = 11.2462
Iteration 190: Best Cost = 50.353 * --> Time = 11.305
Iteration 191: Best Cost = 50.3529 * --> Time = 11.3632
Iteration 192: Best Cost = 50.3529 * --> Time = 11.4219
Iteration 193: Best Cost = 50.3529 * --> Time = 11.48
Iteration 194: Best Cost = 50.3529 * --> Time = 11.5388
Iteration 195: Best Cost = 50.3527 * --> Time = 11.5979
Iteration 196: Best Cost = 50.3527 * --> Time = 11.6545
Iteration 197: Best Cost = 50.3525 * --> Time = 11.7152
Iteration 198: Best Cost = 50.3524 * --> Time = 11.7752
Iteration 199: Best Cost = 50.3524 * --> Time = 11.8341
Iteration 200: Best Cost = 50.3524 * --> Time = 11.8938
fx >>
```

Fig. 5 Variation of the best cost with the increase in iterations (time taken from the start specified in the rightmost column)

4.2 Analysis of Results

The UAV path planning problem studied here is on a two-dimensional cartesian space, and thus, using an optimisation algorithm with an exploration strategy similar to the driver's mindset in a manual situation is appreciable. BA_2 uses an exploration strategy that chooses a random 2D vector on a circular disc, thus not creating any directional bias and leaving the inclusion of bias to the cost computation once the UAV advances to a particular point in the space. BA_2 also uses site abandonment and the stagnation limit to reduce searches in less feasible regions, which reduces the complexity of the program. Neighbourhood shrinking makes the search more precise in the area of interest. The choice of a forager bee as the sequence of planned coordinates of the path is also something which is of interest since choosing a different forager bee formulation such as the very next point in the planned path would have given a different result. All of the above combined algorithms are believed to make BA_2 perform better than the other benchmark algorithms for this particular UAV path planning setup.

The logarithmic recruitment can be seen to be giving a better cost in a smaller amount of time, which is of considerable interest, and the theoretical analysis as to why it happens is something that will be taken up for discussion in future work.

5 Conclusion and Future Work

In this work, we see how BA_2 can be applied for path planning in 2D for UCAVs in the presence of circular obstacles with different radius and respective degrees of threat. We also show that BA_2 can perform better than a few other similar algorithms in terms of the metrics listed earlier. We also studied the effect of choosing a nonlinear distribution function for recruitment and presented its results and comparison with BA_2.

Future work involves solving a similar problem with multiple UCAVs present and extending the same for a 3D case, thus making the problem closer to a real-world scenario and demonstrating the effectiveness of the Bees Algorithm in the optimisation of these kinds of problems. A theoretical analysis of the entire problem with logarithmic recruitment giving a better performance is something that will also be studied.

Acknowledgements We would like to thank Professor Duc Pham for the very insightful discussions and his valuable input while doing this work and while writing it.

References

1. Kennedy J, Eberhart R (1995) Particle swarm optimization. In: Proceedings of ICNN'95-international conference on neural networks, vol 4. IEEE
2. Dorigo M, Di Caro G (1999) Ant colony optimization: a new meta-heuristic. In: Proceedings of the 1999 congress on evolutionary computation-CEC99 (Cat. No. 99TH8406), vol 2. IEEE
3. Pham DT, Ghanbarzadeh A, Koç E, Otri S, Rahim S, Zaidi M (2006) The Bees algorithm—a novel tool for complex optimisation problems. In: Intelligent production machines and systems. Elsevier Science Ltd., pp 454–459
4. Pham DT, Ghanbarzadeh A, Koc E, Otri S, Rahim S, Zaidi M (2005) The Bees algorithm. Technical Note, Manufacturing Engineering Centre, Cardiff University, UK
5. Pham DT, Ghanbarzadeh A, Koç E, Otri S, Rahim S, Zaidi M (2006) The Bees algorithm—a novel tool for complex optimisation problems. Intell Prod Mach Syst 454–459
6. Pham DT, Castellani M, Le Thi H (2014) Nature-inspired intelligent optimisation using the Bees algorithm. Transactions on computational intelligence XIII, pp 38–69
7. Pham DT, Castellani M (2015) A comparative study of the Bees algorithm as a tool for function optimisation. Cogent Eng 2:1091540
8. Pham DT, Otri S, Ghanbarzadeh A, Koc E (2006) Application of the Bees algorithm to the training of learning vector quantisation networks for control chart pattern recognition. In: 2006 2nd International conference on information & communication technologies, vol 1. IEEE, pp 1624–1629
9. Pham DT, Marco C (2009) The Bees algorithm: Modelling foraging behaviour to solve continuous optimization problems. Proc Inst Mech Eng Part C: J Mech Eng Sci 223(12):2919–2938
10. Ismail AH, Ruslan W, Pham DT (2023) A user-friendly Bees algorithm for continuous and combinatorial optimisation. Cogent Eng
11. Zaeri R, Ghanbarzadeh A, Attaran B, Zaeri Z (2011) Fuzzy logic controller based pitch control of aircraft tuned with Bees algorithm. In: The 2nd international conference on control, instrumentation and automation. IEEE
12. Abdul-Razaq TS, Ali FH (2015) Hybrid Bees algorithm to solve aircraft landing problem. J Zankoi Sulaimani JZS 17(1)
13. Hoang VT et al (2018) Angle-encoded swarm optimization for UAV formation path planning. In: 2018 IEEE/RSJ international conference on intelligent robots and systems (IROS). IEEE
14. Tong B, Chen L, Duan H (2021) A path planning method for UAVs based on multi-objective pigeon-inspired optimisation and differential evolution. Int J Bio-Inspired Comput 17(2):105–112
15. Aggarwal S, Kumar N (2020) Path planning techniques for unmanned aerial vehicles: a review, solutions, and challenges. Comput Commun 149:270–299
16. Zhang S, Zhou Y, Li Z, Pan W (2016) Grey wolf optimizer for unmanned combat aerial vehicle path planning. Adv Eng Softw 99:121–136

A Tabu-Based Bees Algorithm for Unmanned Aerial Vehicles in Maritime Search and Rescue Path Planning

Yangyang Guo, Yanjie Song, Jiting Li, Junwei Ou, Lining Xing, and Yue Zhang

Abstract Unmanned aerial vehicles (UAVs) play a vital role in maritime search and rescue (SAR). The ability to find people overboard quickly is closely related to the flight path setting. We construct a mathematical programming model considering the task size of each SAR center and the overall search time as the objective function for the maritime search and rescue path planning (SARPP) problem. Then, a Tabu-based bees algorithm (TBBA) is proposed considering the complex problem space and constraints. The TBBA adopts the idea of the Tabu strategy to guide bees' search and records the improvement of the individual structure of scout bees' search to reduce the possibility of invalid search occurrence. An individual replacement strategy is also used in the algorithm to randomly generate a new site to replace the worst site when the search is unsatisfactory. The effectiveness of the proposed algorithm is verified by various experiments. TBBA can obtain a more ideal SAR solution than the genetic algorithm and neighbourhood search algorithm.

Keywords Bees algorithm · Tabu list · Maritime search and rescue · Path planning · UAV

Y. Guo
School of Systems Science, Beijing Jiaotong University, Beijing, China

Y. Song (✉) · Y. Zhang
School of Reliability and Systems Engineering, Beihang University, Beijing, China
e-mail: songyj_2017@163.com

J. Li
Chinese Academy of Military Science, Beijing, China

J. Ou
School of Computer Science and School of Cyberspace Science, Xiangtan University, Xiangtan, China

L. Xing
Key Laboratory of Collaborative Intelligence Systems, Ministry of Education, Xidian University, Xi'an, China

D. T. Pham and N. Hartono (eds.), *Intelligent Engineering Optimisation with the Bees Algorithm*, Springer Series in Advanced Manufacturing,
https://doi.org/10.1007/978-3-031-64936-3_18

1 Introduction

Maritime safety is an important part of the national safety emergency system. When a person falls into the water or a vessel is damaged, it is necessary to search the marine accident site quickly to ensure the safety of people. However, in the face of the vast sea area, it is difficult to achieve the desired efficiency by relying only on human search and rescue. The cooperation of UAVs in the search can improve search efficiency and enhance the possibility of human rescue. This makes the application of UAVs in maritime rescue receive increasing attention [1, 2]. In overall UAV cooperation to complete maritime search and rescue work, making a reasonable and efficient search and rescue plan is particularly important.

The problem of developing search routes for UAVs is called the UAV maritime search and rescue path planning (SARPP) problem, which requires determining the number of UAVs dispatched by each SAR center and the search tasks for each UAV. The UAV's ability to work strictly limits the number of tasks that can be completed, and each task must be carried out on the premise that the UAV can smoothly return to the SAR center. The SARPP problem is similar to the vehicle routing planning (VRP) problem in that the location of the task areas is known and will not change, and there is no need to explore beyond the task areas. Moreover, each UAV has a range of detection due to power or other constraints, and the task area is discretised according to the range. Therefore, the two problems can be expressed simply in the form of a graph model. However, there are many differences between the two issues. For example, unlike the VRP problem, the UAV in the SARPP problem needs to cover the entire search area, and the failure to cover some areas is not acceptable. In addition, more UAVs can be deployed from search and rescue centers depending on the search process. These factors make the SARPP problem more challenging to solve.

The UAV area coverage path planning problem has recently become a popular research topic. Cho et al. proposed a random search heuristic (RSH) algorithm based on a grid-based region decomposition method to quickly solve the large-scale UAV area coverage path planning problem [3]. Yang et al. combined Markov strategy iteration with reinforcement learning to solve the UAV path planning problem and effectively cover the task area [4]. Based on the genetic algorithm, Li et al. combined the K-means algorithm to solve the optimal path of maritime target search, which improved the convergence speed and global search ability of the algorithm [5]. Zhang et al. applied the ant colony algorithm to the design of the maritime search and rescue route, on which the pheromone concentration update strategy was proposed to improve the performance of the algorithm [6]. The characteristics of large scale and many constraints make heuristic algorithms and metaheuristic algorithms suitable for solving the SARPP problem.

Among many metaheuristic algorithms, the bee algorithm (BA) is popular for its simple structure and excellent global search performance [7–9]. Many studies have solved path-planning problems using BA. Haj et al. proposed a new form of BA to solve the path-planning problem for indoor robots [10]. The initial population is obtained by feasible paths in a static environment, and then the superiority of the algorithm is enhanced by local search to avoid collisions between paths and dynamic obstacles. Xu et al. introduced a coevolution framework into the artificial bee colony algorithm to solve the problem of the algorithm, such as the slow convergence speed and dimension dependence [11]. The effectiveness of the new algorithm is verified in the test of robot path planning. Gu et al. introduced a coevolution strategy into the artificial bee colony algorithm and applied it to the multidepot VRP problem [12]. The simulation experiments proved that the algorithm performed better than the greedy algorithm and genetic algorithm.

In the application, the BA suffers from a deficiency in that the search efficiency is affected by the complexity of the problem and the slow convergence speed [9]. Considering the characteristics of the SARPP problem, we add the Tabu strategy to record invalid search operations in BA [13]. This will prevent similar searches from being conducted within a certain time frame. The Tabu-based bees algorithm (TBBA) can dynamically adjust the search position in the solution space according to the search performance. The main contributions of this chapter are as follows:

1. We construct a mathematical programming model for the SARPP problem. In the model, an optimisation function aiming at minimising the overall search and rescue execution time and balancing the task assignment of each search and rescue center is proposed, and various constraints, such as the UAV working capability and the task area coverage, are considered.
2. We propose a Tabu-based bees algorithm based on the characteristics of the SARPP problem and successfully apply it to simulation experiments. The algorithm uses the Tabu search strategy and the individual replacement strategy to reduce the occurrence of repeated searches and improve the efficiency of optimisation. Simulation experimental results fully verify the effectiveness of TBBA in solving the SARPP problem, which can meet the needs of large-scale scenario applications.

The rest of the chapter is structured as follows. Section 2 introduces the mathematical programming model of the SARPP problem. Section 3 introduces TBBA. Section 4 designs simulation experiments to verify the performance of the proposed algorithm in solving the SARPP. Section 5 summarises the main work of this chapter.

2 Models

2.1 Symbols and Variables

In the SARPP problem, the symbols and variables involved are as follows:

R	Set of SAR centers;
r_k	Search and rescue center;
(lon_k, lat_k)	Coordinate of the SAR center r_k;

U_k	Set of UAVs belonging to the SAR center r_k;
M_k	The number of UAVs contained in the SAR center, $M_k = \|U_k\|$;
u_o	Unmanned aerial vehicle;
v_o	Flight speed of UAV u_o;
e_o	Power consumption per unit time of a single UAV;
E_o	The maximum amount of power that can be stored in the battery of a single UAV;
T	Set of search and rescue tasks;
t_i	Search and rescue task;
(lon_i, lat_i)	Coordinate of task t_i;
D_{ij}	Euclidean distance between task t_i and task t_j;

Decision Variables

x_{ij}^{ko} If UAV u_o from SAR center r_k executes task t_i first and then task t_j, $x_{ij}^{ko} = 1$; otherwise, $x_{ij}^{ko} = 0$;

y_j^{ko} If task t_j is executed by UAV u_o from SAR center r_k, then $y_j^{ko} = 1$; otherwise, $y_j^{ko} = 0$;

t_{ij}^{ko} The flight time of UAV u_o from SAR center r_k between task t_i and task t_j. Its equation is:

$$t_{ij}^{ko} = D_{ij}/v_o \tag{1}$$

2.2 Problem Description

In the SARPP problem, there exists a series of SAR centers R responsible for performing search and rescue tasks at sea, and the number of UAVs that SAR center r_k can dispatch is M_k. The UAVs need to depart from the SAR center and fly to the region of task t_i with coordinate (lon_i, lat_i) in task set T at speed v_o. After completing the search of task t_i, the UAV can continue to search the area where task t_j. is located or return to the SAR center. All task areas need to be detected, and each task only needs to be detected once.

The UAV's capability limits the number of tasks it can complete per flight. The power consumed by the UAV from departure to return to the SAR center cannot exceed the maximum battery capacity. The premise of each UAV task is that the power supply can guarantee its return to the SAR center. Therefore, when the UAV is unable to return to the SAR center because of the next task, it must give up the execution and choose to return directly to the SAR center.

According to the above description, the flight of UAVs in the SARPP can be divided into three phases: departure from the SAR center to the task area, flight between the task areas, and return to the SAR center from the task area. The graph model can represent the relationship between the tasks and the SAR centers. We use the directed graph $G(N, E)$ to represent the path of the UAV from the SAR center to the task area to complete the task and return, where node $N = R \cup T$ denotes the SAR center or the task, and nodes 0 and $|T + 1|$ both denote the SAR center. The edge denotes the execution sequence of task t_j after completing task t_i.

2.3 Mathematical Model

We can construct a mathematical programming model based on the above description of the SARPP problem. The assumptions, objective functions, and constraints of the model are given in this section. The mathematical model of the SARPP problem is built on the following assumptions.

Assumptions

1. The search area has been divided into several search tasks in advance, each of which will be executed successfully with only one search;
2. The flight of a UAV is not subject to mechanical failure and will not be affected by communication factors;
3. A search and rescue center has multiple UAVs to participate in search and rescue tasks;
4. Each UAV carries only one detection load.

Our intended goal for UAV flight path planning is to find the optimal path planning solution that minimises the overall SAR execution time based on a roughly balanced task distribution among SAR centers. Here, we use the number of UAVs dispatched to measure the workload of each SAR center and evaluate whether the task distribution is balanced by calculating the standard deviation of the number of UAVs dispatched by each SAR center. In addition, the final execution time is determined by the maximum flight time of the UAVs dispatched by each SAR center. The objective function is composed in the form of linear weighting through the above two optimisation perspectives.

Objective Function

$$\min \alpha_1 \cdot \left\{ \max_{k \in R} GT_k \right\} + \alpha_2 \cdot SD \tag{2}$$

$$GT_k = \sum_{o \in U_k} \sum_{i \in T \cup \{0\}} \sum_{j \in T} t_{ij}^{ko} \cdot x_{ij}^{ko} \tag{3}$$

$$SD = \sqrt{\frac{1}{|R|} \sum_{k \in R} \left(\sum_{o \in U_k} \sum_{j \in T} x_{oj}^{ko} - \mu \right)^2} \tag{4}$$

$$\mu = \frac{1}{|R|} \cdot \sum_{k \in R} \sum_{o \in U_k} \sum_{j \in T} x_{oj}^{ko} \tag{5}$$

$$\alpha_1 + \alpha_2 = 1 \tag{6}$$

Constraints

$$\sum_{j \in T} x_{0j}^{ko} = 1, \forall k \in R, o \in U_k \tag{7}$$

$$\sum_{i \in T} x_{i(|T|+1)}^{ko} = 1, \forall k \in R, o \in U_k \tag{8}$$

$$\sum_{j \in T} t_{0j}^{ko} \cdot e_o \cdot x_{0j}^{ko} + \sum_{i \in T} \sum_{j \in T} t_{ij}^{ko} \cdot e_o \cdot x_{ij}^{ko} \le E_o, \quad \forall k \in R, o \in U_k \tag{9}$$

$$\sum_{i \in T} \sum_{j \in T} t_{ij}^{ko} \cdot e_o \cdot x_{ij}^{ko} + \sum_{i \in T} t_{i(|T|+1)}^{ko} \cdot e_o \cdot x_{i(|T|+1)}^{ko} \le E_o, \quad \forall k \in R, o \in U_k \tag{10}$$

$$\sum_{k \in R} \sum_{o \in U_k} y_j^{ko} \le 1, \forall j \in T \tag{11}$$

$$\sum_{k \in R} \sum_{o \in U_k} \sum_{j \in T} y_j^{ko} = |T| \tag{12}$$

$$\sum_{i \in T} x_{ij}^{ko} = y_j^{ko}, \forall k \in R, o \in U_k \tag{13}$$

$$\sum_{i \in T \cup \{0\}} x_{ij}^{ko} = \sum_{j' \in T \cup \{|T|+1\}} x_{jj'}^{ko}, \forall k \in R, o \in U_k \tag{14}$$

$$x_{ij}^{ko} \in \{0, 1\} \tag{15}$$

$$y_j^{ko} \in \{0, 1\} \tag{16}$$

Constraint (7) indicates that the UAVs need to depart from all search and rescue centers. Constraint (8) indicates that the UAVs need to return to the search and rescue center from which they started after completing the assigned task. Constraints (9, 10) indicate that the power consumption cannot exceed the upper limit. Constraint (11) indicates that each task needs to be executed only once. Constraint (12) indicates that all the areas need to be searched successfully. Constraints (13, 14) indicate the states that need to be satisfied between nodes in the graph model. Constraints (14, 15) represent the range of values of the variables.

3 Tabu-Based Bees Algorithm

The planning algorithm is extraordinarily important for developing a good UAV search route to quickly rescue crew members who fall into the water. To solve the SARPP problem, we propose a Tabu-based bees algorithm. This algorithm simulates the process of bees searching for food sources and uses a Tabu search strategy to guide the BA search to improve the effectiveness of the algorithm search. In the following, we will introduce the overall framework and main steps of TBBA. The flow chart of the algorithm is shown in Fig. 1.

Fig. 1 Flow chart of TBBA

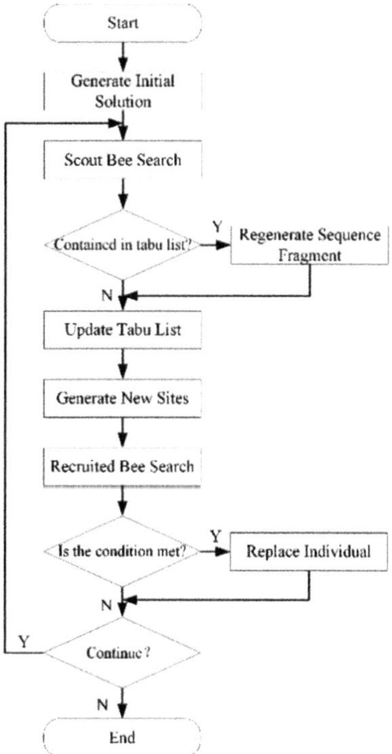

3.1 Framework of TBBA

In this part, the overall framework of TBBA is introduced. We make a series of improvements to the algorithm structure and search strategy based on the BA considering the SARPP characteristics. The TBBA algorithm mainly contains the steps of generating the initial population, the Tabu-based scout bee search stage, the recruited bee search stage, and individual replacement. TBBA randomly generates the initial population, selects the corresponding number of sites, and conducts the subsequent search. The scout bees are mainly responsible for solution space exploration, while the recruited bees are responsible for development. The Tabu search strategy is used in the scout bee search. The search operation will be recorded in the Tabu list when the search performances is unsatisfactory [14, 15]. The individual replacement strategy completes the structural update of the site. The pseudocode of TBBA is shown in Algorithm.

3.2 Tabu-Based Scout Bee Search

Scout bee search is the main search method of TBBA to achieve efficient exploration of the solution space. The scout bee search is carried out based on already found sites and is performed by swapping positions of fragments within individuals, and the length of the swapped fragments is l.

As the search continues, the difficulty of finding higher-quality solutions increases accordingly. To reduce the number of invalid search occurrences, we record the improved sequence fragments of individual structures in the Tabu list. When the scout bee search is performed again, first compare whether the sequence fragment to be used in this search is already included in the Tabu list *Tabu*. If it has already been included in *Tabu*, then the fragment needs to be reselected.

A longer Tabu list length does not mean a more effective algorithm search. Whether the task sequence of the algorithm search is in the Tabu list or not will occupy computing resources and may lead to an insufficient search of the solution space [16]. Therefore, when the length in the Tabu list reaches the set maximum length, it will empty the Tabu list and start the record again.

After completing the search, the scout bees will compare with the original sites to determine the new sites to maintain the same number of sites. The value of the fitness function is the basis for determining whether a new site is formed, and the needed number of sites will be selected in order from lowest to highest.

Algorithm Table 1: Tabu-based Bees Algorithm (TBBA)

Input: search and rescue center set R, task set T, scout bee population size N_s, number of sites N, maximum tabu list length L_{max}, swap fragment length l, the control parameter$Thre$

Output: search and rescue solution S

```
Initialisation algorithm parameters;
Generate the initial population randomly and select sites based on fitness values;
local_best, gobal_best ← inf;
For i = 1 → Gen do
    Generate scout bee population;
    Select the sequence fragment to be updated;
    While the sequence fragment is contained in Tabu do
        Regenerate the sequence fragment to be updated;
    End While
    Scout bees search;
    local_best ← Calculate the fitness function of the new population;
    Merge with the original sites to generate new sites;
    If local_best ≥ gobal_best then
        Tabu ←Add the search records to the tabu list;
        count ← count + 1;
    Else
        gobal_best ← local_best;
    End If
    Recruited bees search;
    If len(Tabu) > L_max then
        Tabu ← [];
    End If
    If count == Thre then
        Generate a new site to replace the worst site randomly;
    End If
End For
```

3.3 Recruited Bees Search

Recruited bee search is also an important search method of TBBA that explores the local solution space to find the optimal solution with better performance [17]. The recruited bee searches on the optimal site, which is done by swapping the positions of two tasks within an individual. If the generated new sequence performs better than the original optimal site, the optimal site is updated; otherwise, the results of this search will be discarded.

3.4 Individual Replacement Strategy

When TBBA often fails to find better UAV search routes, new sites need to be introduced to increase the diversity of the search. When multiple searches are not satisfactory, we use the individual replacement strategy to randomly generate a new site and replace the current worst site with the newly generated site.

4 Simulation Experiments

To verify the effectiveness of the TBBA solution, we conducted a series of simulation experiments on a Core (TM) i5-8265U CPU 1.60 GHz, Windows 10 OS computer with MATLAB R2020b as the coding environment.

We adopted the method of randomly generated scenarios for the experiments. To represent the test set concisely and clearly, we use the form of "T/R" to reflect the task size and the number of SAR centers, where T denotes the number of tasks and R denotes the number of SAR centers. In the scenarios of the simulation experiments, the number of tasks T is 100, 200, 300, and 400, and the number of search and rescue centers R is constant at 3. These values are reflected in the first column of Table 1. The larger the ratio of the number of tasks to the number of SAR centers is, the larger the scale of the problem. Accordingly, the difficulty of planning UAV search and rescue paths will be greater.

To verify the performance of TBBA in solving the problem, we use a genetic algorithm and neighbourhood search algorithm for comparison experiments. The

Table 1 Experimental results for different task sizes (measurement unit: unit time)

Instance	TBBA			GA			NS		
	Min	Ave	Std	Min	Ave	Std	Min	Ave	Std
100/3-1	**6.09**	**6.09**	**0.00**	9.91	10.22	0.18	10.17	10.58	0.26
100/3-2	**6.82**	**6.82**	**0.00**	10.04	10.25	0.13	10.42	10.66	0.17
100/3-3	**5.83**	**5.83**	**0.00**	10.08	10.25	0.13	10.34	10.65	0.15
200/3-1	**4.55**	**4.58**	**0.10**	10.08	10.28	0.11	10.46	10.68	0.19
200/3-2	**3.82**	**3.82**	**0.00**	10.11	10.29	0.13	9.85	10.64	0.31
200/3-3	**5.27**	**5.27**	**0.00**	9.89	10.25	0.16	10.08	10.57	0.27
300/3-1	**4.58**	**4.58**	**0.00**	10.01	10.30	0.18	10.36	10.74	0.24
300/3-2	**4.87**	**4.87**	**0.00**	10.04	10.25	0.14	10.21	10.66	0.22
300/3-3	**4.49**	**4.49**	**0.00**	10.08	10.39	0.22	10.04	10.70	0.32
400/3-1	**5.18**	**5.18**	**0.00**	10.16	10.33	0.16	10.42	10.84	0.29
400/3-2	**4.78**	**4.78**	**0.00**	10.01	10.30	0.23	10.34	10.73	0.30
400/3-3	**4.31**	**4.31**	**0.00**	9.58	10.26	0.25	10.61	10.94	0.22

genetic algorithm (GA) uses the method of population evolution to search the solution space [18]. The neighborhood search algorithm (NS) searches for a better search and rescue solution through the continuous improvement of the neighbourhood structure [19]. We use the algorithms to solve each scenario 10 times and record the minimum value (denoted as Min), the average value (denoted as Ave), and the standard deviation value (denoted as Std) of the results of each algorithm run as indicators to evaluate the search performance of each algorithm.

In TBBA, the maximum number of generations *Gen* is set to 500, the maximum Tabu list length L_{max} is set to 30, the threshold *Thre* is set to 20, and the swap fragment length l is set to 3. Other algorithm parameters are consistent with those in the literature. Both α_1 and α_2 in the problem are set to 0.5.

The experimental results of different task size scenarios are shown in Table 1. As seen from the table, the TBBA proposed by us significantly outperforms the comparison algorithms in all three evaluation indicators for different task scales. This reflects that the Tabu search strategy can help the algorithm continuously explore new solution spaces and find a better UAV flight path. Compared to TBBA, GA and NS are prone to poor search performance due to repeat searches arising from the large problem solution space. The results also clearly show that TBBA can effectively solve large-scale problems, which gives the algorithm strong potential for practical applications.

To see the algorithm programming performance more intuitively, the average value of each algorithm running result of different task size scenarios is given. As shown in Fig. 2, the average value of TBBA in each scenario is significantly lower than that of GA and NS, which indicates that the solution obtained by using TBBA can complete the search of the search area as fast as possible while considering the workload of each SAR center.

The above experiments can effectively demonstrate that the proposed TBBA has a strong ability to solve UAV path planning in the SARPP problem, and the stable performance of the algorithm in large-scale problem scenarios allows the algorithm to be used in task planning systems to provide decision support for maritime search and rescue.

5 Conclusion

We construct a mathematical programming model for the UAV search solution in the SARPP problem considering the search and rescue efficiency and the balanced task allocation among different SAR centers. After that, we propose a TBBA, in which a Tabu list is used to ensure that each search operation does not repeat the previous invalid search. The use of the Tabu search strategy allows TBBA to search quickly to find the ideal UAV search solution in complex problem scenarios.

In future research, we will consider more complex problems, such as different interference events, obstacles, weather factors, and communication restrictions in the actual search and rescue process of UAVs. The model constructed in this chapter

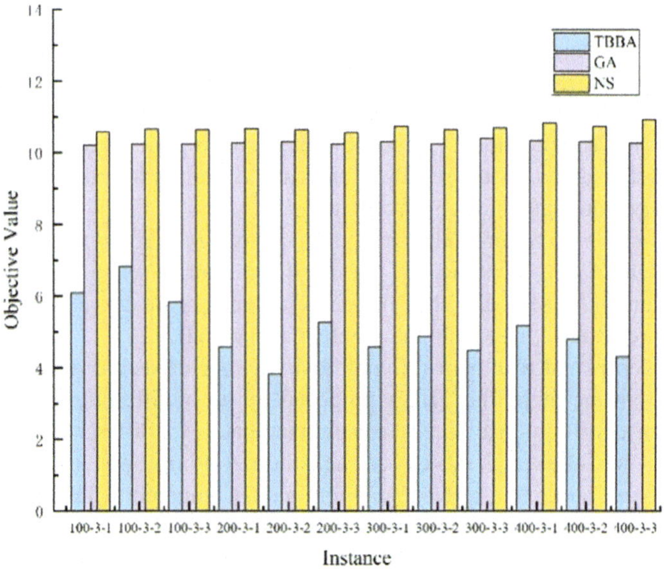

Fig. 2 Comparison of the average values of each algorithm

is somewhat simplified, and it will be further optimised in conjunction with actual maritime SAR operations. In addition, we will also make other improvements to the algorithm and consider other strategies to improve the search efficiency.

References

1. Gallego AJ, Pertusa A, Gil P, Fisher RB (2019) Detection of bodies in maritime rescue operations using unmanned aerial vehicles with multispectral cameras. J Field Robot 36(4):782–796
2. Gonçalves L, Damas B (2022) Automatic detection of rescue targets in maritime search and rescue missions using UAVs. In: 2022 International conference on unmanned aircraft systems (ICUAS). IEEE, pp 1638–1643
3. Cho SW, Park HJ, Lee H, Shim DH, Kim SY (2021) Coverage path planning for multiple unmanned aerial vehicles in maritime search and rescue operations. Comput Ind Eng 161:107612
4. Yang T, Jiang Z, Sun R, Cheng N, Feng H (2020) Maritime search and rescue based on group mobile computing for unmanned aerial vehicles and unmanned surface vehicles. IEEE Trans Industr Inf 16(12):7700–7708
5. Li L, Gu Q, Liu L (2020) Research on path planning algorithm for multi-UAV maritime targets search based on genetic algorithm. In: 2020 IEEE international conference on information technology, big data and artificial intelligence (ICIBA), vol 1. IEEE, pp 840–843
6. Zhang H, Sun J, Yang B, Shi Y, Li Z (2020) Optimal search and rescue route design using an improved ant colony optimization. Inf Technol Control 49(3):438–447
7. Gao H, Shi Y, Pun CM, Kwong S (2018) An improved artificial bee colony algorithm with its application. IEEE Trans Industr Inf 15(4):1853–1865

8. Koc E (2010) Bees algorithm: theory, improvements, and applications. Cardiff University, United Kingdom
9. Pham DT, Ghanbarzadeh A, Koç E, Otri S, Rahim S, Zaidi M (2006) The bees algorithm—a novel tool for complex optimisation problems. In: Intelligent production machines and systems, Elsevier Science Ltd., pp 454–459
10. Haj Darwish A, Joukhadar A, Kashkash M (2018) Using the Bees algorithm for wheeled mobile robot path planning in an indoor dynamic environment. Cogent Eng 5(1):1426539
11. Xu F, Li H, Pun CM, Hu H, Li Y, Song Y, Gao H (2020) A new global best guided artificial bee colony algorithm with application in robot path planning. Appl Soft Comput 88:106037
12. Gu Z, Zhu Y, Wang Y, Du X, Guizani M, Tian Z (2022) Applying artificial bee colony algorithm to the multi depot vehicle routing problem. Softw Pract Experience, 52(3):756–771
13. Imanguliyev A (2013) Enhancements for the Bees algorithm. Doctoral dissertation, Cardiff University
14. Gmira M, Gendreau M, Lodi A, Potvin JY (2021) Tabu search for the time-dependent vehicle routing problem with time windows on a road network. Eur J Oper Res 288(1):129–140
15. Prajapati, VK, Jain M, Chouhan L (2020) Tabu search algorithm (TSA): a comprehensive survey. In: 2020 3rd International conference on emerging technologies in computer engineering: machine learning and internet of things (ICETCE). IEEE, pp 1–8
16. Song Y, Xing L, Chen Y (2023) Application of the dual-population bees algorithm in a parallel machine scheduling problem with a time window. In: Intelligent production and manufacturing optimisation—the bees algorithm approach. Springer, Cham, pp 229–242
17. Wang H, Wang W, Xiao S, Cui Z, Xu M, Zhou X (2020) Improving artificial bee colony algorithm using a new neighborhood selection mechanism. Inf Sci 527:227–240
18. Mirjalili S (2019) Genetic algorithm. In: Evolutionary algorithms and neural networks. Springer, Cham, pp 43–55
19. Chakrabortty RK, Abbasi A, Ryan MJ (2020) Multi-mode resource-constrained project scheduling using modified variable neighborhood search heuristic. Int Trans Oper Res 27(1):138–167

Pedestrian-Aware Cyber-Physical Optimisation of Hybrid Propulsion Systems Using a Fuzzy Adaptive Cost Map and Bees Algorithm

Ji Li, Mingming Liu, Chongming Wang, Yingqi Gu, Quan Zhou, Chengqing Wen, D. T. Pham, and Hongming Xu

Abstract Electrified propulsion systems are a promising way of reducing traffic-related pollution. Because of the characteristics of the exhaust systems of engine-assisted vehicles, it is possible that pedestrians in close proximity to vehicles may encounter situations with high enough concentrations of emissions to cause specific health effects. To decrease the impact of vehicle emissions and pollutants on surrounding pedestrians, this chapter presents a cyber-physical optimisation technique for the pedestrian-aware supervisory control strategy of hybrid propulsion systems. The technique is a combination of the Bees Algorithm and a fuzzy adaptive cost map to optimise the rule-based power-split parameters. It is capable of self-adjusting the intertarget weights of exhaust emissions and fuel with real-time pedestrian density information during the optimisations. To examine the robustness of the hybrid propulsion system optimised by the introduced technique, the effects of communication quality and bootstrap sampling techniques are studied. The results show that applying the developed fuzzy adaptive cost map can reduce emissions to nearby pedestrians by 14.42%.

Keywords Bees algorithm · Cyber-physical optimisation · Fuzzy reasoning · Hybrid propulsion systems · Pedestrian awareness

---

J. Li (✉) · Q. Zhou · C. Wen · D. T. Pham · H. Xu
University of Birmingham, Birmingham, UK
e-mail: J.Li.1@bham.ac.uk

M. Liu · Y. Gu
The Dublin City University, Dublin, Ireland

C. Wang
Coventry University, Coventry, UK

D. T. Pham and N. Hartono (eds.), *Intelligent Engineering Optimisation with the Bees Algorithm*, Springer Series in Advanced Manufacturing,
https://doi.org/10.1007/978-3-031-64936-3_19

1 Introduction

At present, environmental pollution caused by transportation is emerging as a major social problem. Indeed, vehicle mobility has not only contributed to a variety of negative health consequences for city dwellers but has also become the primary contributor to global warming [1, 2]. As a result, new transportation solutions need to be developed to properly handle this dilemma for the sake of long-term social advantages. Plug-in hybrid propulsion systems (PHPSs) are increasingly considered a cost-effective mode of transportation for minimising the rising air quality challenges associated with classic internal combustion engines (ICEs) [3]. In comparison to a full battery electric vehicle, a PHPS typically has greater mileage and greater control over emission actuation [4] by allowing a driver to change to a fully electric mode when the surrounding concern is maximum.

Control systems of heuristic supervision have been widely applied in the industry thanks to their ease of development and high resilience. The rules are developed using human skill, intuition, or mathematical models and, in most cases, without having to know previous knowledge of the driving information in advance [5]. More specifically, charge-sustaining (CS) and charge-depleting (CD) modes are extensively employed in PHPSs [6], allowing for maximum power consumption while the battery is well charged and maintaining a safe level of charge when the battery level is low [7]. To reduce the workload associated with the development of controllers for energy management, Quan et al. investigated a transferable representation modelling routine involving two artificial intelligence technologies, i.e., deep neural networks [8] and Gaussian process regression [9], with an adaptive neuro-fuzzy inference system to facilitate the energy management controller's knowledge transfer. In addition, global optimisation methodologies have been applied to improve the parameters of heuristic control strategies to further increase their efficacy. For instance, Montazeri-Gh et al. [10] first investigated how a genetic algorithm (GA) can be applied to optimise parameters for an electric-assist control strategy and demonstrated that global optimum solutions can be found for predefined deriving cycles. Furthermore, particle swarm optimisation (PSO) is frequently employed in design and control optimisation issues due to its ease of implementation and cheap computing cost [11, 12]. However, Pham et al. demonstrated that when dealing with complicated nonlinear issues, such as PHPS energy management, PSO always performs poorly due to its simplified updating policy of randomness. Shouran et al. successfully used the Bees Algorithm (BA) for parameter optimisation of a control module for a power system and experimentally demonstrated that BA performs significantly better than PSO in this case [13]. Kamaruddin and Latif applied BA to find the optimal solution for the multiple disc clutch design, and experimental results show that BA performs better than adaptive particle swarm optimisation (APSO) since BA introduces both global exploration and local exploitation [14]

Knowing that the management of the energy of PHPSs is an ongoing optimisation issue, the cost function formulation is thus highly dependent on the driving circumstances [15]. Due to the isolation of environmental information in the past, a substantial number of studies have concentrated only on the interests of divers and hence regard subobjective weights in the cost function as fixed, as stated in [16, 17]. Considering the variety of genuine driving conditions, dynamic cost functions might more accurately represent the fitness of holistic objects, yet this remains an open question. Recent work by the authors established a management technique at the system level for a PHPS that responds to pedestrian density along varied routes [18]. This human-centric management approach could effectively help the people who are impacted the most by automobile pollution, namely, pedestrians. This contrasts with the common techniques of vehicle management, which are usually designed to benefit vehicle owners. However, as shown in [19], such a routing challenge makes it very difficult to fully exploit the benefits of hybrid cars alone by shifting traction modes.

The importance of adopting automated and connected vehicle (CAV) technologies for energy and emissions co-optimisation has been recognised [20]. Huang et al. [21] introduce a CAV predictive speed controller that is both emissions and energy conscious, with energy-saving performance based on NOx emission restrictions and the extent of data sharing. Shao et al. proposed a CAV control approach in [22] for optimising the gear position and vehicle speed concurrently in a platooning situation. Through experimental validation, a 10.6% gain in fuel economy over the immediately previous vehicle was realised. Vahidi et al. studied several connection strategies in [23] that contribute to the energy-saving potential of CAVs and the intelligent use of upcoming data. However, communication concerns such as quantisation errors, time delays, and packet loss pose substantial challenges for distributed control in vehicles [24] and even driving safety [25]. Until recently, few notable efforts on EMSs for CAVs that address communication difficulties have been reported.

This chapter proposes a cyber-physical optimisation technique for the pedestrian-aware supervisory control strategy of connected PHPSs to limit pedestrian exposure to nearby vehicle contaminants and emissions. Different from our recent study of engine management at the system level [18], this chapter focuses on the powertrain level, which allows explicit adjustment of the energy flow for less energy consumption and emissions. This is the first study to coevaluate the interests of internal drivers and external pedestrians. The proposed optimisation approach combines the Bees algorithm (BA) and a new fuzzy adaptive cost map to optimise the power split control parameters of the PHPS. This approach allows the self-regulation of the interjective weights of emissions and fuel on the basis of real-time pedestrian density data. The main contributions of our work are summarised as follows:

1. The designs of the supervisory control strategy and power-split hybrid powertrain are investigated, and the associated emission-based optimisation issue is presented.
2. Using the fuzzy satisfactory technique, we create a fuzzy adaptive cost map to capture the link between cost map weights and pedestrian density.
3. The rule-based power-split settings are optimised using the BA in conjunction with the created fuzzy adaptive cost map derived from pedestrian density data.

4. The flexibility of the cost map, the performance of cyber-physical optimisation, the robustness of the vehicle system, and the influence of communication quality are all assessed.

The remainder of the chapter is structured as follows: in section II, we formulate the scientific problem to be studied by modelling a PHPS. Section III describes the pedestrian-aware supervisory control strategy's mechanism and the suggested cyber-physical optimisation approach. Section IV details the collection of real-world driving cycles and the growth of the learning sample utilised in this work. Section V analyses the outcomes of 1) the cost map's flexibility and optimisation performance and 2) the vehicle system's resilience. Finally, section VI concludes our findings in the chapter.

2 Problem Statement

2.1 Configuration of Hybrid Propulsion System

A power-split PHPS propulsion consisting of an integrated starter-generator (ISG), a gasoline engine, two fuel and electric energy sources, and a transmission motor is supervised by the vehicle controller, which is displayed as follows (Fig. 1).

Here, a combination of the power of the internal combustion engine and the trans-motor is achieved by coupling the speed, where the speeds of the power units are

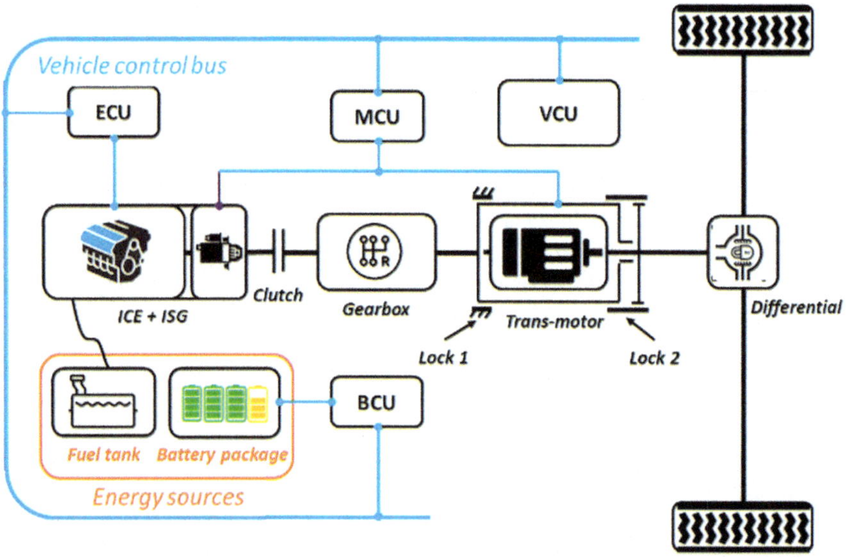

Fig. 1 Architecture of the power-split hybrid propulsion system

Table 1 Main parameters of the PHPS model

Symbol	Parameters	Values
M	Gross mass	1, 500 kg
A_f	Windward area	2 m^2
R_{whtr}	Tire rolling radius	0.3 m
C_{ad}	Air drag coefficient	0.3
i_{0d}	Differential ratio	3.75
i_{gr}	Gearbox ratio	3.55/1.96/1.30/0.89/0.71

decoupled [26]. The maximum power of the trans-motor P_{mot^*} is set to 75 kilowatts, and the maximum power is set to 270 Newton meters (N•m), while for the gasoline engine, these two values are set to $P_{ICE^*} = 63$ kilowatts and 140 N•m. The maximum power of the ISG P_{ISG^*} is set as 32 kilowatts. Medium passenger vehicle data are collected from our earlier work[27], and the suitability has been determined, in which a 2RC electrical model [28] is used to develop a battery pack that includes NCR-18650 series battery cells supplied by Panasonic Automotive & Industrial System Ltd. Table 1 demonstrates the symbol and values of different parameters in the special PHPS model.

A rearward model of the vehicle with longitudinal dynamics is applied. After a bevel gear, which is used to reduce speed, the needed speed n_{dem} and torque T_{dem} are:

$$\left.\begin{aligned} T_{dem} &= \left(\alpha ma + \frac{C_{ad}A_f u^2}{21.15} + mzsin\theta + mzfcos\theta\right) \cdot \frac{R_{whtr}}{i_{0d} \cdot \eta_{i0}} \\ n_{dem} &= \frac{P_{dem}}{T_{dem}} \end{aligned}\right\} \tag{1}$$

where $z = 9.81 m/s^2$ is the general gravitational constant; $\alpha = 1$ is the rolling friction coefficient; $\theta = 0$ represents the slope level; and P_{dem} represents the demand of the propulsion system. The speed of the internal combustion engine after transmission and the speed of the traction motor can both be determined by decoupling the speed, which is

$$n_m + n_e' = n_{dem} \tag{2}$$

in which

$$\begin{cases} n_e = n_e' \cdot i_{gr} \\ T_e = T_e \cdot i_{gr} \end{cases} \tag{3}$$

where n_e and n_e' are the rotation speeds of the internal combustion engine before and after transmission, respectively, and T_e is the needed torque of the internal combustion engine before transmission.

A conventional automatic gearbox with a five-speed transmission model is used to link the rear part of the trans-motor, and it offers the internal combustion engine a suitable gear ratio for efficient operation. To keep from the speed reduction that occurs

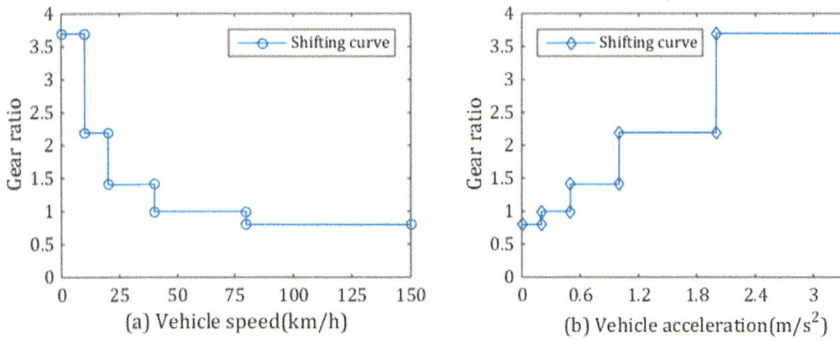

Fig. 2 Maps of shifting logic introduced in the study correspond to **a** vehicle speed and **b** vehicle acceleration

when upshifting, the shifting strategy offers additional limits on vehicle acceleration and speed. When more gear ratio options are available, ensure that the higher gear ratio is always chosen first. Thus, the gear ratio is defined as:

$$i_{gr} = \max(s_{al}(a_{ts}), s_{vl}(v_{ts})) \tag{4}$$

where a_{ts} and v_{ts} are the vehicle acceleration and speed, respectively; s_{al} and s_{vl} are look-up tables introduced to obtain the matching gear ratios; and i_{gr} is the result of the gear ratio (Fig. 2). The shifting curves of the vehicle speed and acceleration maps are displayed as:

2.2 *Rule-Based Control Strategy*

Considering the hybrid propulsion configuration of the vehicle, a classical power-split state machine, as shown in Fig. 3, is used [29] to mean the usage of electricity and fuel according to the states of request demand speed n_{dem}, torque T_{dem}, and status of the battery pack of charge SoC.

In this study, a vector of power-split $\boldsymbol{\Psi}$ is formulated as Eq. (5), with the transmotor speed, n_m; the transmotor torque, T_m; the internal combustion engine speed, n_e; the internal combustion engine torque, T_{ices}; and the ISG power, P_g.

$$\boldsymbol{\Psi} = \left[\, T_m \; n_m \;\; T_e \;\; n_e \; P_g \,\right] \tag{5}$$

In EV mode ($P_d \geq 0 \wedge SoC \leq 0.5$), adequate electricity can be offered to the power needed of the PHPS. Therefore, the ISG and the internal combustion engine can be zero here. The power distribution in EV mode is defined as:

$$\boldsymbol{\Psi}_{EV} = \left[\, T_{dem} \; n_{dem} \;\; 0 \; 0 \; 0 \,\right] \tag{6}$$

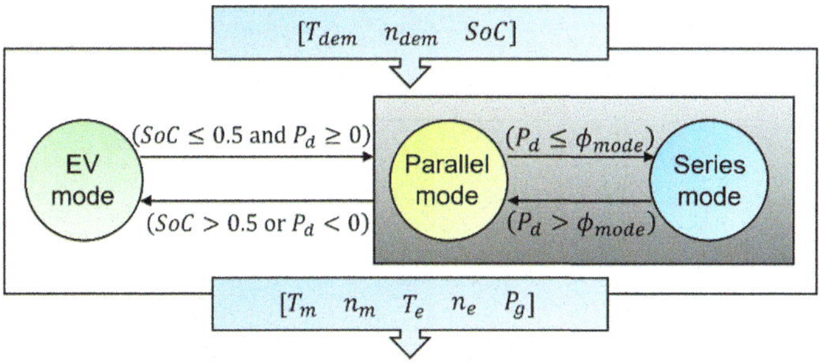

Fig. 3 State machine for mode transition

When under the parallel and series modes ($P_d \langle 0 \vee SoC \rangle 0.5$), switching between parallel and series modes is controlled by the needed power P_d. Additionally, a control index ϕ_{mode} is needed. If $P_d > \phi_{mode}$, the working state will be parallel mode, or it will be series mode. The outputs of the different modes are:

$$\left.\begin{aligned} \boldsymbol{\Psi}_{series} &= \left[\, T_{dem}\ n_{dem} \cdot (1-\chi_1)\ \ T_{dem}\ n_{dem} \cdot \chi_1\ 0 \,\right] \\ \boldsymbol{\Psi}_{parallel} &= \left[\, T_{dem}\ n_{dem}\ \ T_e'(P_g)\ n_e'(P_g)\ P_g^* \cdot \chi_2 \,\right] \end{aligned}\right\} \tag{7}$$

$n_e\prime$ and $T_e\prime$ are the best speed and torque of the internal combustion engine converted based on P_{gen}, the needed power of the ISG; P_g^* is the largest power of the ISG; χ_j (j=1 or 2) is a proportionality factor, which is calculated by SoC as follows: χ_1 and χ_2 are, respectively, for different internal combustion engine control and ISG control [30].

$$\chi_j(SoC) = \begin{pmatrix} 1, & SoC \in (0, 0.2] \\ \left\{1 + \exp\left[\left(\frac{SoC}{SoC^*} + \phi_{i,\beta}\right)\phi_{i,\alpha}\right]\right\}^{-1}, & SoC \in (0.2, 0.5] \\ 0, & SoC \in (0.5, 1] \end{pmatrix} \tag{8}$$

where SoC^* represents a scaling factor of the BP's SoC, and for the control parameters of the internal combustion engine and ISG, there are a total of four control parameters: $\phi_{i,\alpha}(i = 1, 2) \in [0.01, 50]$ and $\phi_{i,\beta}(i = 1, 2) \in [-6, 6]$.

3 Proposed Solution

To decrease pedestrian access to emissions fuel consumption of vehicles and emissions as well as pollutants in the vicinity, this study proposes a cyber-physical approach to develop a pedestrian-aware cyber-physical optimisation, as shown in Fig. 4. This technique for optimisation integrates the Bees algorithm and a new fuzzy adaptive cost map to search for the best solution of the parameters for power

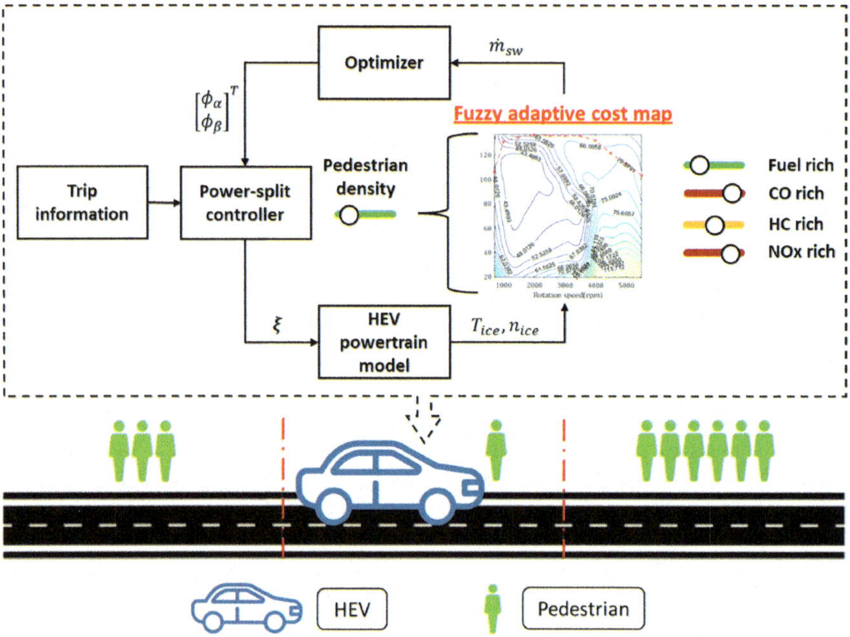

Fig. 4 Diagram of pedestrian-aware cyber-physical optimisation

split control. The proposed optimisation consists of 1). Normalisation and aggregation of cost maps for nitrogen oxides (NOx), fuel, hydrocarbons (HC), and carbon monoxide (CO); 2). Inferring the connection between traveler density and cost map weights by fuzzy satisfaction approaches; 3). Optimisation of power distribution control system parameters by combining BA and fuzzy adaptive cost diagrams.

3.1 Cost Map Standardisation

Initially, the exhaust emission and fuel cost maps for internal combustion engines are standardised and aggregated before self-adapting to pedestrian density. A classic approach in earlier studies [31, 32] was proposed to deduce separate cost maps for the emission components and fuel and perform a multiobjective optimisation. The processing integrates emission components and multiple fuels into a single objective to reduce the calculation burden of several table lookup actions.

Figure 5 shows a quasisteady-state cost map for a specific engine, showing fuel and emissions in units of grams per kWh. In addition, the location of the highest fuel efficiency is not necessarily related to the location of optimal emissions. To simultaneously decrease emissions and fuel consumption, a clear balance of weights between different emissions and fuel efficiency is needed. Considering the various scales of the emission and fuel consumption maps, it is necessary to normalise them, and the equation is defined as

$$\dot{m}_{s,i}(n_e, T_e) = \frac{\dot{m}_i(n_e, T_e) - \min[\dot{m}_i(n_e, T_e)]}{\max[\dot{m}_i(n_e, T_e)] - \min[\dot{m}_i(n_e, T_e)]} \tag{9}$$

where $m_i(n_e, T_e)$ is the initial cost map and $\dot{m}_{s,i}(n_e, T_e)$ is the normalised cost map (with values between 0 and 1). Then, the weighting process is carried out by:

$$\dot{m}_{sw}(n_e, T_e) = \frac{\sum_{i=0}^{i=l} k_i \dot{m}_{s,i}}{\sum_{i=0}^{i=l} k_i} \tag{10}$$

where the k_i factor depends on the components needed to minimise and the number of demands and the value of i is 1, 2, 3, 4 corresponding to HC, fuel, NOx and CO. The value of these factors is usually a fixed set that is determined by expertise from humans. Nevertheless, pedestrians in the vicinity of a PHPS optimised through the

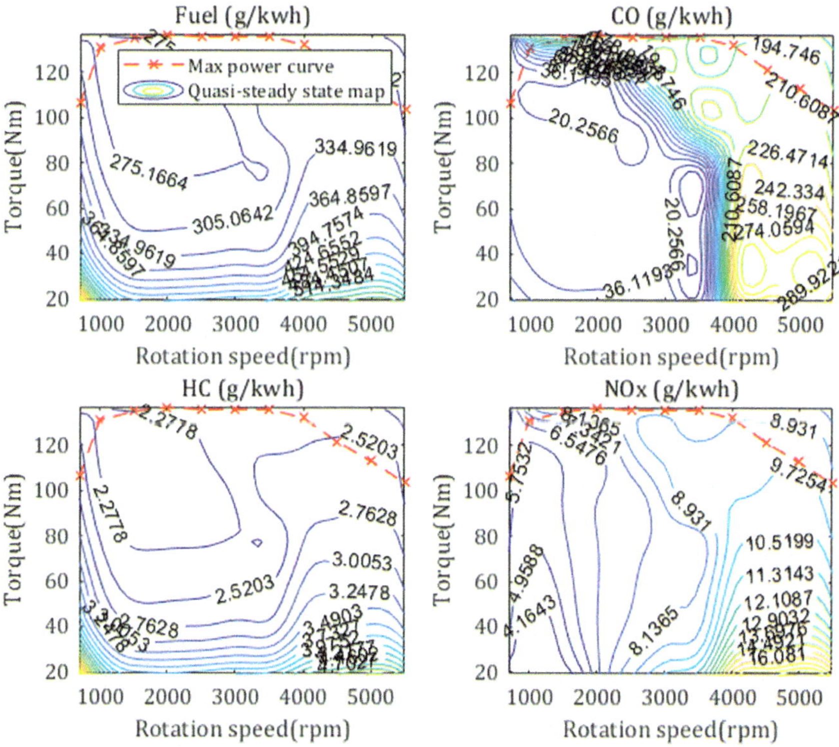

Fig. 5 Quasisteady state cost maps of the internal combustion engine containing fuel, HC, CO, and NOx [33]

use of cost maps could result in situations where the HC/NOx/CO quantity is large enough to lead to a health problem.

Through the introduction of a fuzzy adaptive cost map, the optimal combination of objectives is

$$J = \int_{0}^{t_f} \dot{m}_{sw} dt \tag{11}$$

where t_f is the last timepoint of the driving process cycle and $\dot{m}_{sw}$ represents the flow rate of mass that resulted from the fuzzy adaptive cost map. Thus, the constraints that can be applied in emission-oriented optimal targets are listed as follows.

$$s.t. \begin{cases} SoC_t, \; SoC_t \in \left[SoC^-, SoC^+\right] \\ n_{m,t}, \; n_{m,t} \in [0, n_{m^*}] \\ T_{m,t}, \; T_{m,t} \in [-T_{m^*}, T_{m^*}] \\ P_{e,t}, \; P_{e,t} \in [0, P_{ICE^*}] \\ P_{g,t}, \; P_{g,t} \in [0, -P_{ISG^*}] \end{cases} \tag{12}$$

where SoC^- and SoC^+ are the bottom and top of the value of the battery SoC, respectively.

3.2 Fuzzy Satisfaction for Pedestrian Adaptation

The connection between the density of pedestrians and the cost map weights is reasonably established to give the cost map the ability to fit the density of pedestrians. In consideration of the imprecise nature of the judgment from divers, it is easy to believe that humans cannot consider every objective clearly and precisely when making decisions. A fuzzy satisfaction approach is formulated to explain the connection between 1 input: density of pedestrians and 4 outputs as a result: cost map weights for HC, fuel, NOx, and CO.

The rule base is used to identify the control outputs M, N, Z, and Y with the input states I and A by using a 'if I & A then M & N & Z & Y' policy. A mathematical equation of the 'if I then M & N & Z & Y' policy is

$$[M, N, Z, Y] = (I \times A) \circ X. \tag{13}$$

where 'M', 'N', 'Z', and 'Y' represent the fuzzy sets of cost map weights related to HC, fuel, NOx, and CO, respectively. 'I' represents the reference value of scaled pedestrian density. 'A' represents the crisp of the reference of scaled request power. The values of both 'I' and 'A' are in [0, 1]. 'X' represents the fuzzy relationship matrix consisting of the cross product of one input fuzzy set.

The fuzzy rule was originally designed to reduce the exposure of pedestrians in close proximity of car emissions first and then to protect the interests of vehicle owners. Among the input variables, the density of pedestrians is the primary factor in measuring the weight of the emission component (for example, pedestrian benefit) and fuel consumption (for example, owner benefit). The power requirement of a PHPS is a secondary parameter for adjusting the weight distribution, particularly in the case of very low or very high power demand. The linguistic set of pedestrian densities is assigned to simplify the formulation of the fuzzy logic inference: 1 means 'Very Small; 2 means 'Small; 3 means 'Medium'; 4 means 'Large'; and 5 means 'Very Large'. To linguistic sets of demand power: 1 means 'Very large', 2 means 'Large', 3 means 'Medium', 4 means 'Small', and 5 means 'Very Small'. Thus, the process that is based on Eq. (13) with the fuzzy set can be defined as

$$\left.\begin{array}{l} ifI + A \in [2,3] \\ ifI + A \in [4,5] \\ ifI + A \in [6] \\ ifI + A \in [7,8] \\ ifI + A \in [9,10] \end{array}\right\} thenMis \left\{\begin{array}{c} VeryLarge \\ Large \\ Medium \\ Small \\ VerySmall \end{array}\right\} andN, Z, Yare \left\{\begin{array}{l} VerySmall \\ Small \\ Medium \\ Large \\ VeryLarge \end{array}\right. \tag{14}$$

Within this reasoning mechanism, implicit fuzzy sets are generated by applying max–min combinations. Hidden fuzzy sets are integrated in defuzzification to offer a clear value for the controller output. Some approaches [34] have been proposed to achieve the defuzzification process, among which the area centroid method is chosen as the basis. The final output is the mean value of the individual centroids weighted by the value of the membership, which is defined as

$$Out = \frac{\sum_{i=1}^{n} \pi_i \cdot \varphi_i}{\sum_{i=1}^{n} \varphi_i}, \tag{15}$$

where π_i is the result of rule base i and φ_i represents the centroids of the equation. Here, the weighted summation refers to the result of the adaptation of every index to the density of pedestrians. The functions are treated as a triangular membership equation, which is defined as

$$q_i = \max\left(\min\left(\frac{x - \left(1/4 i - 1/2\right)}{1/4}, \frac{1/4 i - x}{1/4}\right), 0\right), \quad i = 1, 2, 3, 4, 5 \tag{16}$$

3.3 Bees Algorithm for Cyber-Physical Optimisation

Because of the increase in dynamic optimisation goals and external information about pedestrians, the improvement of pedestrian-aware supervision systems will be more challenging. The BA proposed by Pham et al. observes the behavior of bees to obtain the optimal solution [35]. By comparison with classic PSO and GA, the BA is more appropriate for the precise search of holes and narrow valleys, as well as for highly multiple modal equations [36]. The flowchart of BA-driven cyber-physical optimisation with a designed fitness assessment is illustrated as follows (Fig. 6).

Every particle in the search area in the BA is considered a target. 'Scout bees' gain spatial information at random. Through the formulation equation, 'Scout bees' returns the value of accessed locations. Solutions collected are ranked. At the same time, other 'bees' are going to find the landscape about the largest-ranking locations.

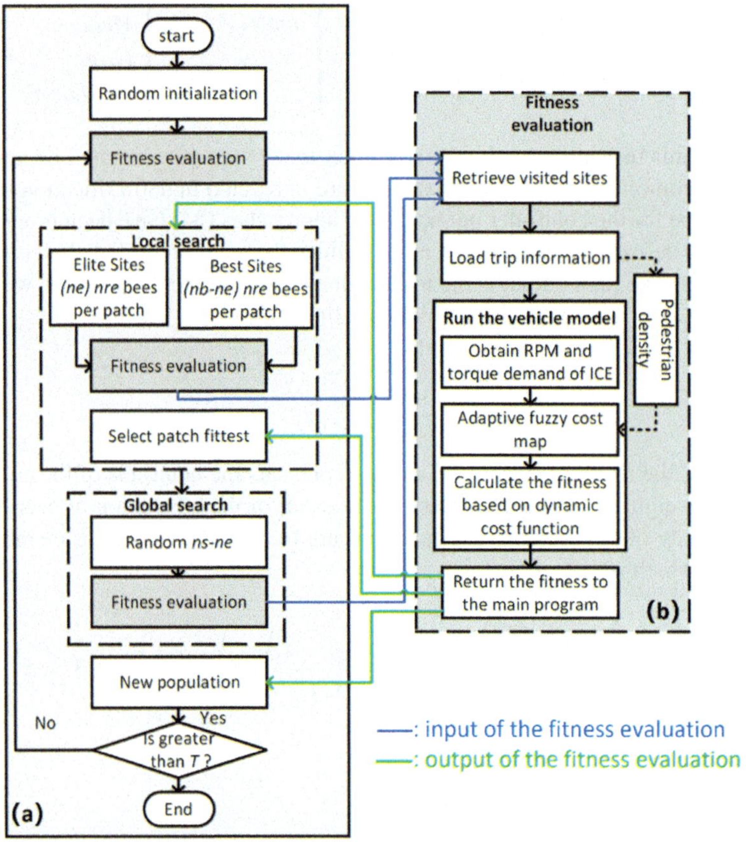

Fig. 6 Cyber-physical optimisation flowchart based on BA including **a** main loop of optimisation and **b** subloop of fitness evaluation

Here, 'flower patch' is defined as the area around that optimal solution. The BA targets the optimal solution and explores its surrounding area to find the global optimal value of the objective function [36]. In this case, the fitness can be calculated based on the pedestrian density and the vehicle model.

First, the visited sites are searched for aptitude assessment. When the PHPS model is operating, the torque and rpm needed of the internal combustion engine are obtained. At the same time, information about the environment with respect to pedestrian density is loaded into the designed fuzzy adaptive cost map to adjust the intertarget weights of emissions and fuel in real time. Then, the fitness can be obtained and sent back to the main program. Based on the calculated fitness, elite bees are used for searching the local optimum, and scout bees are used for the global optimum.

4 The Results and Discussion

The studied real-world driving cycle was collected from the authors' previous work [18], where the route is made up of sidewalks that allow local residents and students to access different facilities through the campus from east to west. The speed limit is 35 km/h for the entire route, with a total length of 1.1 km. The cloud server (free mobile station from OpenCellID) will send specific pedestrian information to the smartphone (model GT-I9500 Samsung Galaxy S IV) when receiving the request. The smartphone allows the data transmitted to be recorded by the application developed in our recent work [18]. The pedestrian density is estimated based on the number of pedestrians walking on a given (indexed) road segment with a fixed length of 100 m. GPS signals are used to determine whether a pedestrian is on the road segment upon receiving the request.

In the real world, the information about routes is basically stored in the vehicle database and can be extracted for analysis when needed. To examine the improvement of vehicle system robustness by the cyber-physical optimisation method, extra trip data are necessary as learning trials to assess the behavior. The existing samples obtained are limited and cannot represent the whole distribution of various traffic scenarios. Therefore, only using these samples for cyber-physical optimisation may reduce the system robustness. Bootstrap (BS) [37], a commonly used resampling technique, is designed to create multiple simulated samples of the same size as the original dataset by randomly sampling the existing data several times. As the sample size increases, the new dataset obtained by the BS converges to the correct sampling distribution. Thus, BS does not need to consider the distribution of the original data to estimate the equations of the sampling distribution [38]. BS is also relatively simple and more accurate than the asymptotic approximation of the standard interval obtained using the sample variance and normality assumptions [39]. Therefore, BS is selected in this study to evaluate the confidence interval of the involved parameters by selecting a random sample with replacement from the sample data.

Figure 7 illustrates the experimental driving profiles applied in the investigation, which include (a) information on real-world trips gathered by using the information on real-time pedestrian and vehicle speed; (b) information on real-time pedestrians resampled; and (c) values of the mean and standard deviation of the 233 bootstrap samples. As can be observed in Figs. 8b and c, the random sample has a wider distribution of bootstrap means than the distribution of bootstrap standard deviations, which is more similar to the scope of the distribution of the real data, as shown in Fig. 8a, [0, 7]. The pedestrian density dataset, resampled via the bootstrap methods described above, is then combined with the Worldwide harmonised Light vehicles Test Cycle. These data are applied as a sample of extra trips to evaluate the robustness of the vehicle system.

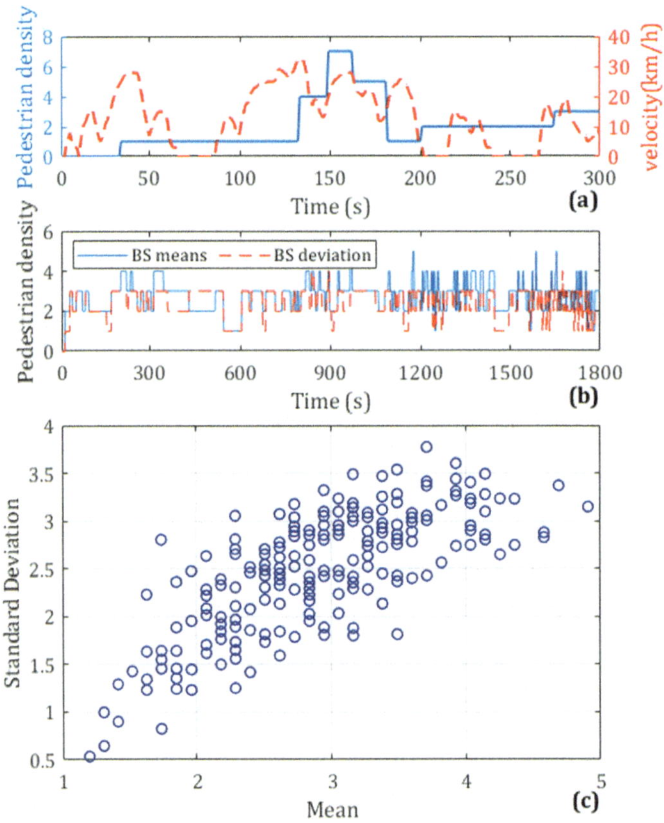

Fig. 7 Overview of test driving, including **a** information collected on real-world trips; **b** information on resampled real-time pedestrians; and **c** a mean and standard deviation pair of 233 estimated bootstraps

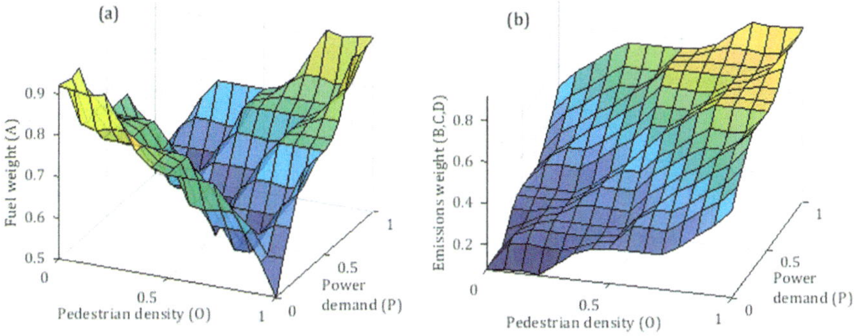

Fig. 8 Fuzzy response surfaces of **a** fuel weight and **b** emissions weight

4.1 *Cost Map Adaptability*

Here, we analyse the efficiency of the designed fuzzy adaptive cost map and compare the optimisation performance of 3 different algorithms. Figure 8 shows the response surface with the fuzzy satisfaction method. Significantly, the weight of fuel will approach the maximum when the density of pedestrians and power demand are both close to the minimum 0 or close to the maximum 1. This design is intended to best improve the experience of vehicle users in situations where there is low pedestrian density or high-power demand. For the weight of emissions, both the density of pedestrians and power demand are positively correlated.

Figure 9 shows the fuzzy adaptive cost maps for different densities of pedestrians, $I = 0, 0.5, 1$. With increasing pedestrian density, the high-efficiency region of special consumption moves from the region of $T_e \in [80, 120]$, $n_e \in [1000, 2500]$ to the region of $T_e \in [40, 120]$, $n_e \in [1000, 2000]$. Combined with Fig. 5, since the high-efficiency region of NOx and CO cost maps fills the whole low-speed region, the peak region shifts to this area when the density of pedestrians rises. However, the efficiency is always low in regions with low torque ($<$ 40 Nm) and high speed ($>$ 4000 rpm) demand. In combination with the two charts in Fig. 9, the transition process from weights of fuel to emission weights gradually stabilises after a quick increase.

4.2 *Cyber-Physical Optimisation Performance*

The GA and PSO, two classical optimisation algorithms, are introduced for comparison to evaluate the optimisation performance of the BA. The parameters of the BA and PSO used here are carefully defined based on the research of Pham et al. [40]. For GA, population sizes are commonly set to 20, 50, 100, and 200 according to the research of Hassanat et al. [41]. Here, for a fair comparison under the same scale as

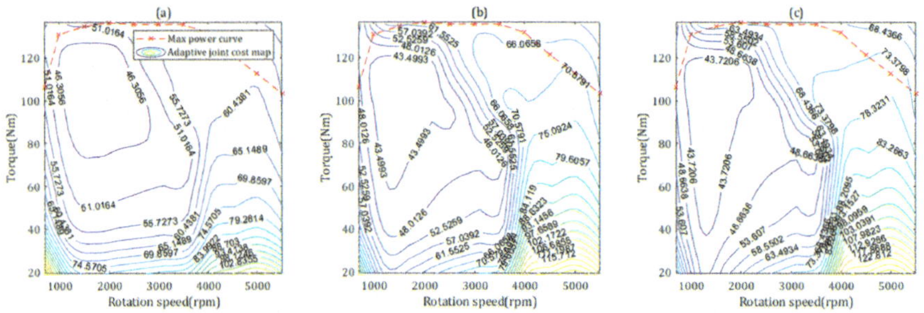

Fig. 9 Comparison of the fuzzy adaptive cost map over different pedestrian densities: **a** $I = 0$; **b** $I = 0.5$; and **c** $I = 1$

the BA and PSO, 50 is selected as the population size. Details of the three algorithms used are shown in the Appendix.

The initial battery SoC is assumed to be 0.5, and each case is run 30 times repeatedly on the real driving collection cycles for 12 rounds. To make a fair comparison, set the three optimisation algorithms to be compared based on the condition of setting the same size of population, i.e., $N = 20, 50, 100$. The maximum number of iterations T is defined as 100 iterations.

Figure 10 shows the mean convergence capability of the cyber-physical optimisation process for PSO, GA, and BA when the population size is set as 20 and 100. In terms of convergence speed, PSO can quickly reach the stable phase (approximately 10 iterations) in cyber-physical optimisation compared to GA (approximately 20 iterations) and BA (approximately 15 iterations). However, PSO has the poorest ability in searching for the global optimal solution, which is probably due to the simplified particle guidance mechanism. Different from the BA, the GA, with faster convergence in cyber-physical optimisation, is more active in searching for the global optimal solution. In addition, the reduction in population size causes a significant decrease in PSO optimisation performance. In summary, in comparison with the other two methods, BA has superior optimisation performance in the cyber-physical optimisation of pedestrian-aware cyber-physical systems.

Table 2 contains a summary of the performance comparison of PSO, GA, and BA. In this case, the situation when the fitness value of a single operation is not larger than 1% of the optimal fitness value achieved by BA is defined as an index'success'. The success rate was determined by applying different sizes of population and different level iterations, and BA showed a completely dramatic advantage over the other two algorithms in cyber-physical optimisation (198/270). When the maximum number of iterations is set to $q = 20$ iterations, the BA has almost twice the success rate of the GA, yet PSO does not succeed in reaching the global optimum. For this target, the PSO has the ability to quickly reach a region (number of iterations to reach 90% success rate:≤ 12) where potential global optimal solutions occur, yet it is extremely difficult to find a global optimal solution. GA and BA work on a more gradual strategy (number of iterations to 90% success rate:≥ 14)). This strategy initially assigns populations to the solution space equally and then makes them gradually approach

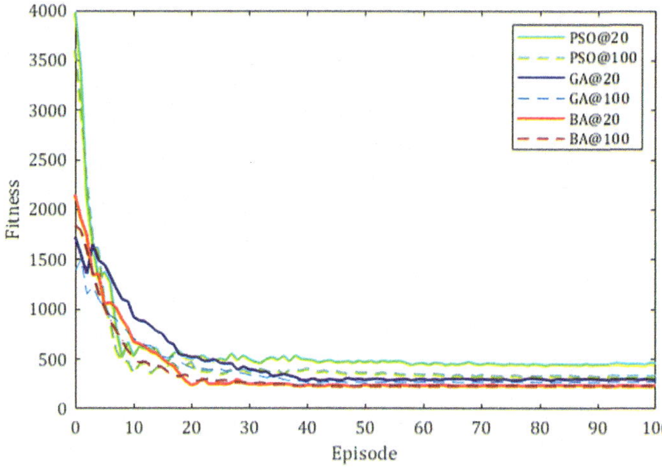

Fig. 10 Convergence process of PSO, GA, and BA during the cyber-physical optimisation

the global optimum. Since the convergence process of GA algorithms is difficult to reverse, the populations are easily trapped in suboptimal performance peaks. In addition, exploration and exploitation in GA and PSO are correlated; thus, resources need to be balanced carefully. Therefore, the performance of PSO is more sensitive to the choice of the optimal set of learning parameters [36]. Through cyber-physical optimisation, the optimal solution is shown as follows (Table 3).

Table 2 Performance comparison of three Optimisation Algorithms

Studied algorithm	Population	Iterations to 90% success	Success rate via q iterations		
			q = 20	q = 50	q = 100
	20	19	9/30	15/30	16/30
GA	50	18	11/30	15/30	21/30
	100	14	12/30	19/30	22/30
	20	12	0/30	1/30	0/30
PSO	50	10	0/30	0/30	5/30
	100	10	0/30	2/30	8/30
	20	16	14/30	20/30	24/30
BA	50	16	18/30	21/30	30/30
	100	15	20/30	21/30	30/30

Table 3 Optimised result of the control parameters

Parameter	$\phi_{ice,\alpha}$	$\phi_{ice,\beta}$	$\phi_{gen,\alpha}$	$\phi_{gen,\alpha}$
Value	0.016	0.277	0.360	0.032

For a sample concept to be implemented in a real-world vehicle application, section B explains how a laboratory cycle (i.e., WLTC) can be used to optimise a pedestrian-aware cyber-physical optimisation for real-world driving. Although the above results are obtained from the investigated theoretical model, the optimisation algorithm proposed in this chapter has the ability to be applied to other situations, including the control systems of real vehicles.

4.3 Robustness and Economy of Vehicle System

This part investigates the robustness and economics of a pedestrian-aware cyber-physical supervised control system with the designed cyber-physical optimisation technique. "Equilibrium" [42], "fuel only", and "CO priority" [43] are taken as control groups, as shown in Table 4.

To calculate the impact of vehicle emissions on nearby pedestrians, the relative emissions mass is determined to assess three mapped treatments and one untreated behavior, and the equation is defined as

$$\begin{cases} m'_{i,t} = (1 - r_i) \int_0^t \dot{m}'_{i,t} dt \\ \dot{m}'_{i,t} = \dot{r}_{i,t} \cdot I \end{cases} \tag{17}$$

where r_i is the efficiency of catalyst removal for HC, NOx, and CO; $\dot{m}_{i,t'}$ and $\dot{m}_{i,t}$ are the absolute and relative mass rates of pollutant i, respectively; and I is in [0, 1], which is a real-time signal from the scaled density of pedestrians. Generally, the catalyst removal efficiency $r_i(\lambda, T_{cat})$ is an equation of the air–fuel ratio λ and catalyst temperature T_{cat}. High-fidelity fuel economy and emission modelling are not considered as the focus of this chapter, as the main consideration of this chapter is the possible and feasibility benefits of the designed optimisation method. The catalyst removal efficiency ratio in this case is defined under an ideal constant scenario with $\lambda = 1$, $T_{cat} = 400°C$. In this case, the efficiency of catalyst removal for HC, NOx, and CO is 88%, 98%, and 94%, respectively, which is applied from our previous work [44].

Table 4 Inter-Objective Weight Comparison of Studied Mapping Treatments

Mapping treatment	Weight distribution [HC Fuel NOx CO]
Fuel only	[0100]
Equalisation	[0.250.250.250.25]
CO Priority	[0.110.110.110.67]
FACM*	Self-adaptive

Note 'FACM' indicates the fuzzy adaptive cost map

Table 5 lists the comparative performance of the vehicle system applying the studied mapping treatment for 12 rounds of actual driving cycles collected when the initial battery SoC is set to 0.5. Compared to the other three methods, the mapping treatment that only considers fuel consumption has the smallest fuel consumption but the largest relative mass of all emissions. This means that pedestrians on the test route are exposed to high levels of HC, NOx, and CO pollution generated by the studied PHPSs. The system of vehicles based on the designed fuzzy adaptive cost map has the second rank of fuel used, yet the largest SoC value. Furthermore, in terms of avoiding pedestrian exposure to near-vehicle emissions, this vehicle system has the ability to decrease the total emissions by 14.42% compared to the CO priority (12.16%) and equalisation (6.13%) systems. This is due to the real-time self-adjustment of the intertarget weights of exhaust emissions and fuel according to the real-time signal from the density of pedestrians during the process of optimisation.

Figure 11 illustrates a comparison of the real-time performance of the different resampling techniques, including bootstrap averaging and standard deviation. To show the result better, an entire cycle [0,300] is chosen as representative for analysis. The results show that the weight of vehicle exhaust emissions increases with higher pedestrian density, while the weight of fuel consumption decreases in regions with higher pedestrian density. The optimised PHPS system using the standard deviation technique is more adaptable to the test cycle, saving more fuel and reducing more exhaust emissions than using the bootstrap averaging technique. Considering Fig. 5, the bootstrap bias methodology applies a conservative policy to distribute the random samples to a higher probability range. The bootstrap averaging methodology will recover the distribution of the original sample as much as possible; on the other hand, several marginal samples will appear, and they cannot be aligned with the timing of the rest of the information in the source sample.

In Fig. 12, further comparisons are made between the overall performance of the vehicle systems optimised by bootstrap averaging and standard deviation. The results are consistent with the expected results; the vehicle system optimised by applying the bootstrap averaging methodology performs poorly in terms of HC, fuel, NOx, and CO. Although a higher SoC value (0.246) can be obtained by using this technique, it does not effectively convert the surplus power into fuel and emission reductions. The bootstrap deviation approach achieves higher fuel economy (0.5%) and lower relative emissions (> 1.1%) at this point compared to the bootstrap average approach. Generally, both resampling techniques studied are alternatives

Table 5 Vehicle system performance comparison of using studied mapping treatments

Mapping treatment	Fuel (g)	SoC	Rel CO mass (g)	Rel HC mass (g)	Rel NOx mass (g)	Total emissions (g)	
						Rel. mass	Decline %
Fuel only	880.4	0.232	0.562	0.400	0.099	1.061	-
Equalisation	884.4	0.221	0.584	0.328	0.086	0.996	6.13%
CO Priority	883.6	0.218	0.496	0.365	0.072	0.932	12.16%
FACM	881.2	0.236	0.529	0.310	0.069	0.908	14.42%

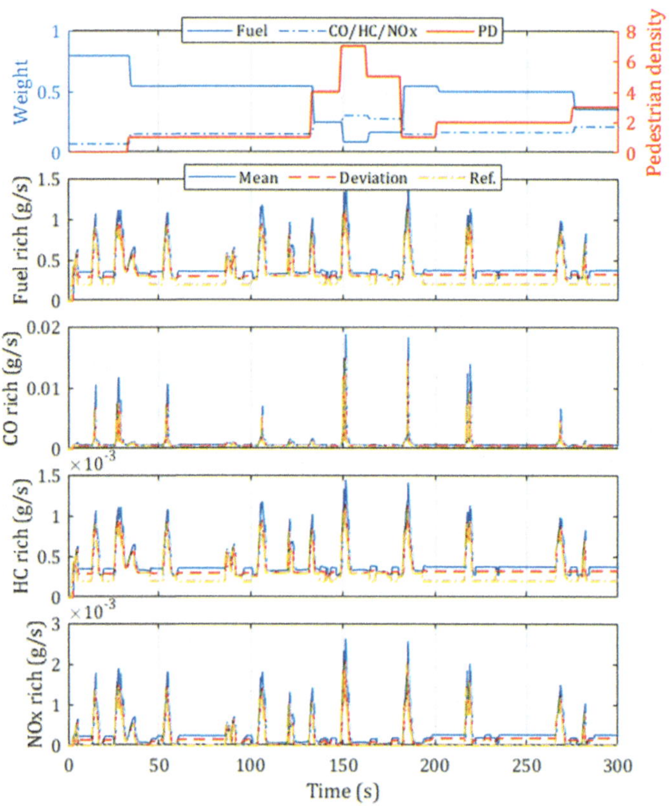

Fig. 11 Using bootstrap averaging and standard deviation for real-time performance comparison

for regenerating information related to the driving cycle, where the deterioration $\leq 5.80\%$. The deterioration can be further reduced to 2.17% using the bootstrap deviation methodology compared to using the bootstrap average methodology.

4.4 *Impact of Communication Quality*

The impact of communication quality is a critical factor influencing the success ratio of distributed systems. On the other hand, signal delays and losses also occur frequently. For CAVs, the quality of the overall transmitted and received signals from the vehicle terminals must be carefully considered [45]. 0–2000 ms is the conventional latency range for dedicated short-range communication and 4G-LTE for connected vehicles [46]. However, the update rate of the density of pedestrian information is low, and it has been investigated that there is a slight reduction ($< 0.1\%$) in fuel-saving performance even with a communication delay of 2000 ms.

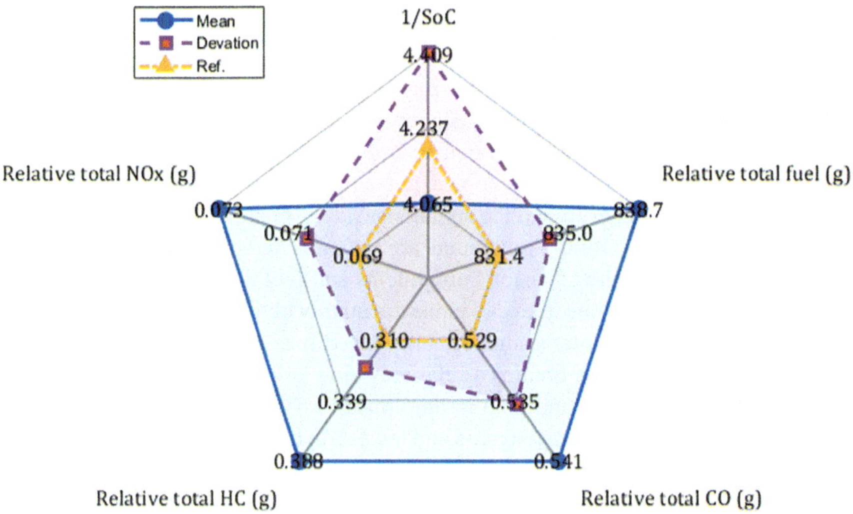

Fig. 12 The overall performance comparison improved by applying two BS techniques

With that in mind, this part investigates the influence of signal loss during pedestrian-aware cyber-physical optimisation (Fig. 13).

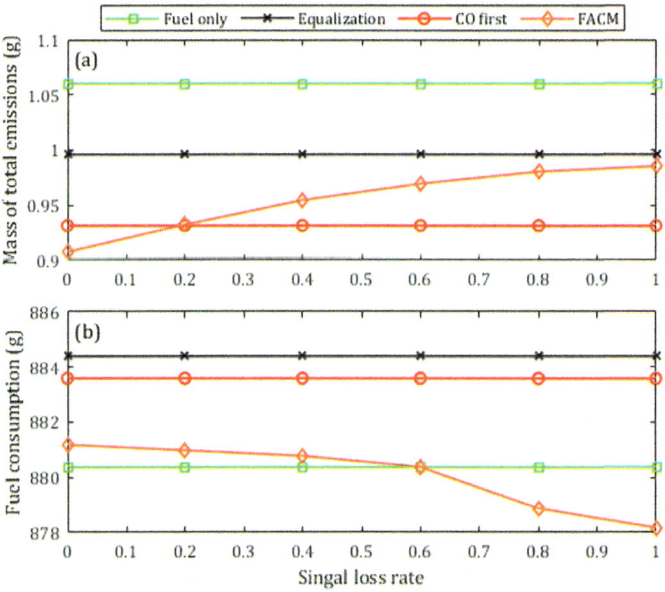

Fig. 13 System robustness performance under different signal loss rates

To simulate the signal loss in V2X networks between roadside devices and vehicles, 6 levels of signal loss rates are considered and applied to evaluate the robustness of the proposed pedestrian-aware cyber-physical optimisation. The design of a triggered square signal to multiply the signal, where every triggered interval is applied to activate the signal of pedestrian density for a fixed time of 30 s. In contrast, a nontriggered section is introduced to deactivate the signal of pedestrian density, the duration of which is according to the signal loss rate. Apparently, the "equalisation", "fuel only", and "CO priority" treatments are not impacted by signal loss since they have fixed weights for the emission components and fuel in real-time driving. With an increasing signal loss rate, the total emission quality obtained by pedestrian-aware cyber-physical optimisation through the introduced fuzzy adaptive cost map gradually increases. This system obtains a relatively slow drop in fuel consumption first to a signal loss rate of 0.6 and then a rapid drop to 878.2 g. The proposed system (with relatively low total emission mass and lowest fuel consumption) remains highly robust under conditions of complete signal loss, where the only change is that the system benefit has shifted from the pedestrian to the vehicle user.

5 Conclusion

This chapter presents a cyber-physical optimisation technique for the pedestrian-aware supervisory control strategy of hybrid propulsion systems to decrease the impact of vehicle emissions and pollutants on surrounding pedestrians. The conclusions of the investigations conducted are listed.

1. The proposed fuzzy adaptive cost map enables the activation of fixed weights for self-regulation and submaps of intertarget weights according to the density of pedestrians in the test scenario.
2. The pedestrians involved can reduce the mass of exhaust emissions by 4.42% with the assistance of the proposed method, compared to only considering fuel consumption in the mapping treatment.
3. Compared to the bootstrapped average, the proposed bootstrapped deviation technology enables lower emissions (>1.1%) and better fuel economy (0.5%).
4. Compared to the 2 baselines of the genetic algorithm (79/90) and particle swarm optimisation (13/90), the adapted BA used for cyber-physical optimisation indicates the ability to obtain the largest success rate (86/90), which shows the superiority of control optimisation for PHPSs.
5. The proposed system (with relatively minimal fuel consumption and low total emissions) remains highly robust in the case of complete signal loss, where the only changing thing is that the benefit of the system has shifted from the pedestrian to the vehicle owner.

Acknowledgements This work has emanated from research supported in part by Science Foundation Ireland under Grant Number 21/FFP-P/10266 and SFI/12/RC/2289_P2 (Insight SFI Research

Centre for Data Analytics), co-funded by the European Regional Development Fund in collaboration with the SFI Insight Centre for Data Analytics at Dublin City University.

Appendix

This section details the parameters of the three applied optimisation algorithms. All parameters of the algorithms are set as the commonly used values under the same scale based on the studies of [40] and [41]. The Tables 6, 7 and 8 show the details of three optimisation algorithms, including BA, PSO, and GA.

Here, C1 and C2 are trust parameters, where C1 expresses how much confidence a particle has, while C2 expresses how much confidence a particle has in its neighbours. w max and w min are weight parameters, which determine the contribution rate of a particle's previous velocity to its velocity at the current time step.

The crossover probability represents the frequency of performing a crossover. If there is no crossover, the offspring is an exact copy of the parents. The mutation probability indicates the frequency of mutations in parts of the chromosome, which are performed to prevent the GA from falling into a local optimum. Based on the

Table 6 The parameters for the BA used in this case [40]

Parameters	Value
Number of Scout Bees in the Selected Patches	50
Number of Selected Sites	15
Number of Selected Elite Sites	3
Number of Recruited Bees for Selected Sites	8
Number of Recruited Bees for Elite Sites	12

Table 7 The parameters for the PSO used in this case [40]

Parameters	Value
population size set	100
C1	2
C2	2
w max	0.9
w min	0.4

Table 8 The parameters for the GA used in this case [41]

Parameters	Value
population size set	50
crossover probability	0.9
mutation probability	0.03

research of Hassanat et al., population sizes are commonly set to 20, 50, 100, and 200 [41]. Here, to make a comparison under the same scale as the BA studied in this chapter, 50 is selected as the population size.

References

1. Shiva Nagendra SM, Schlink U, Dheeraj Alshetty V, Diya M, Menon JS (2020) Traffic-related air pollution, human exposure, and commercially available market solutions: perspectives from the developing nation context. In: Traffic-Related Air Pollution. Elsevier, pp 531–540. https://doi.org/10.1016/B978-0-12-818122-5.00022-3
2. Naoum-Sawaya J, Crisostomi E, Liu M, Gu Y, Shorten R (2017) Smart procurement of naturally generated energy (SPONGE) for plug-in hybrid electric buses. IEEE Trans Autom Sci Eng 14(2):598–607. https://doi.org/10.1109/TASE.2016.2633001
3. Li J, Wu D, Mohammadsami Attar H, Xu H (2022) Geometric neuro-fuzzy transfer learning for in-cylinder pressure modelling of a diesel engine fuelled with raw microalgae oil. Appl Energy 306:118014. https://doi.org/10.1016/j.apenergy.2021.118014
4. Zhou Q et al (2019) Multi-step reinforcement learning for model-free predictive energy management of an electrified off-highway vehicle. Appl Energy 255:113755. https://doi.org/10.1016/j.apenergy.2019.113755
5. Yang C, Zha M, Wang W, Liu K, Xiang C (2020) Efficient energy management strategy for hybrid electric vehicles/plug-in hybrid electric vehicles: review and recent advances under intelligent transportation system. IET Intell Transp Syst 14(7):702–711. https://doi.org/10.1049/iet-its.2019.0606
6. Enang W, Bannister C (2017) Modelling and control of hybrid electric vehicles (A comprehensive review). Renew Sustain Energy Rev 74:1210–1239. https://doi.org/10.1016/j.rser.2017.01.075
7. Li J, Zhou Q, Williams H, Xu H, Du C (2022) Cyber-physical data fusion in surrogate- assisted strength Pareto evolutionary algorithm for PHEV energy management optimization. IEEE Trans Ind Inf 18(6):4107–4117. https://doi.org/10.1109/TII.2021.3121287
8. Zhou Q, Zhao D, Shuai B, Li Y, Williams H, Xu H (2021) Knowledge implementation and transfer with an adaptive learning network for real-time power management of the plug-in hybrid vehicle. IEEE Trans Neural Netw Learn Syst 32(12):5298–5308. https://doi.org/10.1109/TNNLS.2021.3093429
9. Zhou Q et al (2022) Transferable representation modelling for real-time energy management of the plug-in hybrid vehicle based on k-fold fuzzy learning and Gaussian process regression. Appl Energy 305:117853. https://doi.org/10.1016/j.apenergy.2021.117853
10. Montazeri-Gh M, Poursamad A, Ghalichi B (2006) Application of genetic algorithm for optimization of control strategy in parallel hybrid electric vehicles. J Franklin Inst 343(4–5):420–435. https://doi.org/10.1016/j.jfranklin.2006.02.015
11. Li J, Zhou Q, Williams H, Xu H (2020) Back-to-Back competitive learning mechanism for fuzzy logic based supervisory control system of hybrid electric vehicles. IEEE Trans Ind Electron 67(10):8900–8909. https://doi.org/10.1109/TIE.2019.2946571
12. Li J et al (2019) Dual-loop online intelligent programming for driver-oriented predict energy management of plug-in hybrid electric vehicles. Appl Energy 253:113617. https://doi.org/10.1016/j.apenergy.2019.113617
13. Shouran M, Anayi F, Packianather M, Habil M (2021) Load frequency control based on the bees algorithm for the great britain power system. https://doi.org/10.3390/designs5030050.
14. Kamaruddin S, Abd Latif MAH (2019) Application of the bees algorithm for constrained mechanical design optimisation problem. Int J Eng Mater Manuf 4(1):27–32. https://doi.org/10.26776/ijemm.04.01.2019.04

15. Zhang L, Zhang Z, Wang Z, Deng J, Dorrell DG (2021) Chassis coordinated control for full x-by-wire vehicles-a review. Chinese J Mech Eng 34(1):42. https://doi.org/10.1186/s10033-021-00555-6
16. Rezaei A, Burl JB, Zhou B, Rezaei M (2019) A new real-time optimal energy management strategy for parallel hybrid electric vehicles. IEEE Trans Control Syst Technol 27(2):830–837. https://doi.org/10.1109/TCST.2017.2775184
17. Qi X, Luo Y, Wu G, Boriboonsomsin K, Barth M (2019) Deep reinforcement learning enabled self-learning control for energy efficient driving. Transp Res Part C Emerg Technol 99:67–81. https://doi.org/10.1016/j.trc.2018.12.018
18. Gu Y, Liu M, Naoum-Sawaya J, Crisostomi E, Russo G, Shorten R (2018) Pedestrian-aware engine management strategies for plug-in hybrid electric vehicles; pedestrian-aware engine management strategies for plug-in hybrid electric vehicles. IEEE Trans Intell Transp Syst 19(1). https://doi.org/10.1109/TITS.2017.2697044
19. Bahrami S, Nourinejad M, Amirjamshidi G, Roorda MJ (2020) The plugin hybrid electric vehicle routing problem: a power-management strategy model. Transp Res Part C Emerg Technol 111:318–333. https://doi.org/10.1016/j.trc.2019.12.006
20. Kopelias P, Demiridi E, Vogiatzis K, Skabardonis A, Zafiropoulou V (2020) Connected & autonomous vehicles—environmental impacts—a review. Sci Total Environ 712:135237. https://doi.org/10.1016/j.scitotenv.2019.135237
21. Huang C, Salehi R, Ersal T, Stefanopoulou AG (2020) An energy and emission conscious adaptive cruise controller for a connected automated diesel truck. Veh Syst Dyn 58(5):805–825. https://doi.org/10.1080/00423114.2020.1740283
22. Shao Y, Sun Z (2021) Vehicle speed and gear position co-optimization for energy-efficient connected and autonomous vehicles. IEEE Trans Control Syst Technol 29(4):1721–1732. https://doi.org/10.1109/TCST.2020.3019808
23. Vahidi A, Sciarretta A (2018) Energy saving potentials of connected and automated vehicles. Transp Res Part C Emerg Technol 95:822–843. https://doi.org/10.1016/j.trc.2018.09.001
24. Li SE et al (2017) Dynamical modeling and distributed control of connected and automated vehicles: challenges and opportunities. IEEE Intell Transp Syst Mag 9(3):46–58. https://doi.org/10.1109/MITS.2017.2709781
25. Zhang L, Wang Z, Ding X, Li S, Wang Z (2021) Fault-tolerant control for intelligent electrified vehicles against front wheel steering angle sensor faults during trajectory tracking. IEEE Access 9:65174–65186. https://doi.org/10.1109/ACCESS.2021.3075325
26. Modern Electric, Hybrid Electric, and Fuel Cell Vehicles (2018) Third Edition. CRC Press. https://doi.org/10.1201/9780429504884
27. Li J, Zhou Q, He Y, Williams H, Xu H (2020) Driver-identified supervisory control system of hybrid electric vehicles based on spectrum-guided fuzzy feature extraction. IEEE Trans Fuzzy Syst 28(11):2691–2701. https://doi.org/10.1109/TFUZZ.2020.2972843
28. Zhou Q, Zhang Y, Li Z, Li J, Xu H, Olatunbosun O (2018) Cyber-physical energy-saving control for hybrid aircraft-towing tractor based on online swarm intelligent programming. IEEE Trans Ind Inform 14(9):4149–4158. https://doi.org/10.1109/TII.2017.2781230
29. Peng J, He H, Xiong R (2017) Rule based energy management strategy for a series–parallel plug-in hybrid electric bus optimized by dynamic programming. Appl Energy 185:1633–1643. https://doi.org/10.1016/j.apenergy.2015.12.031
30. Chen Z, Xiong R, Wang C, Cao J (2017) An on-line predictive energy management strategy for plug-in hybrid electric vehicles to counter the uncertain prediction of the driving cycle. Appl Energy 185:1663–1672. https://doi.org/10.1016/j.apenergy.2016.01.071
31. Deb M, Debbarma B, Majumder A, Banerjee R (2016) Performance—emission optimization of a diesel-hydrogen dual fuel operation: a NSGA II coupled TOPSIS MADM approach. Energy 117:281–290. https://doi.org/10.1016/j.energy.2016.10.088
32. Michel P, Charlet A, Colin G, Chamaillard Y, Bloch G, Nouillant C (2017) Optimizing fuel consumption and pollutant emissions of gasoline-HEV with catalytic converter. Control Eng Pract 61:198–205. https://doi.org/10.1016/j.conengprac.2015.12.010

33. Markel T et al (2002) ADVISOR: a systems analysis tool for advanced vehicle modeling. J Power Sources 110(2):255–266. https://doi.org/10.1016/S0378-7753(02)00189-1
34. Zadeh LA (1976) A fuzzy-algorithmic approach to the definition of complex or imprecise concepts. Int J Man Mach Stud 8(3):249–291. https://doi.org/10.1016/S0020-7373(76)80001-6
35. Pham DT, Ghanbarzadeh A, Koç E, Otri S, Rahim S, Zaidi M (2006) The Bees algorithm—a novel tool for complex optimisation problems. In: Intelligent production machines and systems—2nd I*PROMS virtual international conference, 3–14 July 2006, pp 454–459. https://doi.org/10.1016/B978-008045157-2/50081-X
36. Pham DT, Castellani M (2009) The bees algorithm: modelling foraging behaviour to solve continuous optimization problems. Proc Inst Mech Eng Part C J Mech Eng Sci 223(12):2919–2938. https://doi.org/10.1243/09544062JMES1494
37. Efron B, Tibshirani RJ (1994) An introduction to the bootstrap
38. DiCiccio TJ, Efron B (1996) Bootstrap confidence intervals. Stat Sci 11(3). https://doi.org/10.1214/ss/1032280214
39. Xu L, Gotwalt C, Hong Y, King CB, Meeker WQ (2020) Applications of the fractional-random-weight bootstrap view supplementary material applications of the fractional-random-weight bootstrap. Am Stat 74(4):345–358. https://doi.org/10.1080/00031305.2020.1731599
40. Yuce B, Packianather MS, Mastrocinque E, Pham DT, Lambiase A (2013) Honey bees inspired optimization method: the bees algorithm. Insects 4:646–662. https://doi.org/10.3390/insects4040646
41. Hassanat A, Almohammadi K, Abunawas E, Hammouri A, Surya Prasath VB (2019) Choosing mutation and crossover ratios for genetic algorithms-a review with a new dynamic approach. Information 10:390. https://doi.org/10.3390/info10120390
42. Sezer V, Gokasan M, Bogosyan S (2011) A novel ECMS and combined cost map approach for high-efficiency series hybrid electric vehicles. IEEE Trans Veh Technol 60(8):3557–3570. https://doi.org/10.1109/TVT.2011.2166981
43. Wang Y, Shen Y, Yuan X, Yang Y (2015) Operating point optimization of auxiliary power unit based on dynamic combined cost map and particle swarm optimization. IEEE Trans Power Electron 30(12):7038–7050. https://doi.org/10.1109/TPEL.2014.2383443
44. Hasan AO, Abu-Jrai A, Al-Muhtaseb AH, Tsolakis A, Xu H (2016) HC, CO and NOx emissions reduction efficiency of a prototype catalyst in gasoline bi-mode SI/HCCI engine. J Environ Chem Eng 4(2):2410–2416. https://doi.org/10.1016/j.jece.2016.04.015
45. He J, Yang K, Chen H-H (2021) 6G Cellular networks and connected autonomous vehicles. IEEE Netw 35(4):255–261. https://doi.org/10.1109/MNET.011.2000541
46. Dey KC, Rayamajhi A, Chowdhury M, Bhavsar P, Martin J (2016) Vehicle-to-vehicle (V2V) and vehicle-to-infrastructure (V2I) communication in a heterogeneous wireless network—performance evaluation. Transp Res Part C Emerg Technol 68:168–184. https://doi.org/10.1016/j.trc.2016.03.008

Surrogate Model-Assisted Bees Algorithm for Global Optimisation of Microwave Filters

Feiying Lan, Lu Qian, Marco Castellani, Yi Wang, D. T. Pham, and Yongjing Wang

Abstract Microwave filter optimisation is an important example of black-box optimisation, where the objective function is unknown and requires full-wave electromagnetic (EM) simulations. This problem is challenging and even computationally intractable for commonly used global optimisation techniques due to the multimodal and computationally expensive nature of its objective function. This chapter proposes the surrogate-model-assisted Bees Algorithm. Gaussian process regression is used to model the unknown objective function and prescreen promising candidates for expensive EM simulations. In this scheme, the Bees algorithm is used to perform a global search and intelligent sampling for surrogate modelling. This method was evaluated on 7 benchmark functions and compared with the standard Bees Algorithm. Mann–Whitney U tests indicated the statistical significance of the results. A case study involving a microwave dielectric filter demonstrated the significant advantages of using the proposed method in terms of high-quality design and a reduced number of EM simulation-based evaluations.

F. Lan · M. Castellani · D. T. Pham · Y. Wang
Department of Mechanical Engineering, University of Birmingham, Birmingham, UK
e-mail: fxl655@student.bham.ac.uk

M. Castellani
e-mail: m.castellani@bham.ac.uk

D. T. Pham
e-mail: d.t.pham@bham.ac.uk

Y. Wang
e-mail: y.wang@bham.ac.uk

L. Qian (✉) · Y. Wang
Department of Electronic and Electric System Engineering, University of Birmingham, Birmingham, UK
e-mail: l.qian.1@bham.ac.uk

Y. Wang
e-mail: y.wang.1@bham.ac.uk

D. T. Pham and N. Hartono (eds.), *Intelligent Engineering Optimisation with the Bees Algorithm*, Springer Series in Advanced Manufacturing,
https://doi.org/10.1007/978-3-031-64936-3_20

Keywords Bees algorithm · Surrogate model · Global optimisation · Gaussian process · Black-box optimisation · Microwave filter

1 Introduction

Microwave filters are an important class of electronic components in modern wireless communication and radar systems, where they are used to select, separate, or combine different electromagnetic spectra. The simulation-driven shape optimisation process is a critical step in the design of microwave filters. Given an initial filter model, the geometric parameters are optimised to satisfy the designated specifications via full-wave electromagnetic (EM) simulations. This step is challenging due to the multimodal and time-consuming nature of the objective function. Common local optimisation methods (e.g., interpolated quasi-Newton [1] and the Nelder–Mead simplex method [2]) are employed by popular commercial EM simulation software (e.g., CST Microwave Studio), but they are susceptible to local minima. To mitigate this problem, problem-specific techniques such as space mapping [3, 4] and port tuning [5, 6] were introduced to assist local optimisation. These techniques require knowledge of the transcendental mapping functions (approximate analytical expressions or equivalent circuit models) during the optimisation process. Moreover, these methods only improve the efficiency of local optimisation procedures rather than strengthening the global search capacity.

Global optimisation techniques have been increasingly applied to microwave circuit design [7–9]. Examples are Particle Swam Optimisation (PSO) and Evolutionary Algorithms (EA). Directly adopting standard metaheuristic algorithms in microwave filter design is impossible because the full-wave EM simulation is very computationally intensive and time-consuming, especially when the complexity of the filter model increases. To address this challenge, surrogate model-assisted global optimisation methods have been introduced for microwave circuit design [10–12]. The computational cost is reduced because a surrogate model is trained with the EM simulation results and partially replaces full-wave EM simulation with low-cost predictions of the surrogate model.

The Bees Algorithm (BA) is a nature-inspired algorithm for global optimisation inspired by the foraging behaviour of biological bees [13, 14]. The main motivation for adopting the BA is to obtain a nearly optimal solution within a reasonable time for practical engineering applications. The BA conducts exploitative neighbourhood search combined with random explorative search. Via a mechanism simulating the waggle dance of biological bees, the BA exchanges global information on the most promising regions in the solution space and distributes the local exploitative search effort accordingly. Neighbourhood shrinking and site abandonment are two key strategies that intensify the search in local regions and prevent the algorithm from becoming stuck in local optima. The BA literature includes theoretical analysis of its search mechanisms [15] as well as applications to engineering optimisation problems [16].

This chapter proposes the Gaussian process (GP)-assisted Bees Algorithm for microwave filter optimisation. The strategy consists of using BA intelligent global sampling for surrogate modelling, while GP modelling is used to prescreen promising candidate solutions before they are tested via expensive EM simulations. For each iteration, the GP model is built using the observations from previous EM simulations. Namely, in the local and global search procedures of the Bees Algorithm, the foragers and scouts visit locations (candidate solutions) whose fitness is predicted by the low-cost GP model, and only the best candidates are selected for expensive full objective function evaluation. Unlike the differential evolutionary algorithm reported in [17], which requires a post-Gaussian local search, the advantage of the BA is that it is designed to balance explorative and exploitative search. The proposed method follows Bayes' intuition that the computational search effort should be made in promising areas of the solution space, based on previous evaluations using the black-box objective function.

This chapter is structured as follows. Section 2 surveys the literature on the state-of-the-art methods of surrogate model-assisted global optimisation for microwave filter design. Section 3 introduces the proposed GP-assisted BA. Section 4 presents the experimental results of the comparison of the proposed method with the standard BA on 7 benchmark functions. These functions include 2 unimodal and 5 multimodal fitness landscapes and were compared. Section 5 reports the results of a test on a microwave dielectric filter as a case study and demonstrates the compelling performance of the proposed algorithm in terms of design results and economy of EM simulations.

2 Related Work

Evolutionary Algorithms (EAs) and Particle Swarm Optimisation (PSO) are popular techniques for microwave filter design due to their generality and global search capabilities [18–21]. However, as the dimensionality of the solution space increases, using a sufficiently large population size becomes unfeasible due to the high computational cost of the EM simulations. Surrogate model-assisted evolutionary algorithms adjust the search effort to favour promising areas of the solution space based on a computationally light model of the objective function. The surrogate model is usually implemented using a data-driven technique, such as a Support Vector Machine (SVM) [22–24], Artificial Neural Network (ANN) [25–27], or Gaussian Process (GP) [28, 29]. The first two learning models (SVM and ANN) attempt to identify the objective function under a certain metric. Using an SVM poses the challenge of selecting an appropriate kernel function. In addition, the computational complexity and memory consumption of the model grow with the size of the training dataset. ANNs have gained large popularity in recent years [30]. However, architecture design [31], overfitting [32], and generalisation [33] from small training datasets using overparameterised ANNs are the major issues. Gaussian process modelling is a technique that identifies the distribution of the objective function instead of fitting it. This approach can predict Gaussian uncertainties in unseen areas and mitigate the overfitting problem, even when the dataset is very small.

In recent years, a few surrogate model-assisted methods for microwave circuit optimisation have been reported. Liu et al. [17] proposed a surrogate model-assisted evolutionary algorithm for the global optimisation of microwave filters. In this scheme, the EA was combined with the GP model to balance the exploration ability and the surrogate model quality. This first phase is followed by a hybrid technique to improve the optimisation efficiency [34]. Torun and Swaminathan [35] combined deep partitioning tree (DPT) and additive Gaussian process (ADD-GP) modelling for high-dimensional global optimisation in high-frequency electronic design. Akinsolu et al. [36] addressed the goal of parallel computation and proposed a parallel surrogate model-assisted evolutionary algorithm for electromagnetic design optimisation. Parallel computing and surrogate model-assisted evolutionary algorithms are integrated to improve efficiency and optimisation capacity. Nguyen et al. [37] conducted a comparative study on surrogate modelling methods for microwave circuit applications, where the performance of the nonparametric Gaussian process was compared with other common modelling techniques.

3 Methodology

In this section, the GP model and the BA are introduced in Sect. 3.1 and Sect. 3.2, respectively. They are followed by the description of the proposed GA-assisted BA in Sect. 3.3.

3.1 Gaussian Process Regression

GP regression models the probabilistic distribution of the possible predictive function across the solution domain. It is a non-parametric learning model and can predict the uncertainty in unseen areas based on observations. The response y at observation point $\boldsymbol{x} \in \boldsymbol{R}^d$ is considered a random variable given by Eq. (1):

$$y = f(x) + \omega, \omega \in N\left(0, \sigma^2\right) \tag{1}$$

where $f(x)$ is the predictive model, and ω is the Gaussian uncertainty on the variance σ^2.

Given a set of observations $\mathcal{X} = \{\boldsymbol{x}_1, \boldsymbol{x}_2, \ldots, \boldsymbol{x}_n\}$ and responses $\mathcal{Y} = \{y_1, y_2, \ldots, y_n\}$, the marginal distribution of $\boldsymbol{f} = \left[f_1, f_2, \ldots, f_n\right]^T$ is depicted by a multivariate Gaussian $p(\boldsymbol{f}) = N(0, \boldsymbol{K})$, where n is the number of data points, and $\boldsymbol{K}$ is the Gram matrix determined by a kernel function $\kappa\left(\boldsymbol{x}_i, \boldsymbol{x}_j\right)$ such that $\boldsymbol{K}_{ij} = \kappa\left(\boldsymbol{x}_i, \boldsymbol{x}_j\right)$. The conditional distribution of $\boldsymbol{y} = \left[y_1, y_2, \ldots, y_n\right]^T$ on $\boldsymbol{f}$ is given by $p(\mathbf{y}|\mathbf{f}) = N(\mathbf{y}|\mathbf{f}, \sigma^2 \mathbf{I})$, where $\mathbf{I}$ is the $n \times n$ identity matrix. Thus, the distribution of $p(\mathbf{y})$ is marginalised as follows:

$$p(\mathbf{y}) = \int p(\mathbf{y}|\mathbf{f}) p(\mathbf{f}) d\mathbf{f} = N(\mathbf{y}|0, \mathbf{\Sigma}) \tag{2}$$

where $\mathbf{\Sigma}$ is the covariance of $\mathbf{\Sigma} = \mathbf{K} + \sigma^2 \mathbf{I}$, i.e., $\mathbf{\Sigma}_{ij} = \kappa\left(\mathbf{x}_i, \mathbf{x}_j\right) + \sigma^2 \mathbf{I}_{ij}$.

For predicting a new observation $\hat{\mathbf{x}}$, the joint distribution of $\hat{y}$ and $\mathbf{y}$ is given by:

$$\begin{bmatrix} \mathbf{y} \\ \hat{y} \end{bmatrix} \sim N(0, \begin{bmatrix} \mathbf{K} + \sigma^2 \mathbf{I} & \kappa(\hat{\mathbf{x}}, \mathbf{x}) \\ \kappa(\hat{\mathbf{x}}, \mathbf{x})^T & \kappa(\hat{\mathbf{x}}, \hat{\mathbf{x}}) \end{bmatrix} \tag{3}$$

where $\kappa(\hat{\mathbf{x}}, \mathbf{x})$ is a column vector generated by $[\kappa(\hat{\mathbf{x}}, \mathbf{x}_1)\kappa(\hat{\mathbf{x}}, \mathbf{x}_2) \ldots \kappa(\hat{\mathbf{x}}, \mathbf{x}_n)]^T$.

The response $\hat{y}$ is predicted by the marginal distribution below:

$$p\left(\hat{y}|\mathcal{X}, \mathcal{Y}, \hat{\mathbf{x}}\right) = N(\hat{\mu}, \hat{\sigma}^2) \tag{4}$$

where:

$$\begin{cases} \hat{\mu} = \kappa(\hat{\mathbf{x}}, \mathbf{x})^T \left[\mathbf{K} + \sigma^2 \mathbf{I}\right]^{-1} \mathbf{y} \\ \hat{\sigma}^2 = \kappa(\hat{\mathbf{x}}, \hat{\mathbf{x}}) - \kappa(\hat{\mathbf{x}}, \mathbf{x})^T \left[\mathbf{K} + \sigma^2 \mathbf{I}\right]^{-1} \kappa(\hat{\mathbf{x}}, \mathbf{x}) \end{cases} \tag{5}$$

A popular instance for the kernel function is the exponential quadratic kernel:

$$\kappa\left(\mathbf{x}_i, \mathbf{x}_j\right) = \sigma'^2 \exp\left(-\frac{\mathbf{x}_i - \mathbf{x}_j^2}{2\ell^2}\right) \tag{6}$$

where σ' is the amplitude and ℓ is the length scale. The hyperparameters can be optimised, and the details are enclosed in [38].

3.2 *Bees Algorithm for Global Search and Intelligent Sampling*

The Bees Algorithm (BA) simulates the foraging behaviour of honeybee colonies. During the food harvesting process, a bee colony employs a given number of scout bees to explore the area near the hive and randomly search for food. These scouts exchange and propagate information on the quality and location of their findings to idle mates via a ritual known as the '*waggle dance*'. Informed of the location of a rich food source by a dancer scout, onlookers of a waggle dance proceed to harvest it and eventually waggle the dance themselves to pass the information to other mates. The nature-inspired BA is elaborated below.

At the beginning of the BA, the bee colony is initialised by scattering *ns* scouts onto the solution space randomly and uniformly (explorative search). Each scout evaluates the cost of the solution where it landed via the objective function. The solutions found by the scouts are then ranked based on their cost (minimisation problem) and receive a number of foragers based on their rank (*waggle dance*). That is, the scouts that found the top *nb* solutions are called the best bees. Among them, the scouts that found the top *ne* bees are called elite bees ($ne \leq nb$). The *ne* elite bees recruit *nre* foragers each, and the $(nb - ne)$ remaining best bees recruit *nrb* foragers ($nrb \geq nre$) each for local exploitative search.

During the local search phase, each forager is randomly placed in proximity to the solution found by the scout. The area where foragers are placed is called a '*flower patch*'. Each forager assesses the goodness of the solution where it landed, and the forager that found the solution of lower cost becomes the new scout. In the next BA cycle, this scout will perform the waggle dance. Each flower patch has a '*stagnation count*' to accumulate the number of cycles in which the local best is not improved. If all foragers in a flower patch fail to land on a solution of lower cost than the scout, the stagnation count for this patch is incremented, and the local neighbourhood size is shrunk. This latter procedure is called '*neighbourhood shrinking*' and aims to progressively focus and refine the search in a flower patch. The stagnation count will be reset to zero if an improvement occurs in a flower patch. If the stagnation count reaches a limit, this flower patch will be abandoned, and the scout will be sent to a random site. This latter procedure is known as *site abandonment*. This latter strategy [14] aims to prevent the local search from being stuck in local minima. The remaining $(ns - nb)$ scouts perform random explorative search in the solution space. The bee colony size n is calculated as $n = ne \times nre + (nb - ne) \times nrb + ns$.

The BA repeats cycles of local and explorative search until a stopping criterion is met, e.g., the maximum number of iterations has elapsed, a given number of cost function evaluations have been performed, or a solution of desired cost has been found.

3.3 Surrogate Model-Assisted Bees Algorithm

The GP surrogate model prescreens candidate solutions for the global search mechanism of the BA. At each cycle of the BA, the solutions evaluated via EM simulation are collected as observations and fed into the training dataset for surrogate modelling. Namely, a number of $c \cdot n$ candidate solutions is generated at each BA cycle, where c is the over generating factor, and n is the size of the bee colony. The cost of these solutions is predicted using the low-cost GP model, and the top n solutions are visited by scouts or foragers for expensive EM simulation. This strategy enhances the information exchange mechanism of the BA, that is, the waggle dance and local search only send the bees to promising locations based on machine learning of the previously explored history. Additionally, the prescreeing phase reduces the number of expensive EM simulations for less valuable bees, thus improving the BA search efficiency and convergence speed. The global search mechanism of the BA with the GP surrogate model is presented in Fig. 1.

In Fig. 1, the BA initialises *ns* random scouts for cost function evaluation, followed by the waggle dance. The solutions evaluated via EM simulation are collected as observations for GP surrogate modelling. Then, the BA enters the local and explorative search phases with GP-assisted prescreening, as shown in Fig. 2.

In the local search phase, for each best site (including the elite sites), for each forager, c candidate solutions are generated, where c is the factor for prescreening. The GP model predicts their cost via a Gaussian distribution $y \sim N(\mu, \sigma^2)$. In this

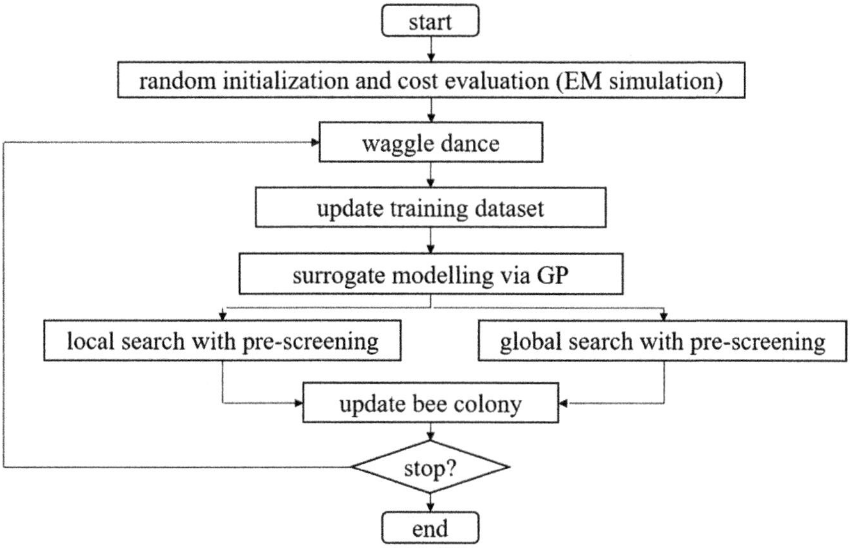

Fig. 1 Global search of GP-assisted BA for microwave filter optimisation

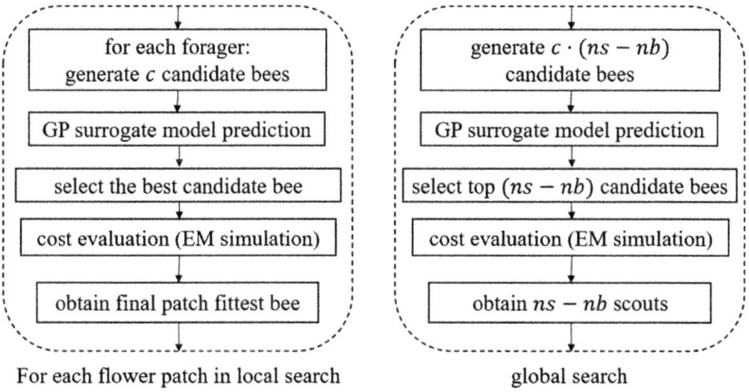

Fig. 2 Local and explorative search of BA with GP surrogate model prescreening

study, the $1 - \sigma$ rule was adopted, i.e., the predicted cost value for the candidate solution is the lower-bound cost value with a 68% confidence level:

$$\hat{y} = \mu - \sigma \tag{7}$$

The generated solutions are prescreened by selecting the best solution for computationally expensive EM simulation and cost evaluation. This step is repeated for all recruited foragers, and the final fittest scout bee is obtained.

In the explorative search phase, $c(\text{ns} - \text{nb})$ solutions are generated randomly and uniformly in the whole solution space. The GP model predicts their cost value for prescreening. The top (ns − nb) candidates are used for EM simulation and cost evaluation.

The bee colony is updated with the results of the local and explorative search efforts. The samples evaluated via EM simulation are appended to the dataset for improved surrogate modelling in the next cycle. The BA repeats the procedure until the stopping criterion is met.

4 Experiments on Benchmark Functions

This section presents the results of the experimental evaluation of the proposed method on 7 benchmark functions. These benchmarks include two unimodal functions (ellipsoid and sphere) and five multimodal functions (Ackley, Griewank, Rastrigin, Rosenbrock, and Weierstrass) [39]. The proposed GP-assisted BA was compared with the standard BA. The results were analysed using the Mann–Whitney U test of statistical significance.

4.1 Benchmark Functions and Parameter Settings

The 7 benchmark functions are detailed in Table 1. For the function $y = f(\boldsymbol{x}), x \in \boldsymbol{R}^d$, the optimal solution is $x^* = 0^d$, and the global minimal value is 0. The parameterisation of the algorithms is presented in Table 2.

4.2 Results and Discussion

The experiments were repeated 20 times for each algorithm on each benchmark function. Since the proposed optimisation method is a stochastic process, the results were statistically analysed, and the average values are presented in Table 3. Figure 3 visualises the summary of the statistical analysis. For each algorithm and each benchmark, the summary of the statistical analysis includes the minimum, median, maximum, and

Table 1 Seven benchmark functions and their domains

	Expression	domain
Ellipsoid	$y = \sum_i^d i \cdot x_i^2$	$[-5.12, 5.12]^d$
Sphere	$y = \sum_i^d x_i^2$	$[-5.12, 5.12]^d$
Ackley	$y = -20\exp\left(-0.2\sqrt{1/d}\sum_i^d x_i^2\right) + 20 - \exp\left((1/d)\sum_i^d \cos(2\pi x_i)\right) + e$	$[-32, 32]^d$
Griewank	$y = \sum_i^d x_i^2/4000 - \prod_i^d \cos\left(x_i/\sqrt{i}\right) + 1$	$[-600, 600]^d$
Rastrigin	$y = 10d + \sum_i^d [x_i^2 - 10\cos(2\pi x_i)]$	$[-5.12, 5.12]^d$
Rosenbrock	$y = \sum_i^{d-1}\left[100(x_{i+1} - x_i^2)^2 + (x_i - 1)^2\right]$	$[-5, 10]^d$
Weierstrass	$y = \sum_{i=1}^d \left\{\sum_{k=0}^{20}\left[0.5^k \cos\left(2\pi \cdot 3^k(x_i + 0.5)\right)\right]\right\} - d\sum_{k=0}^{20}\left[0.5^k \cos\left(2\pi \cdot 3^k \cdot 0.5\right)\right]$	$[-0.5, 0.5]^d$

Table 2 Parameterisation of the algorithms

Bees Algorithm	# Maximum iterations	100	*n*	20
	# Stopping criteria	0	*ne*	1
	# Population size	20	*nb*	4
	Stagnation limit (*stlim*)	10	*nre*	4
	Neighbourhood shrinking rate	0.75	*nrb*	4
	Prescreening factor	5	*ns*	4
Gaussian Process	Initial σ	σ lower bound	Standardise	
	$0.5 \cdot std(\mathbf{y})$	$0.1 \cdot std(\mathbf{y})$	Yes	

interquartile range (IQR, the height of the boxes) of the results of the 20 optimisation trials.

Figure 3 shows significant improvements in the BA results when using surrogate modelling via GP. The improvements were particularly compelling on the two unimodal functions (ellipsoid and sphere), where the median result decreased from

Table 3 Summary of the results obtained by the standard BA and GP-enhanced BA on the 7 benchmark functions

		Min	Median	Max	IQR
Ellipsoid	BA	21.4445	44.7673	59.8379	21.4978
	BA-GP ($\times 10^{-4}$)	**0.0825**	**0.2758**	**0.9553**	**0.2558**
Sphere	BA	6.3630	8.9481	10.4196	2.1092
	BA-GP ($\times 10^{-4}$)	**0.0613**	**0.1163**	**0.1684**	**0.0463**
Ackley	BA	12.8136	15.2451	16.8305	1.6362
	BA-GP	**0.7048**	**1.5511**	**2.2482**	**0.7724**
Griewank	BA	12.3704	28.9428	43.8403	12.1590
	BA-GP	**0.0005**	**0.3751**	**0.6879**	**0.2238**
Rosenbrock	BA	1571.1	3769.6	6990.5	3148.8
	BA-GP	**13.7250**	**70.6708**	**151.4737**	**51.2624**
Rastrigin	BA	51.4245	68.5635	79.8734	10.7945
	BA-GP	**4.2119**	**13.7270**	**36.8412**	**6.9039**
Weierstrass	BA	8.3068	9.7411	10.4101	0.8198
	BA-GP	**1.4389**	**2.5178**	**4.8052**	**0.7989**

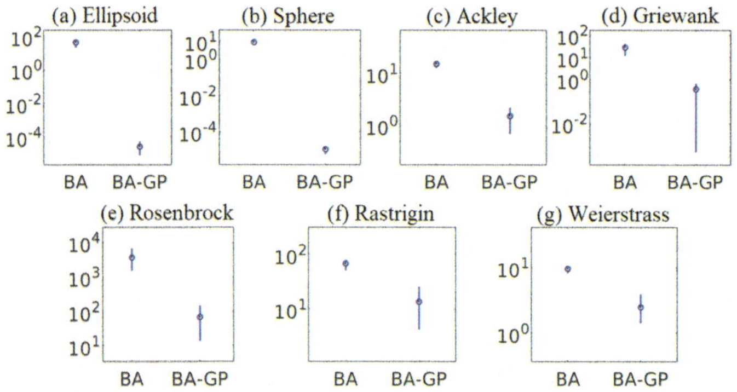

Fig. 3 Boxplots of the optimisation results on the 7 benchmark functions. The y axis is plotted on a logarithmic scale

44.7673 and 8.9481×10^{-4} to 0.2758 and 0.1163×10^{-4}, respectively. The surrogate model also reduced the spread of the results: the IQR dropped from 21.4978 on the ellipsoid and 2.1092 on the sphere to below 10^{-5}. These improvements are clearly visible in the boxplots of Fig. 3.

For the remaining 5 multimodal functions, the surrogate model implemented by GP improved both the median cost and the IQR of the standard BA by at least one order of magnitude (two orders of magnitude for the Rosenbrock and Griewank functions). Thus, the surrogate model excelled in increasing the global search capacity of

Table 4 Mann–Whitney U test results. The null hypothesis is that there is no difference in performance between the standard and the GP-enhanced BA

Ellipsoid	Sphere	Ackley	Griewank	Rosenbrock	Rastrigin	Weierstrass
0	10^{-4}	0	0.0007	0	0	0

the BA owing to the prescreening strategy but also helped to produce high-precision solutions.

The statistical significance of the differences between the results obtained by the standard and the GP-enhanced BA was tested via pairwise nonparametric Mann–Whitney U tests [40]. The results of the tests are summarised in Table 4.

In Table 4, the p values are less than the standard 0.05 threshold for all the benchmark functions where the two algorithms were compared. That is, the Mann–Whitney U test suggests that the differences in the results obtained by the standard BA and GP-assisted BA are statistically significant.

5 Case Study on a Microwave Dielectric Filter

This section presents a case study of microwave dielectric filter optimisation using the GP-assisted Bees Algorithm, and compares its performance with that of the standard BA and a Genetic Algorithm in the commercial EM full-wave simulation tool CST Studio [41].

5.1 Dielectric Filter Model and Parameter Settings

The design example adopted in this case study is a fourth-order dielectric waveguide filter. It is composed of four coupled dielectric waveguide resonators operating in TE_{101} resonance mode and the input/output ports. Figure 4a shows the configuration of this filter and six key dimensional parameters. From the design point of view, these parameters are necessary to provide full control of the filter performance. Therefore, they are all selected for optimisation. The initial value of these design parameters is obtained by dimensioning using the coupling matrix method and has been presented in Table 5.

Figure 4b shows a comparison between the simulated initial filter response and the ideal filter response from the coupling matrix. The initial response in Fig. 4b is far from the ideal response and the design specification. It should be noted that in microwave circuit design, the design specification is described by scattering parameters (S parameters). The S parameters are defined in terms of the wave variables of different ports of the microwave network, and they are usually expressed in decibels (dB). The sample filter used in this study is specified in terms of a centre frequency of

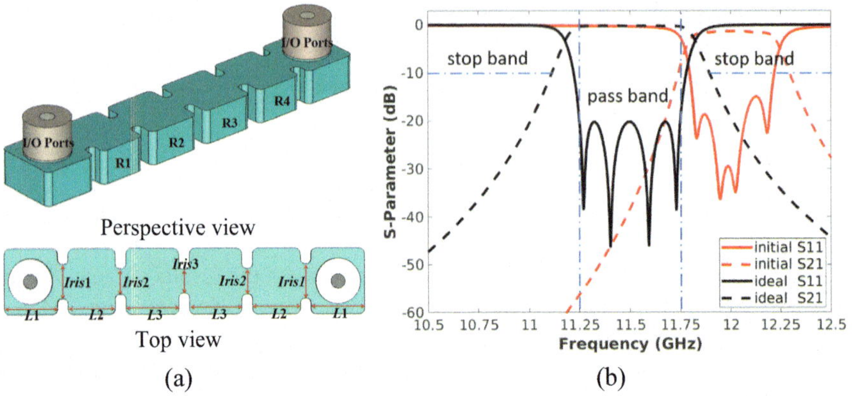

Fig. 4 Dielectric waveguide filter model and its S-parameters. **a**: Conceptual drawing of the filter model, where R1-R4 are TE101 dielectric resonators, and I/O ports are input/output ports. **b**: Comparison between the initial filter response and ideal response synthesised from the coupling matrix

Table 5 Initial parameters for the dielectric filter

Initial Dimension Parameters (mm)					
Iris1	Iris2	Iris3	L1	L2	L3
3.0	2.1	2.0	4.8	4.1	4.6

11.5 GHz and a bandwidth of 0.5 GHz. The $max(S11)$ within the passband should be at least less than − 20 dB, and the stopbands are from 10.5 GHz to 11 GHz and from 12 GHz to 12.5 GHz GHz, where $max(S21)$ should be lower than $-10dB$. Therefore, the cost function for this optimisation problem can be defined below.

$$x^* = \underset{x}{\operatorname{argmin}} \max S_{11}(f), 11.25GHz \leq f \leq 11.75GHz \tag{8}$$

$$\text{s.t.} \begin{cases} \max S_{21}(f) \leq -10, f \leq 11GHz \\ \max S_{21}(f) \leq -10, f \geq 12GHz \end{cases}$$

The constraints in Eq. (8) are interpreted as a regularisation term in the cost function:

$$\mathcal{F} = E_0 + \lambda_1 E_1 + \lambda_2 E_2 \tag{9}$$

where:

$$
\begin{aligned}
E_0 &= \begin{cases} [\max S_{11}(f)] - (-20), & \text{if} \max S_{11}(f) \geq -20 \\ 0, & \text{otherwise} \end{cases}, f \in [11.25, 11.75] \\
E_1 &= \begin{cases} [\max S_{21}(f)] - (-10), & \text{if} \max S_{21}(f) \geq -10 \\ 0 & \text{otherwise} \end{cases}, f \leq 11 \\
E_2 &= \begin{cases} [\max S_{21}(f)] - (-10) & \text{if} \max S_{21}(f) \geq -10 \\ 0 & \text{otherwise} \end{cases}, f \geq 12
\end{aligned} \tag{10}
$$

λ_1 and λ_2 are penalty factors set to be 25 empirically in this study. The parameterisation for BA and GP was set consistent with that in Sect. 4.

5.2 *Results and Discussion*

The optimisation process for the dielectric filter is time-consuming. In this case study, the GP-assisted BA was tested, and its performance was compared with that of the standard BA and the GA in the commercial software CST Studio. The final cost values are shown below:

The GP-assisted BA method obtained the desired solution within 84 cycles, which consumed 1680 EM simulations and 8400 GP surrogate model predictions. The GA and the standard BA terminated after 100 cycles performing 2000 EM simulations. The cost curves of the methods during the optimisation process and the final responses are plotted below:

As shown in the numeric results in Table 6, the standard BA obtains the solution with a lower cost value than GA in CST Studio. Both the GA in CST Studio and the standard BA failed to obtain a solution that satisfies the design specification since the search is not sufficient and thorough within the limitation of a small number of function evaluations. The GP-assisted BA successfully minimised the cost value to below $-20dB$ with a relatively low number of cycles (i.e., 84 cycles were used), which fulfilled the specifications for the dielectric filter. The optimisation curve during iterations in Fig. 5a visualised the advantages of the GP surrogate model. The optimisation curve of GA is not available since CST Studio is a closed source, but it is still possible to confirm that the standard BA outperforms the GA from their final optimisation results. For the GP-assisted BA, the cost curve drops rapidly and returns the desired solution with 84 iterations. Figure 5b confirms that the performance of the solution obtained using the GP-assisted BA meets the design specifications.

Table 6 Final solutions and cost values obtained by the three methods

	Iris1	Iris2	Iris3	L1	L2	L3	Cost
GA	3.1005	2.3202	2.1421	4.9741	4.3707	4.9220	-8.0077
BA	3.4296	2.4996	2.2617	4.4527	3.9575	4.7733	-12.138
BA-GP	3.5058	2.5187	2.3477	4.2576	3.9000	4.7392	-20.137

Fig. 5 Cost curves during optimisation (**a**) and their final S-Parameters (**b**)

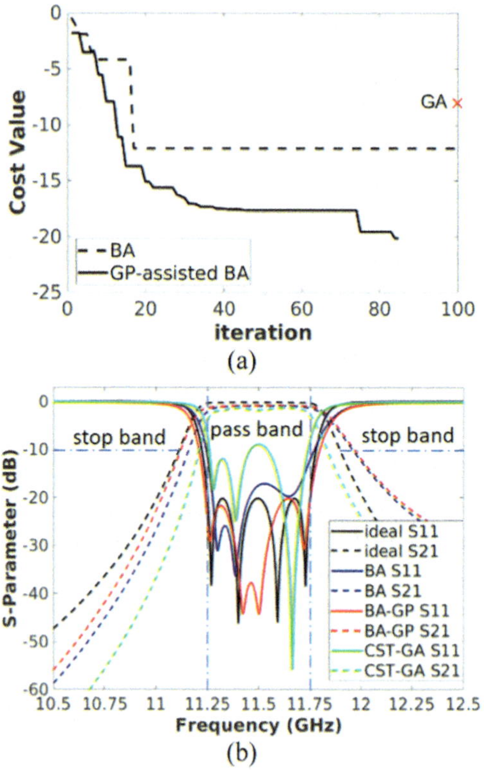

6 Conclusion

Microwave filter optimisation is an important instance of black-box optimisation and is challenging due to the time-consuming evaluation of the multimodal cost function. This chapter proposed the Gaussian Process assisted Bees Algorithm, which efficiently reduces the computational overhead by prescreening promising candidate solutions for evaluation via expensive EM simulations. Gaussian process regression excels in modelling functions using only a handful of data samples and estimating the uncertainties of the model. The combined exploitative search and explorative search conducted by the BA works as an intelligent sampling strategy for the surrogate model, while the BA exploitative search enhances the precision of the surrogate model in the most promising regions, and the explorative search continues to probe the entire solution landscape for further coverage. Experimental results showed the compelling performance of the GP-assisted BA on 7 benchmark functions, and a case study on a dielectric microwave filter demonstrated the effectiveness and efficiency of the proposed method for EM optimisation.

References

1. Gill PE, Murray W (1972) Quasi-Newton methods for unconstrained optimization. IMA J Appl Math 9:91–108
2. Olsson DM, Nelson LS (1975) The Nelder-Mead simplex procedure for function minimization. Technometrics 17:45–51
3. Koziel S (2017) Space mapping: performance, reliability, open problems and perspectives. In: 2017 IEEE MTT-S international microwave symposium (IMS), pp 1512–1514
4. Melgarejo JC, Ossorio J, Cogollos S, Guglielmi M, Boria VE, Bandler JW (2019) On space mapping techniques for microwave filter tuning. IEEE Trans Microw Theory Tech 67:4860–4870
5. Swanson DG (2007) Narrow-band microwave filter design. IEEE Microw Mag 8:105–114
6. Rautio JC (2017) Tuning ports in the middle of resonators. In: 2017 IEEE MTT-S international microwave symposium (IMS), pp 1509–1511
7. Fakhfakh M, Tlelo-Cuautle E, Siarry P (eds) (2015) Computational intelligence in analog and mixed-signal (AMS) and radio-frequency (RF) circuit design. Springer International Publishing, Cham
8. He N, Xu D, Huang L (2009) The application of particle swarm optimization to passive and hybrid active power filter design. IEEE Trans Ind Electron 56:2841–2851. https://doi.org/10.1109/TIE.2009.2020739
9. Deb A, Roy JS, Gupta B (2014) Performance comparison of differential evolution, particle swarm optimization and genetic algorithm in the design of circularly polarized microstrip antennas. IEEE Trans Antennas Propag 62:3920–3928. https://doi.org/10.1109/TAP.2014.2322880
10. Rayas-Sánchez JE (2004) EM-based optimization of microwave circuits using artificial neural networks: The state-of-the-art. IEEE Trans Microw Theory Tech 52:420–435. https://doi.org/10.1109/TMTT.2003.820897
11. Hawe GI, Sykulski JK (2008) A scalarizing one-stage algorithm for efficient multi-objective optimization. IEEE Trans Magn 44:1094–1097. https://doi.org/10.1109/TMAG.2007.915977
12. Couckuyt I, Declercq F, Dhaene T, Rogier H, Knockaert L (2010) Surrogate-based infill optimization applied to electromagnetic problems. Int J RF Microw Comput Eng 20:492–501. https://doi.org/10.1002/mmce.20455
13. Pham DT, Ghanbarzadeh A, Koç E, Otri S, Rahim S, Zaidi M (2006) The bees algorithm—a novel tool for complex optimisation problems. In: Intelligent production machines and systems. Elsevier, pp 454–459
14. Pham DT, Castellani M (2009) The bees algorithm: modelling foraging behaviour to solve continuous optimization problems. Proc Inst Mech Eng Part C J Mech Eng Sci 223:2919–2938
15. Baronti L, Castellani M, Pham DT (2020) An analysis of the search mechanisms of the bees algorithm. Swarm Evol Comput 59:100746
16. Pham DT, Baronti L, Zhang B, Castellani M (2018) Optimisation of engineering systems with the bees algorithm. Int J Artif Life Res 8:1–15
17. Liu B, Yang H, Lancaster MJ (2017) Global optimization of microwave filters based on a surrogate model-assisted evolutionary algorithm. IEEE Trans Microw Theory Tech 65:1976–1985. https://doi.org/10.1109/TMTT.2017.2661739
18. Zaharis ZD, Gravas IP, Yioultsis TV, Lazaridis PI, Glover IA, Skeberis C, Xenos TD (2017) Exponential log-periodic antenna design using improved particle swarm optimization with velocity mutation. IEEE Trans Magn 53:2–5. https://doi.org/10.1109/TMAG.2017.2660061
19. Kahng S, Khattak MK, Kim H-S, Oh S-H, Kim H-S (2015) A PSO-optimized, compact UWB bandpass filter suppressing the WLAN and the spurious harmonics. In: 2015 Asia-Pacific microwave conference (APMC). IEEE, pp 1–4
20. Rocca P, Oliveri G, Massa A (2011) Differential evolution as applied to electromagnetics. IEEE Antennas Propag Mag 53:38–49. https://doi.org/10.1109/MAP.2011.5773566

21. Goudos SK, Sahalos JN (2010) Pareto optimal microwave filter design using multiobjective differential evolution. IEEE Trans Antennas Propag 58:132–144. https://doi.org/10.1109/TAP.2009.2032100
22. Trinchero R, Manfredi P, Stievano IS, Canavero FG (2018) Machine learning for the performance assessment of high-speed links. IEEE Trans Electromagn Compat 60:1627–1634. https://doi.org/10.1109/TEMC.2018.2797481
23. Ciccazzo A, Di Pillo G, Latorre V (2015) A SVM surrogate model-based method for parametric yield optimization. IEEE Trans. Comput. Des Integr Circuits Syst 35:1224–1228
24. Lal A, Datta B (2018) Development and implementation of support vector machine regression surrogate models for predicting groundwater pumping-induced saltwater intrusion into coastal aquifers. Water Resour Manag 32:2405–2419
25. Feng F, Zhang C, Ma J, Zhang QJ (2016) Parametric modeling of em behavior of microwave components using combined neural networks and pole-residue-based transfer functions. IEEE Trans Microw Theory Tech 64:60–77. https://doi.org/10.1109/TMTT.2015.2504099
26. Dong J, Li Y, Wang M (2019) Fast multi-objective antenna optimization based on RBF neural network surrogate model optimized by improved PSO algorithm. Appl Sci 9:2589
27. Sun G, Wang S (2019) A review of the artificial neural network surrogate modeling in aerodynamic design. Proc Inst Mech Eng Part G J Aerosp Eng 233:5863–5872
28. Koziel S, Sigurðsson ATA, Vidarsson FV (2018) Accurate design-oriented modeling of compact microwave couplers in constrained domains. In: 2018 IEEE MTT-S International conference on numerical electromagnetic and multiphysics modeling and optimization NEMO 2018, pp 1–3. https://doi.org/10.1109/NEMO.2018.8503172
29. Torun HM, Hejase JA, Tang J, Beckert WD, Swaminathan M (2018) Bayesian active learning for uncertainty quantification of high speed channel signaling. EPEPS 2018 - IEEE 27th Conference on electrical performance of electronic packaging and systems, pp 311–313. https://doi.org/10.1109/EPEPS.2018.8534251
30. Krizhevsky A, Sutskever I, Hinton GE (2012) Imagenet classification with deep convolutional neural networks. Adv Neural Inf Process Syst 25
31. Castellani M (2013) Evolutionary generation of neural network classifiers—an empirical comparison. Neurocomputing 99:214–229
32. Srivastava N, Hinton G, Krizhevsky A, Sutskever I, Salakhutdinov R (2014) Dropout: a simple way to prevent neural networks from overfitting. J Mach Learn Res 15:1929–1958
33. Feng X, Wu QMJ, Yang Y, Cao L (2020) An autuencoder-based data augmentation strategy for generalization improvement of dcnns. Neurocomputing 402:283–297
34. Liu B, Grout V, Nikolaeva A (2018) Efficient global optimization of actuator based on a surrogate model assisted hybrid algorithm. IEEE Trans Ind Electron 65:5712–5721. https://doi.org/10.1109/TIE.2017.2782203
35. Torun HM, Swaminathan M (2019) High-dimensional global optimization method for high-frequency electronic design. IEEE Trans Microw Theory Tech 67:2128–2142. https://doi.org/10.1109/TMTT.2019.2915298
36. Akinsolu MO, Liu B, Grout V, Lazaridis PI, Mognaschi ME, Barba PD (2019) A parallel surrogate model assisted evolutionary algorithm for electromagnetic design optimization. IEEE Trans Emerg Top Comput Intell 3:93–105. https://doi.org/10.1109/TETCI.2018.2864747
37. Nguyen T, Shi B, Ma H, Li EP, Chen X, Cangellaris AC, Schutt-Aine J (2021) Comparative study of surrogate modeling methods for signal integrity and microwave circuit applications. IEEE Trans Components, Packag Manuf Technol 11:1369–1379. https://doi.org/10.1109/TCPMT.2021.3098666
38. Bishop CM, Nasrabadi NM (2006) Pattern recognition and machine learning. Springer
39. Suganthan PN, Hansen N, Liang JJ, Deb K, Chen Y-P, Auger A, Tiwari S (2005) Problem definitions and evaluation criteria for the CEC 2005 special session on real-parameter optimization. KanGAL Rep. 2005005:2005
40. MacFarland TW, Yates JM (2016) Mann—whitney u test. In: Introduction to nonparametric statistics for the biological sciences using R. Springer, pp 103–132
41. STudio CSTM (2014) CST Studio Suite 2011. Comput Simul Technol AG

Index

D. T. Pham and N. Hartono (eds.), *Intelligent Engineering Optimisation with the Bees Algorithm*, Springer Series in Advanced Manufacturing,
https://doi.org/10.1007/978-3-031-64936-3

The manufacturer's authorised representative in the EU is Springer
Nature Customer Service Centre GmbH, Europaplatz 3, 69115 Heidelberg,
Germany. If you have any concerns regarding our products, please
contact ProductSafety@springernature.com

Printed and bound by CPI Group (UK) Ltd, Croydon, CR0 4YY
30/09/2026
02225586-0014